Sustainable Artificial Intelligence-Powered Applications

IEREK Interdisciplinary Series for Sustainable Development

The interdisciplinary series "Sustainable Artificial Intelligence Powered Applications," (SAIPA) aims to cover a wide range of Artificial Intelligence applications with a specific focus on sustainability to further support the sustainable development goals. AI systems are able to perform a wide range of tasks, including high-level image and video recognition, perspective modeling, smart automation, advanced simulation, and complex analytics among many others, and has high learning and improvement potential through machine learning and deep learning. These capabilities have many uses and provide many benefits in all industries applications. SAIPA includes a diverse range of these applications, perspectives, and expertise, where it seeks to advance the understanding and implementation of sustainable AI-powered solutions across various scientific disciplines. Its aim is to become a comprehensive guide to different experimental sciences. The "Sustainable AI-Powered Applications" publishes conference proceedings, monographs, and textbooks that address the intersection of artificial intelligence, sustainability, and advanced technologies. Authors, researchers, contributors, and scientists from all scientific disciplines related to AI and advanced technologies, are all invited to contribute to this series to shape the future of AI in a sustainable and responsible manner. Thus, the series represents an excellent endeavor that promotes a collaborative and interdisciplinary strategy for the development of sustainable AI-driven applications. The scope of this series includes, but is not limited to: Promoting sustainability and human well-being, health and medical use, optimizing smart cities planning, transportation systems management, infrastructure development, creating livable urban environments, renewable energy, agricultural Innovations, precision agriculture, weather agriculture, sustainable food production, crop & animal monitoring, and agricultural robotics, Socio-Economic Impacts, optimizing energy consumption in buildings, factories in energy usage and lower carbon, smart grids applications, smart grids management, climate modelling, environmental monitoring, waste management, Internet of things (IOT) and other domains related to the use of AI applications that improve overall environmental impact.

Series Editor: Mourad Amer **editor@saipa-series.com**

Karthika Subbaraj · Balamurugan Balusamy ·
Chithirai Pon Selvan M. · Feslin Anish Mon ·
Chander Prabha
Editors

Computational Bioacoustic Artificial Intelligence

Springer

Editors
Karthika Subbaraj
Department of Information Technology
Sri Sivasubramaniya Nadar College
of Engineering
Kalavakkam, Tamil Nadu, India

Chithirai Pon Selvan M.
School of Science and Engineering
Curtin University Dubai
Dubai, United Arab Emirates

Chander Prabha
Department of Computer Science
and Engineering
Chitkara University Institute of Engineering
and Technology
Rajpura, Punjab, India

Balamurugan Balusamy
School of Engineering and IT, Manipal Academy
of Higher Education, Dubai Campus
Dubai, Dubai, United Arab Emirates

Feslin Anish Mon
Department of Information Technology
University of Technology
Musact, Oman

ISSN 3005-1762 ISSN 3005-1770 (electronic)
Sustainable Artificial Intelligence-Powered Applications
ISBN 978-3-032-06087-7 ISBN 978-3-032-06088-4 (eBook)
https://doi.org/10.1007/978-3-032-06088-4

This Springer imprint is published by the registered company Springer Nature Switzerland AG
The registered company address is: Gewerbestrasse 11, 6330 Cham, Switzerland

Contents

About the Editors

Dr. S. Karthika Associate Professor, in the Department of Information Technology has over 16 years of teaching and 1.5 years of exclusive research experience. To her credit she has 80+ research articles and delivered 20+ keynote talks in her research domain of computational science using AI. She heads the AI for Social Good (AISG) Research Lab and Cyber Security Working Group (CSWG). She is currently executing DST funded CSRI project for worth approximately 20 lakhs and has also completed projects worth 4.1 lakhs from TNSCST and seed funding of 14 lakhs from SSN Trust. She has executed 2 consultancies worth 7 lakhs and has published 1 US patent. She has organized 3 international conferences, ATAL FDP, ACM Summer School, and 25+ other activities. She has played many vitals roles as an academician and researcher by serving as Board of Studies (BoS), doctoral committee member, and reviewer for reputed international and national journals and conferences. She is a recipient of UGC BSR Fellowship, Teacher Par Excellence awards, and Intel oneAPI Innovator. She is an active ACM and Senior IEEE member. She is an ardent admirer of Indian dance and music art forms.

Prof. Balamurugan Balusamy is Professor and Chairperson of the School of Engineering and IT at Manipal University, Dubai. He previously served as Professor and Associate Dean (Academics) at the Shiv Nadar Institution of Eminence, Delhi-NCR, and held positions as Professor, Director (International Relations), and earlier Associate Professor during his 15-year tenure at VIT University, Vellore. He has also been Professor in the School of Computing Sciences and Engineering at Galgotias University, Greater Noida, and is currently Adjunct Professor in the Department of Computer Science and Information Engineering at Taylor University, Malaysia.

He was listed among the Top 2% Scientists Worldwide 2023 by Stanford University in the field of data science/AI/ML. His work spans engineering education, blockchain, and data science.

Prof. Balusamy has published 200+ high-impact journal papers (Springer, Elsevier, IEEE) and has edited or authored over 200 books, collaborating with leading scholars from top QS-ranked universities. He has delivered over 210 talks at international events and has traveled to more than 15 countries for academic and research engagements. He has also organized and hosted over 10 IEEE and ACM international conferences.

His teaching philosophy emphasizes design thinking and learner-centered approaches.

Dr. Balamurugan Balusamy, Professor and Chairperson, School of Engineering and IT, Manipal Academy of Higher Education, Dubai Campus, Dubai, UAE kadavulai@gmail.com

Prof. Chithirai Pon Selvan has extensive experience in teaching engineering students and has worked in academia for over twenty-five years. He has published/presented more than 150 research articles in journals and conferences. He has been invited and honored as a keynote speaker, session chair, resource person, and technical committee member of various conferences held in UAE, India, Thailand, Malaysia, Germany, Italy, Australia, Oman, Qatar, and the UK. His research interests are in the areas of machine design, optimization techniques, manufacturing practices, renewable energy, and engineering sustainable development. He is a well-known researcher in the field of Abrasive Waterjet Cutting Technology and has evaluated several PhD theses in Mechanical Engineering from various universities. He is the approved supervisor of many universities including Curtin University, Australia, to guide PhD scholars and has produced many PhDs. He is a member of many professional societies including SAE, ASHRAE, IMechE, ASME, EI, ASQ, IEEE, and ISTE. He is also a senior fellow of the Higher Education Academy (SFHEA), UK. Before joining Curtin University Dubai in July 2018, he served for 10 years in the UAE with Manipal University Dubai and Amity University Dubai.

Pon has received several prestigious awards including the following.

- "Teaching Excellence Award (2014)" from Manipal University, Dubai.
- "Dubai Award for Sustainable Transport (2017)" from Road Transport Authority (RTA), Govt. of Dubai.
- "Dr. Kalam's International Excellence Award for Education (2017)" from Dr. APJ Abdul Kalam Lovers Foundation, UAE.
- "Distinguished Conservation Project Award (2018)" from Dubai Electricity and Water Authority (DEWA), Govt. of Dubai.
- "Alleem Sustainability Researcher Award (2018)" from Alleem Research and Development Centre, Sharjah.
- "Sustainability Ambassador Award (2019)" from Road Transport Authority (RTA), Govt. of Dubai.
- "Research Excellence Award (2020)" from Curtin University Dubai.
- "TAG Founders Award - Academic of the Year (2021)" from Transnational Academic Group (TAG), Dubai.
- "Curtin Global Award (2022)" from Curtin Global, Curtin University, Australia
- "Ezhumin Award for Education - Dubai (2023)" from News 7 Tamil TV Channel.
- "Curtin Award for Excellence in Innovation and Teaching (2023)" from Curtin University, Australia.

Dr. Feslin Anish Mon a researcher specializing in Trust-based Security in IoT Networks, holds a Ph.D. in Information and Communication Engineering. With good publications in reputed journals and a patented contribution to his name, Dr. Feslin has notable contribution in academia as well. Currently, he serves in the Department of Information Technology at the University of Technology and Applied Sciences, Ibri Branch, where he actively contributes to research and inspires students and colleagues alike. His work reflects a commitment to advancing knowledge and innovation in the dynamic field of information technology.

Dr. Chander Prabha is working as a professor in the Department of Computer Science and Engineering at Chitkara University Institute of Engineering and Technology, Chitkara University, Punjab, India. She had done her Ph.D. from M.M. University in Computer Science Engineering in 2017 and her master's in Engineering from Punjab Engineering College (2004), PEC, Chandigarh, India. She has a total of 21+ years of academic/research experience with more than 110+ publications in various national, international conferences, and international journals. She has published many book chapters in CRC press. She is the guest editor journal of Artificial Intelligence and Internet of Things, Near East University, Turkey. Also, she is a reviewer of many international conferences and journals of repute. Many online certifications are also there in her credit. She is a lifetime member of educational societies like ISTE, IAENG, SCIEI, IIRJC, IACSIT, etc.

Advanced Bioacoustic Signal Processing and Classification

Advanced Signal Processing Techniques in Bioacoustics by Elevating Wildlife Monitoring

T. Velmurugan, V. Manimala, V. Parameshwari, S. Mahendrakumar, and R. Gopal

Abstract

Bioacoustics is a vital field that significantly contributes to our understanding and monitoring of wildlife, offering a non-invasive approach to studying animal behavior, communication, and overall ecosystem health. By analyzing sound patterns animals produce, researchers can gain valuable insights into their ecological roles, interactions, and responses to environmental changes. However, as the volume and complexity of bioacoustics data continue to increase, the need for advanced signal processing techniques has become imperative. These techniques are essential for improving the accuracy, efficiency, and scope of wildlife monitoring efforts. This chapter explores various innovative signal processing methods specifically designed for bioacoustics applications, including spectral analysis, machine learning algorithms, and time–frequency techniques. In spectral analysis, the complex sound signals are broken into their essential frequencies, which help to provide a clearer picture of the sound of wild animal species. Meanwhile, machine learning techniques are used for the automatic classification of the sound, and time–frequency analysis is used to enrich the behavioral patterns, and communication strategies to change over time to time. The bioacoustics methods used to develop the accuracy of spotting and classifying animal sounds to support the investigation of intricate acoustic environments. In real-time monitoring, these methods are used to assist in responding to ecological changes that threaten the species' natural habitats. Studies of the environmental benefit in the long term, it is used to help datasets for better understanding of species connections over time. The bioacoustics systems, integrating those refined methods permits canvassers to improve species detection, difference overlapping signals, and automated monitoring systems. Eventually, using advancements in signal processing technologies significantly improve our monitor systems in biodiversity and wildlife inhabitants. In the evolution of signal processing in bioacoustics methodology to expose new frontiers in the wildlife species to provide more effect and understanding of the complex relationship within ecosystem.

Keywords

Bioacoustics · Signal processing · Filter · Machine learning algorithms · Ecosystem

T. Velmurugan (✉)
Department of Computer Science and Design, Kongu Engineering College, Perundurai, Erode 638060, Tamilnadu, India
e-mail: ecevel@gmail.com

V. Manimala
Department of ECE, Builders Engineering College, Nathakadaiyur, Kangeyam, Tirupur 638108, India
e-mail: malakrishnasamy@gmail.com

V. Parameshwari
Department of ECE, Nandha Engineering College, Main Road, Vaikkaalmedu, Erode 638052, India
e-mail: parameshwari.velumani@nandhaengg.org

S. Mahendrakumar
Department of ECE, Velalar College of Engineering and Technology, Thindal, Erode 638012, India
e-mail: mahendrakumar@velalarengg.ac.in

R. Gopal
Department of Information and Communication Engineering, College of Engineering, University of Buraimi, Al Buraimi, Oman
e-mail: rgopalkarur@gmail.com

K. Subbaraj et al. (eds.), *Computational Bioacoustic Artificial Intelligence*, Sustainable Artificial Intelligence-Powered Applications, https://doi.org/10.1007/978-3-032-06088-4_1

1 Introduction

1.1 Importance of Wildlife Monitoring in Biodiversity Conservation

Wild fauna observation is therefore considered to be part of the conservation process of biodiversity, which offers key information in understanding and protecting ecosystems. Biodiversity is essential for ecological balance and facilitates a wide variety of essential ecosystem services, including pollination, nutrient cycling, and climate regulation. Thus, wildlife monitoring will therefore enable scientists and policymakers to understand the health status of ecosystems, monitor changes in populations of species, and discover new threats to take necessary responses.

One of the key benefits of wildlife monitoring is related to assessments of population dynamics among several species, particularly those declared as endangered or vulnerable.

Researchers can identify extinct species and start conservation initiatives like habitat restoration and breeding programs to slow reductions as a result of population shifts. For instance, monitoring tiger populations in India has sparked anti-poaching initiatives and the potential for habitat restoration, both of which will eventually aid in the long-term recovery of the species (Kerry et al. 2022).

Wildlife monitoring is a precursor alarm system for changes in environmental conditions. Many species are used as bioindicators, that is, they reflect the condition of their ecosystems. For example, amphibian population declines often indicate water pollution or habitat degradation. Monitoring these species allows for preventive measures to take place before fundamental environmental issues become worse.

Surveillance also allows the management of protected areas and conservation plans. Knowledge about species distribution and migration patterns is used to improve the design of wildlife corridors and reserves so that critical habitats are connected and accessible. It also helps in assessing the effectiveness of conservation measures, thus allowing adaptive management based on scientific information.

Monitoring wildlife also prevents illicit activities like trafficking and poaching. Camera traps, drones, and acoustic sensors are examples of advanced technologies that can remotely and in real-time monitor wildlife, boosting the detection and prevention of illicit activity.

Given the global environmental issues posed by climate change, wildlife monitoring can be a valuable tool for understanding how animals respond to changing conditions. Changes in mating seasons, migration patterns, or habitat preferences that occur overnight may have broader ecological ramifications and impact global conservation objectives.

In conclusion, wildlife monitoring is critical to maintaining biodiversity because it offers the information and instruments required to safeguard ecosystems and the numerous species that depend on them. Its integration with modern technologies and scientific methodology enhances its effectiveness, so it becomes an integral component of international efforts to conserve our natural heritage.

1.2 Bioacoustics Roles in Non-invasive Monitoring

Bioacoustics, which concerns the production, transmission, and reception of sound in animal populations, has transformed the range of wildlife monitoring techniques by providing a non-invasive way to study and then conserve biodiversity. Compared with methods such as trapping, tagging, or direct observation, bioacoustics causes little disturbance to animals as well as their habitats, thus being both ethically and practically beneficial in assessing wildlife (Kerry et al. 2022).

The most important advantage of bioacoustics is that it may successfully detect cryptic or nocturnal species, generally difficult to observe. Many animals, including birds, marine mammals, bats, and frogs, rely on vocalizations for communication, navigation, and mate-finding. It is possible to monitor species without physically capturing or seeing them by recording and examining these auditory signals. It has proved very useful especially when it comes to the study of endangered species or reclusive ones that are sensitive to human presence.

Bioacoustics facilitates the ongoing and prolonged observation of environmental conditions. Acoustic sensors are capable of being installed in natural settings for significant durations, autonomously documenting auditory landscapes continuously. This method creates a large dataset that captures seasonal or daily variations in animal behavior as well as temporal trends. To improve understanding of migratory ecological processes, e.g., the bioacoustic evaluation of migratory bird species provides information on their arrival and departure times.

A significant role that bioacoustics plays in the use of comprehensive monitoring projects lies in the fact that advances in technology now allow acoustic sensors to be installed over wide land or sea areas with the possibility of simultaneously monitoring different species and ecosystems (Kerry et al. 2022). This scaling makes bioacoustics a critical tool in biodiversity surveys and habitat assessments, especially in inaccessible areas.

Bioacoustics also proves effective for species presence/absence detections, which are important for assessing the health of habitat. Acoustic indices, extracted from soundscapes recordings, serve as a proxy for levels of biodiversity.

Any changes in soundscapes may be a reflection of habitat deterioration or the invasion of non-native species, and high auditory variety frequently indicates abundant biodiversity.

Modern analytical methods like artificial intelligence and machine learning are seamlessly included into the study of bioacoustics. The advancements in technology have made it simpler to automatically categorize and recognize species-specific calls without relying on human analysis for extended periods of time. For instance, computers can quickly perform assessments of biodiversity in a tropical rainforest by easily differentiating the sounds of various frog species.

To sum up, bioacoustics is essential to non-invasive wildlife monitoring since it offers a morally sound, practical, and expandable method of examining animal behavior, population dynamics, and ecosystem health. It is a fundamental component of modern conservation science because of its ability to provide vast amounts of data continuously while causing little ecological disturbance.

1.3 Current Challenges in Bioacoustic Data Collection and Analysis

Despite the fact that bioacoustics presents many prospects for wildlife monitoring and conservation, there are some obstacles in data collecting and analysis that must be addressed to improve its effectiveness. These challenges are due to environmental influences, technological constraints, and the complexities associated with data processing.

1.4 Environmental Noise and Signal Interference

The problem most prominently exists in the form of both environmental and anthropogenic noise. Such sounds like wind, rain, running water, and rustling foliage can mask the calls of the animals, making it tough to separate the target sounds. It is more difficult in the vicinity of human habitation due to noises produced by humans, such as machinery, automobile traffic, and industrial operations. The sounds of marine mammals and other aquatic species are also muffled by ship noise, sonar, and undersea construction activities in marine habitats (Xie et al. 2020).

Bioacoustic monitoring is often composed of long term, continuous recording, and hence gigantic datasets. Recording audio 24/7 using a single acoustic sensor can result in hundreds of hours of audio within just a few weeks. It is quite impractical to analyze large amounts of data manually. Hence, the requirement for automated processing tools and computational infrastructure exists. Storage and management of datasets remain logistical and financial challenges in many research projects.

1.5 Complexity of Species Vocalizations

The various calls from different species occur in such close frequencies that it is hard to differentiate between one call from another, especially in a noisy or biodiverse recording place. For instance, if many species call at one time in tropical rainforests, it becomes difficult to untangle the sounds and distinguish between individual calls. Overlap between calls and patterns of change depending on individuals, regions, and seasons make the analysis a challenging process (Xie et al. 2020).

1.6 Limited Labeled Datasets

Machine learning and artificial intelligence are increasingly applied in bioacoustics for automatic species identification and analysis. These approaches, however, depend on labeled datasets that require annotation of species-specific calls within recordings. Such datasets are often scarce or lacking for most species, especially those poorly studied or newly discovered, limiting the accuracy and applicability of models.

1.7 Technological Limitations

Field recording equipment is prone to durability and battery life issues, especially when used in remote or extreme environments. The devices have to withstand harsh weather conditions, humidity, and physical degradation while operating for a long time. Moreover, sensor networks in inaccessible areas require more labor and cost. Interdisciplinary Expertise Requirement.

Bioacoustics necessitates knowledge that spans numerous disciplines, such as computational science, signal processing, and biology (Xie et al. 2020). Due to variations in research objectives, nomenclature, and techniques, interdisciplinary collaboration in these fields can be difficult. This disparity frequently prevents sophisticated technology from being used effectively in bioacoustic research.

1.8 Data Interpretation Challenges

Bioacoustic data analysis goes beyond species identification. Understanding behavioral contexts, like those related to mating, foraging, or distress sounds, frequently requires combining ethological and ecological knowledge. Inaccurate judgments about the presence, abundance, or preferred habitats of species may arise from incorrect interpretations.

2 Overview of Signal Processing in Bioacoustics

2.1 Basic Principles of Signal Processing

The process of evaluating, altering, and deciphering signals in order to derive valuable information from them is known as signal processing. Bioacoustics is the study of animal signals, which are mainly sound waves, such as whale sounds, bat echolocation clicks, or bird songs. It is necessary to understand the fundamentals of signal processing in order to effectively interpret those acoustic signals.

2.2 Signals and Their Representation

A signal is a time course of physical events, such as sound waves, and is characterized by such properties as amplitude, frequency, and phase. Typically, in bioacoustics, sound signals are recorded as waveforms in the time domain, with time plotted on the *x*-axis and amplitude, or loudness, on the *y*-axis (Kvsn et al. 2020). However, the analysis of the frequency content of these signals often reveals far greater insight, and this is done through the conversion of the signal into the frequency domain.

2.3 Time and Frequency Domains

The time domain represents how a signal changes with the time factor, while the frequency domain simply describes its structure in terms of different frequencies. For example, the call of the bird can be expressed by a sequence of pulses in terms of the time domain but can reveal certain harmonic types in the frequency domain. Methods in signal processing involving the Fourier Transform make moving between these domains possible hence enabling more detailed analysis of auditory properties.

2.4 Fourier Transform and Spectrograms

The Fourier Transform is a mathematical technique that decomposes the time-domain signal into its constituent frequency components. It has become one of the most common techniques applied in bioacoustics to generate spectrograms, graphical representations of sound that depict how frequency changes over time (Kvsn et al. 2020). This is very important for identifying the patterns in animal vocalizations such as the frequency sweeps of bat echolocation or the rhythmic patterns in frog calls.

2.5 Filters and Noise Reduction

Filters isolate specific frequencies or remove unwanted noise from a signal. High-pass filters will eliminate low-frequency background noise, while a band-pass filter will isolate the frequency spectrum of interest in a study, say, the vocal range of a specific bird species. Adaptive filtering, spectral subtraction, and other noise reduction techniques improve the signal-to-noise ratio, which lowers interference.

2.6 Feature Extraction

Finding a signal's conspicuous features that enable classification or analysis is known as feature extraction. Energy, perceived loudness, zero-crossing rate the frequency at which the waveform shifts from positive to negative and vice versa and the spectral centroid, or "center of mass," within the frequency spectrum are typical characteristics (Kvsn et al. 2020). These are some of the features important to distinguish between species or behaviors in bioacoustics.

2.7 Digitization and Sampling

Analog waveforms are typically recorded in bioacoustics and then sampled for examination. This would involve sampling the signal—that is, measuring it at discrete time intervals—and the representation of amplitude levels using numerals, or quantizing. The sampling rate gives the frequency range accessible with the sample; more elaborate sampling captures more detail at the expense of requiring storage to accommodate more samples (Fig. 1).

2.7.1 Traditional Versus Advanced Techniques

The field of signal processing within bioacoustics has significantly advanced from traditional methods relying heavily on manual analysis to modern sophisticated techniques supported by contemporary computational resources. Each methodology has specific advantages and disadvantages, thereby influencing its relevance in the contexts of wildlife monitoring and conservation.

2.7.2 Traditional Methods

Conventional signal processing methods were the backbone of the initial bioacoustic studies. These mainly included field recording, graphical representations, and general analyses of animal sounds.

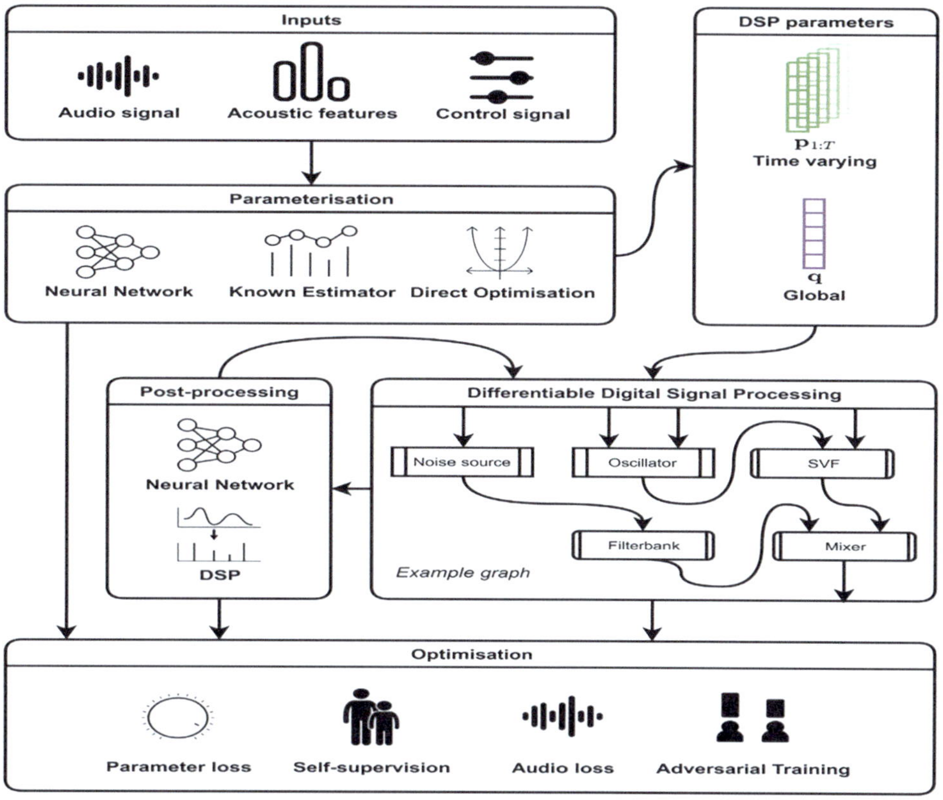

Fig. 1 High level overview of the general structure of a typical DDSP synthesis system (Hayes et al. 2024)

Manual Examination

Researchers used analog or early digital recording equipment to capture sounds, which were then analyzed either graphically (spectrograms) or audibly (playback). By examining patterns like frequency sweeps or harmonics, experts were able to identify species or behaviors thanks to the representation of time and frequency provided by spectrograms.

Sieve Method

Basic filtering methods were used to eliminate unwanted noise or separate important signals. For example, band-pass filters focused on the voice frequency ranges of the target species, whereas high-pass filters eliminated low-frequency background noise.

Expert Knowledge Dependency

In order to understand vocal acoustics, the conventional methods mostly depended on the judgment and intuition of scientists. For example, it was frequently necessary to have a detailed knowledge of the characteristic cries of two bird species in order to differentiate them.

These conventional techniques were labor-intensive, time-consuming, and prone to human error, even though they were helpful for small-scale research (Kvsn et al. 2020). They struggled to scale in noisy situations and with multipath signals or huge datasets.

2.7.3 Sophisticated Methodologies

In order to understand vocal acoustics, the conventional methods mostly depended on the judgment and intuition of scientists. For example, it was frequently necessary to have a detailed knowledge of the characteristic cries of two bird species in order to differentiate them.

These conventional techniques were labor-intensive, time-consuming, and prone to human error, even though they were helpful for small-scale research. They struggled to scale in noisy situations and with multipath signals or huge datasets.

Automated Examination

Tasks like segmentation, call detection, and species identification can now be automated thanks to modern signal processing methods. SVMs, on one hand, and deep learning architectures for a specific type that includes CNNs, which are driven by the convolutional principle, may put sounds into their respective categories.

Complex Noise Reduction

Adaptive filters, including Wiener and Kalman filters, and wavelet-based strategies are also sophisticated filtering methodologies. These techniques can be applied to real-time audio processing in a range of settings, including forests and oceans, because they successfully separate the target signals from noise.

Feature Extraction and Recognition

Complex properties that earlier techniques were unable to directly address, such as spectral entropy or Mel-frequency Cepstral Coefficients, can now be extracted from signals using sophisticated techniques. Such features are critical in distinguishing between closely related calls or analyzing subtle behavioral nuances.

Scalability with Big Data

It has also managed large collections of data from sensor networks as well as from long-time monitoring programs. Tools including cloud computing and parallel processing can scan over thousands of hours of recording, hence whole ecosystems can be monitored.

AI and IoT: Integrations

The monitoring of biacoustics has been greatly enhanced by the use of artificial intelligence and the Internet of Things (IoT). For example, intelligent sensors that process and classify data locally eliminate the need for human system monitoring or sending massive volumes of data.

2.8 Applications of Signal Processing in Wildlife Monitoring

Since signal processing makes it possible to analyze and understand acoustic data—which is essential for researching animal behavior, population dynamics, and the general health of ecosystems—it plays a critical role in wildlife monitoring. This field's application of signal processing has transformed non-invasive wildlife observation techniques, allowing for the tracking of species across a range of habitats.

Taxonomy and Categorization of Species

Techniques in signal processing are widely applied in species identification and classification using the analysis of their calls. Analyzing particular characteristics of animal calls, including frequency, duration, and harmonic patterns, helps identify species, even in different ecosystems. For instance:

Bird Calls

Spectrograms and Mel-Frequency Cepstral Coefficients (MFCCs) enable the classification of bird species (Kerry et al. 2022).

Marine Mammals

Frequency-based analysis differs between whale songs and dolphin clicks.

The process is enhanced by machine learning models to auto-identify species with high accuracy.

Population Estimation

Acoustic monitoring is a reliable method for assessing the amounts of animals, especially among vocalizing species (MacCurdy et al. 2011). Such techniques as estimation of call density, and spatial audio analysis, make it possible to count the individuals in the certain area. For instance:

- If frogs are calling in wetlands, then their population size may be forecast.
- Bat echolocation data gathered with ultrasonic detectors enable population monitoring in caves or forests.

Behavioral Surveys

Signal processing helps in the study of animal communication patterns, social behaviors, and responses to environmental stimuli. Examples include:

Mating Calls

Monitoring the frequency and timing of mating calls in amphibians and avians aids in comprehending reproductive behaviors.

Predator-Prey Dynamic

Acoustic cues between predators and prey reveal ecological interactions.

Habitat Quality Assessment

Acoustic indices obtained from signal processing techniques, such as acoustic diversity index and acoustic entropy, will be used as indicators of ecosystem health.

Rich Soundscapes

A habitat with a high level of auditory variety is one that is biodiverse and healthy.

Environmental Stress

A decline in the quality of habitats may be indicated by a decrease in avian vocalizations.

Tracking Migration Trends

Use signal processing to follow a migratory species' sounds.

Bird Migration

The time and method of migration are revealed by a temporal analysis of calls.

Marine Animals

Certain migratory routes and breeding areas are identified through the analysis of whale songs.

Threat Detection and Conservation

Signal processing can identify threats like habitat encroachment or poaching.

Acoustic sensors detect gunshots in forests, thus helping in combating poaching activities.

Invasive Species Analysis of non-native species' calls helps monitor and control their spread.

Change Studies of the Climate

Changes in vocalization patterns may also be used to understand the effects of climate change.

Variations in bird movement or vocalizations could be a sign of shifting habitat or temperature.

3 Advanced Techniques in Acoustic Signal Processing

3.1 Noise Reduction and Signal Enhancement

The potential for unwanted noise to obscure or alter the intended signals is a major obstacle in bioacoustic monitoring. These noises may originate from human activity, such as automobile traffic, industrial machinery, or building operations; environmental conditions, such as wind, rain, or water movements; or equipment constraints (Kahrs et al. 1998). To extract the pertinent animal vocalizations and guarantee precise analysis and reliable results, noise reduction and signal enhancement techniques must be implemented.

Adaptive Filtering Techniques

Adaptive filtering methods are engineered to eradicate noise that may have temporal fluctuations. Unlike a fixed method of static filters, adaptive filters change their operation based on the changing signal. Algorithms used in these filters evolve its parameters minimizing noise interference and regulating dynamic settings whose characteristics are liable to evolve (Kahrs et al. 1998).

Wiener Filter

An adaptive filter which makes the signal estimate on the basis of its statistical properties and reduces the noise by minimizing mean square error between the filtered and the true signal. It is effective in removal of random noise from a signal.

Kalman Filter

It uses real-time updates based on incoming data to follow a model of the signal and noise. In systems where there is a temporal correlation between signal and noise, such as underwater acoustics, this filter is quite effective.

These adaptive filtering techniques enable noise reduction in real time and without disruption; thus, they are quite reliable for extended reconnaissance under changing conditions.

Spectral Subtraction Techniques

One of the most common methods of noise reduction when noise is stationary or temporally predictable is known as spectral subtraction. This methodology presupposes that noise spectra could be estimated during silent periods or other phases of low activity (Kahrs et al. 1998). The estimated noise spectrum is then subtracted from the total signal, hopefully retrieving the original clean signal and removing noise components effectively within the frequency domain.

Application

Bioacoustics uses this spectral subtraction to remove unwanted noise, for instance a running engine or bird vocalization, to enhance the recognition and identification of animal calls.

Challenges

The greatest challenge with spectral subtraction is distortion in the signal itself, mainly when noise and signal have almost the same spectral properties. The algorithm parameters should therefore be optimized to reduce as much as possible the incidence of artifacts.

Despite these disadvantages, spectral subtraction is well used in time-domain applications due to its simplicity and effectiveness.

Wavelet Transform for Denoising

Wavelet transforms are very useful in the reduction of noise, especially for signals that have sharp transients or otherwise irregular patterns, such as those common in animal vocalizations (Kahrs et al. 1998). Wavelet transformations represent the signal in both the frequency and time domains, in contrast to Fourier transforms, which only work on the frequency domain.

Multiresolution Analysis

By enabling analysis at many scales, the wavelet transformation is useful for identifying both low-frequency traits, such as whale sounds, and high-frequency elements, such as bat echolocation clicks.

Denoising

Denoising of the wavelet involves decomposition of a signal into several frequency bands and filtering out noise within the higher frequencies bands; afterward, reconstruct the signal with these components. Denoising by the method above may effectively eliminate broad-band noise without altering the integrity of the original vocalizations.

For bioacoustic data collected in challenging settings like woods, where high-frequency interference and wind noise are common, wavelet denoising is very helpful.

Signal Enhancement

The practice of enhancing recorded signals to make them more noticeable than background noise and therefore simpler to analyze is known as signal augmentation. Techniques such as gain control, dynamic range compression, and high-pass or band-pass filtering are used in combination with noise reduction techniques (Kahrs et al. 1998).

Gain

The system affects the gain or amplification of the signal to make those specific sounds come out while quieting background noise. In recordings in which animal voices are too soft or swamped out by ambient sound, it is very helpful.

Dynamic Range Compression

This processing technique compresses variance in level between loud and soft sounds, resulting in the weak animal vocalizations of more audible value and thus lesser loud noise within a signal.

Band-Pass Filtering

This technique reduces out-of-band noise (such as distant thunder or human talk) by choosing or isolating the frequency range of interest, which is often the vocal range of the species being studied.

When used in conjunction with noise reduction techniques, these strategies increase signals and maintain the essential qualities of animal vocalizations, enabling more accurate identification and analysis.

3.2 Feature Extraction

An essential component of the study of the bioacoustic data is feature extraction. This entails locating and removing significant characteristics or features from unprocessed audio signals, which can then be applied to additional classification, pattern recognition, etc. (Kahrs et al. 1998). In order to convert complex audio data into useful information, bioacoustics—whose goals frequently include species identification, animal activity monitoring, or environmental condition evaluation—needs effective feature extraction.

3.2.1 Features Typing in Bioacoustics

These features can be obtained across the various domains, including time domain, frequency domain, and the time–frequency domain. The features to extract depend on the nature of the sound and the goals of the analysis.

Time-Domain Characteristics

These characteristics depict the waveform of a signal directly within the time domain. The common time-domain characteristics include:

Zero-Crossing Rate (ZCR)

The number of times the signal crosses the axis of zero amplitude; information on noisiness and periodicity of the sound.

Root Mean Square (RMS) Energy

This quantifies the signal's energy content or loudness, which is correlated with an animal's activity level or vocalization strength.

Autocorrelation

Assists in detecting periodicity or recurring patterns in the signal, such recurring songs or calls.

Frequency Domain Features

The frequency composition of the signal is described by its frequency domain characteristics. They are very helpful in determining the vowel-like characteristics of vocalizations, which vary greatly among species (Kahrs et al. 1998).

Spectral Centroid

The spectral centroid, which indicates the brightness or timbre of the sound, is the "center of mass" of the spectrum. While low values typically correlate with deeper, bass-heavy sounds, large spectral centroids which are frequently linked to high-pitched or sharp sounds.

Spectral Flux

The measurement of spectral flux, which can be helpful in recognizing abrupt shifts or changes in auditory signals, may be helpful in differentiating between animal calls or pinpointing particular vocalization occurrences.

Spectral Roll off

The frequency below which a certain percentage of total spectral energy lies. A feature which can be important in discriminating between harmonic and percussive sounds, for example, possibly characteristic of different species.

Time–Frequency Domain Features

These are characteristics arising from methods that inspect the signal along both axes in time and frequency axes such as STFT and wavelet transforms. It proves very useful for handling signals that are not stationary because they change with time, like animal calls.

Mel-Frequency Cepstral Coefficients

It are one of the most widely used features in bioacoustics. Coefficients are obtained through the log transformation of the power spectrum of the signal and are used for numerous applications in speech and sound recognition. They are successful in representing the timbral properties of the sound and are very useful for species identification and the analysis of vocalizations.

Chroma Features

These features take care of the harmonic and pitch information-related ones and are often used in identifying musical elements in bioacoustic sounds. They are particularly useful in research involving animals with melodic, tonal cries, such as whales or birds.

3.2.2 Feature Extraction Applications in Bioacoustics

A key element of many significant wildlife monitoring programs is feature extraction.

Identification of Species

Scientists can distinguish between species by examining particular aspects of a species' vocalization, such as rhythm, spectral patterns, or frequency range. Whale species may vary depending on the distinctive patterns of their songs, just as bird species can be distinguished by their distinct frequency bands.

Analysis of Behavior

These include of energy levels, frequencies, and rhythmic patterns that may indicate specific actions like feeding, territoriality, or attracting mates. Scientists using this character analysis can untangle the ecological and behavioral dimensions of wildlife.

Environmental Monitoring

The extraction of features was used in assessing the health of a particular area's environment. For instance, different frequencies, timing, or intensities of animal calls may reflect some changes within the ecosystem such as habitation destruction by humans, climate change, etc.

Automated Detection Systems

Following feature extraction, the detection and classification of species, events, or behavioral states can be automated using machine learning techniques. For instance, a machine learning model may automatically identify bat echolocation calls or categorize various frog species using MFCCs and spectral properties.

3.2.3 Difficulty of Feature Extraction

Automated Detection Systems

Following feature extraction, the detection and classification of species, events, or behavioral states can be automated using machine learning techniques. For instance, a machine learning model may automatically identify bat echolocation calls or categorize various frog species using MFCCs and spectral properties.

Noise Interference

Environmental background noise, such as wind, rain, or human activity, can skew the signal and make feature extraction more difficult. Prior to feature extraction, the data must be cleaned using sophisticated signal augmentation and noise reduction algorithms.

Absence of Annotated Data

Using machine learning-based feature extraction, it is challenging to obtain a large enough dataset for training, in which the features have previously been annotated.

3.3 Signal Segmentation

One of the key elements in the analysis of bioacoustic data is signal segmentation, which separates a continuous acoustic signal into segments or pieces that correspond to specific properties or events. Certain calls, actions, or interactions with their surroundings could be among the occurrences. An efficient execution of segmentation would allow researchers to focus on the relevant segments of the signal, which is very crucial for accurate species identification, behavioral analysis, and ecological research. Since signals can be complex and noisy in bioacoustics, segmentation helps isolate signals of interest from background noise and overlapping sounds (Ramli and Jaafar 2016).

3.3.1 Energy-Based Segmentation

One of the simplest and most used techniques for separating bioacoustic signals is energy-based segmentation. It is based on the idea that animal sounds frequently have more energy than the background noise or quiet in the environment. This can be achieved by identifying and segmenting those vocalization-indicative bursts of high signal energy into discrete occurrences.

Procedure

Determine the signal's energy over small windows; typically, a sliding window is used to average the energy. At this point, a vocalization is presumed to have occurred if the energy over a certain threshold, and the signal is regarded as an additional event.

Applications

This technique works well for identifying distinct vocalizations, such frog croaks or bird songs, when the energy of the vocalization is significantly higher than the background.

Challenges

Very quiet voices or sounds that are nearly background noise are not picked up by energy-based segmentation. Furthermore, it frequently misclassifies high-energy, non-vocal stimuli, such as wind gusts or passing cars.

3.3.2 Dynamic Thresholding

Dynamic thresholding is a sophisticated energy-based segmentation technique that modifies the threshold for segmentation over time in response to the signal's properties (Ramli and Jaafar 2016). Dynamic thresholding adjusts to shifting auditory environment conditions, in contrast to fixed thresholding, which employs a fixed value for segmentation.

Procedure

Method for adaptive threshold update as the signal continues, the process adjusts the threshold according to the moving average of its energy or amplitude. Any value of the threshold depends on time of day, environmental conditions, or animal behavior under study.

Applications

It is very useful in those environments where the background noise is changing with time, such as a forest or towns. It is also helpful whenever the vocalizations vary in loudness because threshold can shift appropriately.

Problems

While dynamic thresholding will enhance the separation of objects in noisy environments, the approach is still prone to very variable signals or complicated, superimposed calls and will necessarily require extra post-processing techniques to finesse the results.

3.3.3 Machine Learning Algorithm-Based Automated Segmentation

The most advanced machine learning approaches, which have been widely applied to signal segmentation, are in learning complex patterns and handling big data at very high accuracy. Such methods are most useful when traditional techniques of segmentation fail to introduce differences between overlapping or similar sounding vocalizations (Nieto-Mora et al. 2023).

Supervised Learning

Labeled training data is used in supervised learning to teach the model to identify and separate certain signal events. The signal's properties, like MFCCs or spectral features, can be utilized to train neural networks, random forests, or support vector machines, which can categorize and divide the signal into useful segments.

Unsupervised Learning

Another application of clustering algorithms, such k-means or hierarchical clustering algorithms, is in unsupervised learning, where signals are separated solely based on their inherent characteristics rather than labeled data. By combining related auditory events, it will be useful while working with unmigrated or unknown species.

Deep Learning

CNNs and RNNs are two of the most recent deep learning techniques used in bioacoustic signal segmentation. These models are especially effective at identifying complex or delicate vocalizations against noisy backgrounds because they can capture nuanced patterns in the time–frequency domain.

Applications

Machine learning-based segmentation proves to be very effective even when the background noise is the highest intensity, such as in underwater or dense forests. It provides valuable benefits where calls overlap or vary a lot, such as in a few species or in complex social interactions.

Challenges

The key challenge for the machine learning-based process of segmentation is the huge demand for labeled data. Also, these models are extremely computationally costly and, hence, demand a strong processing environment, mainly when dealing with large-scale monitoring activities.

3.3.4 Hybrid Approaches Enhancing Segmentation Hybrid Approaches

In practice, combining different methods of segmentation usually provides better results than relying on a single technique. The energy-based method of segmentation may be used as an initial detection of segments of vocalizations, which then further improved and classified by using machine learning methods. Using supervised learning, dynamic thresholding can also remain accurate while adjusting to changing environmental conditions.

4 Machine Learning and Artificial Intelligence in Bioacoustics

4.1 Automatic Species Identification

In bioacoustics, automated species identification is crucial because it reports on species based on their voices without physically listening to or evaluating them. This is especially useful in places that are too large, too time-consuming, or in challenging situations where human presence is impaired (Nieto-Mora et al. 2023). Automated species identification systems can process large streams of acoustic data in real time with the help of sophisticated signal processing, machine learning, and artificial intelligence. This allows for timely and high-quality insights into wildlife populations, their behaviors, and ecosystems (Fig. 2).

4.1.1 Acoustic Features for Species Identification

Identifications often usually consider several special characteristics of animal voice; therefore, the presence in the voice of all animals is unique. A crucial feature extracted from these acoustical signals includes

Mel-Frequency Cepstral Coefficients (MFCCs)

These features capture the timbral characteristics of sounds very close to human perception of speech. These are useful when classifying calls from birds, mammals and even frogs' croaks.

Spectral Features

The spectral features include spectral centroid, spectral flux, and spectral roll-off, which help in the differentiation of species based on frequency distribution and changes in the harmonic structure of the sound.

Temporal Features

Zero-crossing rate, root mean square energy can be used to identify rhythms as well as energy patterns; thus, it may be pretty different between species.

Such features can be extracted to enable the researchers in making detailed acoustic fingerprints for different species and distinguishing between them.

4.1.2 Machine Learning Methods for Automatic Recognition

The foundation of automated species identification is machine learning algorithms, which learn to categorize species based on patterns found in data derived from features collected from acoustic signals (Nieto-Mora et al. 2023). The following are the main machine learning methods used in automatic species identification:

Supervised Learning

This makes use of a large collection of acoustic recordings with labels. It then discovers the connection between the species label and the auditory characteristics. Common algorithms that can be used with this approach are:

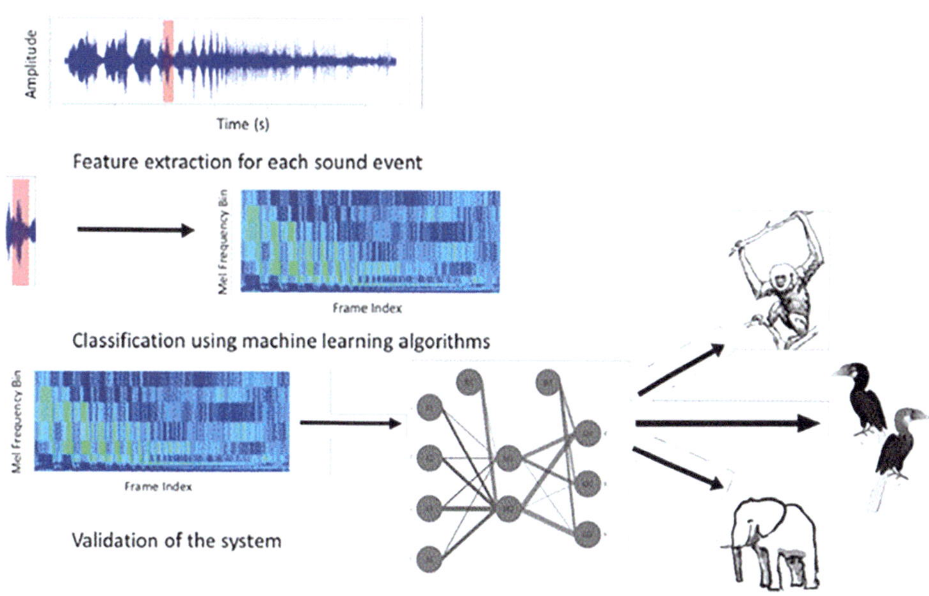

Fig. 2 Schematic of automated detection workflow (Clink and Klinck 2019)

Support Vector Machines (SVM)

A decision boundary that categorizes the different species according to their auditory characteristics is found using SVMs.

Random forests

Combination of decision trees used for processing large and complex datasets; thus, good in using multiple features for species classification.

Neural Networks

If the underlying data is high-dimensional, such as spectrograms or raw audio signal, then techniques like CNN for feature extraction along with classification come into usage.

Unsupervised Learning

Unsupervised learning techniques include clustering algorithms, like k-means, would combine comparable vocalizations based on acoustic characteristics if labels were unavailable. Without any prior knowledge, this could potentially identify unidentified species or vocalization kinds. To map the clusters into particular species, additional validation or post-processing will be necessary.

Deep Learning

The more complex deep learning architectures such as RNNs and LSTM networks can be trained to analyze temporal sequences in vocalizations. In this way, it becomes possible to classify calls that have more complex temporal or sequential patterns, like those of whales or elephants.

4.1.3 Use of Automatic Species Identification

Wildlife Surveys

Auto-identification allows for continuous time-exposure to species in difficult or inaccessible environments, especially inside wetlands and forests and inside all aquatic systems. Acoustic observation instruments with auto-identification software, therefore, generate insights useful for inferring species availability, density, and density spread over longer periods, especially for the long-term efficacy of conservation.

Biodiversity Assessment

The automated systems quickly process large amounts of acoustic data, which enables researchers to rapidly assess biodiversity in large landscapes. It is helpful in identifying species calls within the acoustic data to enable the identification of species richness and ecosystem health.

Behavioral Studies

To understand animal behavior, species need to be identified. Through timing and location tracking of specific species vociferations, scientists may determine periods of reproduction, territorial behaviors, patterns of migration, and sociability.

Conservation and Management

Accurate and automated species identification helps determine critical habitats, monitor endangered species, and detect the occurrence of invasive species. This procedure directs activities to lessen the impact of humans on animals and informs conservation efforts.

4.1.4 Challenges in Automated Species Identification

While automating the identification of species has many benefits, several challenges remain:

Overlapping Vocalizations

Some species vocalize simultaneously, or many species share a common sound. It can be difficult to spot overlap, thus methods like source separation or multi-label categorization may be required.

Environmental Noise

Automated identification methods can become less accurate when background noise from man-made (such as traffic, machinery) or natural (such as wind, water) sources obstructs the quality of animal vocalizations. Before identification, the data must frequently be cleaned using noise reduction and signal enhancement techniques.

Data Labeling and Training

Large, labeled datasets of acoustic recordings are necessary for supervised learning in order to train models. For less-studied or secretive species, it can occasionally be challenging to gather high-quality, labeled data, which limits the system's efficacy.

Generalization

It is possible that models developed on particular species or situations won't translate effectively to novel or diverse settings. To deal with the variations in acoustic circumstances and species vocalizations between regions and seasons, models must be continuously adjusted and retrained.

4.2 Anomaly Detection in Wildlife Soundscapes

Finding odd or unexpected acoustic phenomena occurring within the natural soundscape is the goal of anomaly detection in wildlife soundscapes. Therefore, such a process is of fundamental importance in detecting rare and anomalous events, which can be non-native species invasions, human impacts, environmental disturbances, or changes in the behavior of animals, symbolizing ecological stress (Marcot 2022). Bioacoustic monitoring simply means all the soundscapes, forming a blend of all sounds in an environment, such as sounds from animals, weather, vegetation, and human activities, where the anomalies could then be detected and provide much more critical information about health or ecosystem status, species behavior, and threats to biodiversity.

4.2.1 Value of Anomaly Detection

Anomalies in wildlife soundscapes help to identify anomalies for the purpose of

Ecosystem Monitoring for Health

Unusual noises might imply increase levels of anthropogenic noise (logging, traffic, or construction) that represent environmental stressors affecting habitat quality in species.

New or Invasive Species

Sounds that are unusual for the ecology may be produced by the new or invasive species. Early identification of these unusual noises may lessen their negative ecological effects.

Animal Voice Patterns

It can be used to identify behavioral changes brought on by food scarcity, pressure from predators, or even the start of the reproductive season.

Human Impact Assessment

The acoustic profiles of wildlife habitats are often disturbed by high levels of human activity or illegal activities, such as poaching and deforestation. Quick action and conservation efforts can be facilitated by early detection of these issues.

4.2.2 Methods for Anomaly Detection

Numerous techniques, ranging from traditional statistical methods to more current sophisticated machine learning approaches, have been developed for the purpose of detecting anomalies in wildlife soundscapes (Marcot 2022).

Statistical Methods

Threshold-Based Detection

Quite basic anomaly detection techniques depend on certain sound energy or intensity thresholds. Based on that, when the sound exceeds certain energy threshold values or does not stay within parameters of expected frequency and duration, it is termed as an anomaly. While this approach works well with controlled environments where baseline values are known, it often fails to cope with varying or more complex environments.

Spectral Analysis

Changes in the spectrum of sound can be noted by the presence of aberrations that occur due to shifting in ordinary spectral features, such as the sudden onset of high-frequency noise or unusual shapes in vocalizations.

Machine Learning Approaches

Unsupervised Learning

Techniques in unsupervised machine learning, including clustering algorithms like k-means or DBSCAN, facilitate the categorization of analogous auditory patterns, thereby enabling the identification of outliers or atypical sound patterns that diverge from the usual auditory environment. These methodologies operate without the necessity of labeled data and are effective in uncovering previously unidentified anomalies.

Autoencoders

In the context of deep learning, autoencoders are used to learn a concise representation of the usual acoustic environment, and anomalies are defined as events that do not fit well with the learned model. These models are very good at capturing detailed, high-dimensional acoustic information and detecting slight deviations in sound patterns.

Isolation Forests

This algorithm detects anomalies by creating random decision trees, which isolate anomalous points. It is very efficient for anomaly detection tasks on large datasets, making them useful for processing large numbers of acoustic data collected over long periods of soundscape monitoring.

Time-Series Analysis

Seasonal or periodic rhythms can be heard in many animal vocalizations. The detection of deviations from predicted time-series patterns in call frequencies or sound amplitudes over time can be used to identify anomalies. Time-dependent qualities can be analyzed and anomalous changes can be found using methods like wavelet analysis and Fourier transformations.

4.2.3 Applications of Anomaly Detection

Detection of Environmental Changes

Environmental changes like weather shifts, natural disasters like earthquakes or wildfires, and human activities like urbanization and noise pollution can all be linked to changes in soundscapes (Marcot 2022). Quick detection of such occurrences can help determine their effect on animals and then take corrective measures.

Conservation Monitoring

Animal calls may exhibit irregularities that can be interpreted by anomaly detection systems as a sign of stress, a shift in social dynamics, or even a possible threat. For instance, an unexpected lack of noise in places where animals are known to call could indicate habitat loss or disturbance.

Detection of invasive Species

Many alien sounds are introduced into the system by invasive or alien species. This would enable them to identify it before it destroys the current equilibrium of the environment.

Detection of Illegal Operations

Unusual patterns of sound, e.g., unfamiliar machineries or sounds with signs indicating illegal logging, could be human operations like poaching or deforestation. Anomaly monitoring systems could assist park rangers and conservationists in tracking inaccessible areas for illegal operations.

4.2.4 Difficulty in Anomaly Identification

While anomaly detection is a very powerful tool, it comes with several challenges:

Noise Interference

Wildlife soundscapes are naturally noisy, comprising a mix of biological noises, ecological noises, and anthropogenic noise. In dynamically charged ecosystems, it can be hard to distinguish between the normal fluctuations and real anomalies.

High-Dimensional Data of Acoustics

The bioacoustic dataset typically has high dimension and variability in time (Marcot 2022). Consequently, it becomes hard to detect anomalies without sophisticated algorithms that are efficient for very extensive and complex datasets.

Dynamic and Seasonal Changes

Animal's sounds and environmental noises change according to seasons or natural cycles such as mating seasons. The ability to tell a change from the normal seasonal one requires models that are adaptive for learning dynamic changes.

Data Annotation

In supervised learning methods of machines, appropriate training of models is obtained when there is a decent amount of labeled data available. Finding sufficient labeled data in situations where anomalies occur rarely or are unknown is impossible.

4.3 Using Edge Computing for Instantaneous Assessment

Instead of transmitting such calculations to faraway cloud servers for deliberate review, edge computing is a data processing technology that makes such computations locally, around the source of data production. The edge computing technique thus offers numerous benefits, particularly real-time data analysis and informed decision-making, in the context of bioacoustic monitoring, where vast amounts of acoustic data are continuously being gathered and used in remote or otherwise resource-constrained situations (Kadish and Stoy 2022). This technological method guarantees that timely insights are shown for timely actionability, enhances processing capabilities, lowers latency, and uses less bandwidth.

4.3.1 Edge Computing's Significance in Bioacoustic Surveying

In the majority of conventional bioacoustic monitoring systems, data is usually sent to cloud-based servers for further processing and analysis from remote acoustic sensors or microphones. Significant latency may arise from such transmission, especially when processing huge volumes of audio data. Data processing at or close to the sensor is possible with edge computing.

Instantaneous Analytical Insights

A device's edge computing will allow for local analysis that might quickly identify species, behavior, or abnormalities, enabling quick reactions to environmental concerns or alterations in wildlife behavior.

Reduced Latency

The capacity to process data quickly is essential in situations when prompt responses are required (such as identifying ecological issues or detecting unlawful hunting practices). Without waiting for the data to be transferred and processed at a distant site, edge computing guarantees that real-time insights are accessible.

Bandwidth Efficiency

Wildlife monitoring systems frequently operate in isolated locations where network connectivity may be erratic or unavailable. By enabling local processing, edge computing eliminates the need to send lengthy audio files to central servers, allowing for the transmission of only the most interesting or condensed results species identification and event detection, for example. This implies that data loss can be prevented and bandwidth may be used effectively.

Autonomy in Distant Areas

Edge computing makes it possible for monitoring systems to function autonomously in areas with limited or nonexistent internet connectivity (Kadish and Stoy 2022). Without a dependable network connection, the system can function continuously, offering real-time insights. When connectivity is restored, it can synchronize with central servers.

4.3.2 Edge Computing Architecture for Bioacoustic Systems

Local sensors, edge devices (such as microcontrollers, gateways, or field computers), and a central cloud or server for long-term data storage and additional processing make up a typical edge computing architecture in bioacoustic monitoring (Kadish and Stoy 2022). Important parts consist of:

Local Acoustic Sensors

Sound information from their surroundings can be detected by the local acoustic sensors. The sensors, which capture the environment's soundscape constantly or intermittently, might be placed in woods, marshes, or coastal environments for wildlife monitoring.

Edge Devices

Typically, edge devices are IoT gateways or field computers that preprocess acoustic data. On the recorded audio signals, they perform tasks including event detection, species identification, noise reduction, and feature extraction. These gadgets are typically equipped with strong CPUs or real-time algorithms that are optimized. Despite the fact that the processing is carried out locally on edge devices, a cloud or central server can be used for comparison with larger datasets, long-term data storage, and additional analysis. Additionally, model retraining, system updates, and cooperative research across several monitoring sites can all be done via the cloud.

4.3.3 Edge Computing Uses in Bioacoustic Monitoring

The use of edge computing in bioacoustic monitoring creates new opportunities for animal research and conservation:

Real-Time Species Detection

Edge devices are able to recognize animal vocalizations during real time, such as bird calls, whale songs, or amphibian croaks, and notify researchers or conservationists when a certain species is present. This is particularly helpful for monitoring migrations, tracking endangered species, and identifying species that are hard to see with the naked eye (Ulusar et al. 2019).

Anomaly Detection and Real-Time Alerts

Edge computing allows the system to promptly notify researchers, park rangers, or conservation managers when anomalous occurrences are observed (e.g., the presence of invasive species, environmental disruptions, or illegal actions like poaching). This facilitates prompt intervention and response.

Environmental Monitoring

In addition to biological sounds, edge devices track environmental factors like temperature, humidity, and noise levels. In addition to providing valuable information for habitat management and conservation, a local analysis of this combined data will assist in identifying any changes in habitat conditions that may have an impact on wildlife his is especially useful for tracking endangered species, monitoring migrations, or detecting species that are difficult to observe visually (Ulusar et al. 2019).

Anomaly Detection and Real-Time Alerts

When abnormal events are detected (e.g., the presence of invasive species, environmental disturbances, or illegal activities like poaching), edge computing enables the system to immediately alert researchers, park rangers, or conservation managers. This helps in a quick response and intervention.

Environmental Monitoring

Edge devices monitor environmental parameters such as temperature, humidity, or noise levels besides biological sounds. Analyzing this combined data locally will help detect any changes in habitat conditions, which might affect wildlife and provide useful information for habitat management and conservation.

Wildlife Behavioral Analysis

Continuous real-time analysis of animal vocalizations and patterns of behavior could provide knowledge on social dynamics, mating behaviors, or territory, among different species. These are key issues in determining ecological relationships and supporting programs for the management of species.

4.3.4 Benefits of Edge Computing for Instantaneous Bioacoustic Evaluation

Efficiency

Edge computing only sends the most pertinent information to central servers, such as the presence of species or anomalous occurrences (Ulusar et al. 2019). The efficiency of the entire monitoring system can be raised by reducing the amount of data that needs to be transferred and kept.

Cost savings

The edge computing technique minimizes the need to frequently transfer data to remote servers. It saves on communication expenses, which is especially a factor in areas where connectivity is limited.

Scalability

Edge computing enables scalable deployment of monitoring networks that can be dispersed across entire ecosystems without central servers' overload. This is to say that, with multiple sensors, every sensor and edge device could process independently without overloading the computing resources of centralized systems.

Improved Decision-Making

Real-time data analysis helps in better decision-making by researchers, conservationists, and wildlife managers. For instance, if the invasive species' call is being detected in real time, it could prompt immediate action to stop its spread (Ulusar et al. 2019).

5 Case Studies and Applications

5.1 Case Study: Using Bioacoustics to Monitor Endangered Animals

A number of reasons contribute to the extinction of endangered species, including habitat loss, poaching, climate change, and conflict between humans and wildlife. Therefore, in order to maintain biodiversity, it is important to employ efficient conservation techniques to guarantee the survival of these species. Traditional monitoring techniques, like as video traps or direct observation, can be costly, time-consuming, and impractical in dense or distant settings. A novel non-intrusive technique for observing endangered species' vocalizations and environmental soundscapes is bioacoustic monitoring. This study examines the use of bioacoustic technology for endangered species surveillance, which provides a wealth of data on the species' habitat utilization, population dynamics, and behavioral patterns.

5.2 Case Study: Monitoring Vulnerable Whales in the Pacific Ocean

One of the most endangered species of whales, along with being one of the rarest, is the North Pacific right whale, studied by researchers with bioacoustic technology in the Pacific Ocean. With fewer than a hundred in existence, this species cannot be traced and studied easily by normal tracking methods. The very useful information regarding their movement, health, and behavior comes through the vocalizations of these whales, which include low-frequency calls used for communication and navigation (Ziegenhorn 2022).

5.2.1 Acoustic Monitoring Setup

Researchers placed underwater microphones in important spots along known whale paths and feeding grounds. These microphones listen for the sounds around it throughout the day, continuously catching the sounds of those whales, other sea species, and even the vessels navigating through the waters from various distances. This works to continuously note sounds happening in our real-life ambient. It works year-round so collecting data through all seasons and therefore do not miss other moments like migration or mating moments.

5.2.2 Gather and Analyze Data

All audio data obtained is analyzed with high-powered methods of sound analysis. A particularly important aspect is spectrogram analysis, where people are able to visually see and identify specific patterns in the whale calls. This lets researchers be able to isolate the right whales' sounds from the rest of the ambient noise, such as ships or other sea animals. Scientists can monitor how many times, how long, and how frequently whales call in large oceanic regions to understand where they are and how they travel (Ziegenhorn 2022).

Furthermore, machine learning algorithms automatically pick up the sound of a right whale within a big collection of data. The system learns to locate patterns of sound frequencies with right whale calls that differentiate them from all the other sounds around marine life.

5.2.3 Perspectives and Applications in Conservation

Scientists were able to follow the seasonal migrations of whales and understand better their patterns of migration using bioacoustic data. Such information proves useful in the identification of critical feeding and breeding habitats to be conserved. Knowing the distribution of whales during certain periods helps conservationists argue for measures that reduce the flow of vessels in these important areas, thereby reducing the risk of ship strikes-one of the primary causes of mortality in this species.

Bioacoustic monitoring has added vital information about the species' social organization. Take, for instance, information on communication patterns between the whales, especially when mating: more frequent vocalizations during those periods. This adds to further understanding of their social activities and may contribute to effective plan development for recovery.

5.2.4 Instant Action and Real-Time Alerts

The whales' existence may also be tracked in real time thanks to bioacoustic technology. The monitoring system can produce instant notifications in the event of an uncommon or hazardous circumstance, such as increasing ship traffic in important whale habitats. By sending these notifications to ship operators and conservation authorities, they can promptly take corrective measures, such rerouting ships or temporarily blocking off areas to lessen the impact on people.

5.2.5 Issues and the Plan for Future Action

Although bioacoustic monitoring is quite effective, it has drawbacks. Whale sounds are sometimes obscured and made much more difficult to hear by background noises, such as those made by other marine life or by human activities like fishing or navy operations. In fact, some advanced techniques in signal processing and even noise reduction have helped somewhat solve these problems.

The future of the use of sound to monitor endangered species looks hopeful. Bioacoustic systems, which are getting smarter with machine learning, easier to expand with edge computing, and more precise with real-time data processing, contribute to better and more informed conservation methods over bigger areas.

5.3 Observe the Health of the Amazon Rainforest

In this case study, bioacoustic monitoring was used to check the health of a protected area in the Amazon Rainforest. Its focus was on how it affects bird populations and overall biodiversity (Teixeira et al. 2019). Researchers set up a network of passive acoustic recorders across the region. These recorders captured sounds from different species, environmental factors, and human-made noise.

5.3.1 Acoustic Monitoring Setup

More than 20 recorders were placed at a number of locations in the rainforest, both in and not so disturbed ones. The recorders kept working throughout the year; in addition to environmental noises, such as rain and wind, they recorded animal calls-including bird songs and calls of primates-in addition to insect noises. Therefore, the entire record so obtained gave a holistic representation of the sounds found at different habitats in the rainforest.

5.3.2 Data Collection and Data Analysis

The collected audio data was then analyzed by the specialized software used in bioacoustic analysis. The scientists zeroed in on the calls of specific species, like those of endangered bird species-the Harpy eagle and Scarlet macaw, which are highly sensitive indicators of habitat condition. The Acoustic Diversity Index (ADI) measured the richness and diversity of sounds in the rainforest (Teixeira et al. 2019). A high ADI represented a healthy and diverse ecosystem, whereas a low ADI represented degradation in species presence.

The amount of human-induced noise, including logging equipment, chainsaws, and nearby roads, was quantified to assess noise pollution. Higher levels of human-induced noise in rainforest areas reflected greater human disturbance, which has been known to negatively impact the behavior of wildlife in such aspects as communication, breeding behaviors, and avoidance of predation (Teixeira et al. 2019).

5.3.3 Results and Perspectives

The investigation revealed important differences between the damaged and pristine regions of the rainforest:

Decreased Species Diversity

Areas with strong logging or other anthropogenic activity levels have a reduced aggregate vocalization diversity. Some bird and mammal species either disappeared or hardly called at all, indicating that they have changed or are not able to tolerate the loss of their habitat.

Noise Pollution

Anthropogenic noise levels were considerably higher in the disturbed areas, and this was associated with a lower frequency of animal cries. Species that are sensitive to noise, such as the Harpy eagle, either stopped vocalizing or did not exist in the areas that were disturbed.

Impact on Avian Species

Researchers that were surveying in areas less impacted by disturbance reported observations of more native birds, including several species that rely extensively upon their calls to communicate while defending territories and other associated behaviors. These are readily accepted as indicators that are consistent with an active and undisturbed system.

5.3.4 Implications and Measures for Conservation

The bioacoustic data gathered for this experiment revealed important information on the condition of the Amazonian ecosystem. Conservationists may focus on areas that have already been negatively affected and have fewer types of life in order to focus their efforts on protecting certain locations. Stronger restrictions against logging sensitive locations and establishing buffer zones around critical habitats of endangered species were called for by the data.

5.4 Case Study: Saving Bats Using Sound Monitoring

Bats are very important for ecosystems. They help with pollination, controlling insects, and spreading seeds. But in many areas, bat numbers are going down because of losing their homes, climate change, and diseases like White-Nose Syndrome. Keeping track of bat populations is important to understand how they behave, where they live, and why their numbers are decreasing. Traditional ways to monitor bats, such as using mist nets or radio tracking, can take a lot of time and money. To solve these problems, conservationists in the United Kingdom started a citizen science program to get the public involved in monitoring bats, especially using sound technology (Stahlschmidt and Brühl 2012).

5.4.1 Project Overview: Bat Detectives

A group called the Bat Conservation Trust (BCT) that runs a "Bat Detectives" project got the people of the UK watching bats. The main idea is to get some information about which bat species are in the area and where using portable bat detectors. They capture the sounds that the bats make in navigating, and these sounds vary between the species, thus helping in identifying the various bat species.

5.4.2 Citizen Scientist Involvement

Volunteers were offered bat detectors, and based on the acoustics data retrieved, further training in identifying bat calls was provided to them. Volunteers were instructed to record bat activity within scheduled seasons, say, in summer months, during their peak activity time within their settings. Then they submit the data and recordings over an online interface to a set of specialists who analyzed such information to determine the species of bat.

Besides collecting bat calls, citizen scientists were encouraged to report on local environmental conditions, including the presence of trees, water sources, or agricultural areas, all known to affect bat populations. This information was used to analyze habitat affinities and potential threats to bat populations.

5.4.3 Data Collection and Data Analysis

Volunteers participating in the Bat Detectors over a period of few years recorded a significant acoustical data. Experts used the bioacoustical information gathered with this type of bat detectors toward detection of species-specific call by specialized software packages that processed the data to bring it to a cartography kind of representation, namely plotting out a distribution map throughout the UK.

Apart from data collection, the citizens contributed to filling the knowledge gaps where bat populations were found, something which scientists could not study suitably either in rural or in an urban setting. Large numbers of participation among citizens guaranteed extensive and spatially comprehensive datasets that could be increased toward insightful trends in bat populations.

5.4.4 Outcome and Impact

Data collected from the citizen science project in the field was useful in determining important details on bat populations:

Species Distribution

The study has discovered bat colonies previously unknown in places. In addition, the research has been useful in the study of population health and spread of species that are hard to trace using traditional methods such as the Bechstein's bat.

Habitat Preferences

Based on habitat data collected by volunteers, researchers identified what environmental factors seemed to be definitely required by specific species of bats. Bat-friendly habitats and measures of protection were subsequently erected based on that information.

Public Awareness and Involvement

Increased public awareness of bat needs and management; volunteers reported an attachment to local community wildlife and a disposition to manage habitats for those bats in each community.

5.4.5 Conservation Impact

This citizen science involvement in bat monitoring has been followed by a series of important conservation impacts.

Data-Driven Conservation Planning Data provided from Bat Detectors led to developing more focused-conversation activities, such as better placement of bat corridors and locations along with the effective protection of roosting areas.

This project raised public awareness toward the protection of bats, hence advancing public opinion in affirmation of policies that aim to conserve habitats for bats and reduce the impacts of urban development.

There is long-term monitoring that has been observed after successful citizen science campaigns. This long-term monitoring entails continued public involvement in wildlife conservation, and therefore longer terms to the sustainability of bat populations.

5.4.6 Challenges and Lessons Learnt

Although the Bat Detectives project was created effectively, many problems were found.

Training Volunteers

Citizen Scientists required proper training so that they were well equipped with bat detectors and also could understand how to identify calls. Volunteers must be made aware of the complexity of bat acoustics. Calls, by species, are highly variable among bats.

Data Quality Control

The data collected from volunteers was not of good quality and was quite variable. There was a need for proper validation through expertise. However, appropriate training and guidelines helped garner excellent data from the citizen scientists.

The study further revealed the need for long-term bioacoustic monitoring of overall conservation plans. This provides continuous, live data on how healthy a habitat may be and does not depend on people to stay frequent in remote locations.

6 Future Directions and Challenges in Bioacoustic Monitoring for Conservation of Wildlife

Bioacoustic monitoring is one of the important tools for wildlife conservation, through which species and ecosystems can be monitored in a non-invasive and wide-ranging way. It provides very valuable information about biodiversity, habitat conditions, and species' behaviors. However, despite all this, several challenges need to be overcome so that it becomes even more effective and its potential better realized within wildlife conservation. With advancements in technology, new ways to the future of bioacoustic monitoring are being discovered, which could bring the possibility of better conservation activities.

6.1 Interference of Bioacoustic Monitoring

6.1.1 Data Management and Analysis

One of the major challenges dealing with bioacoustic monitoring is its volume of data. They can create terabytes of data, more so when they are used in extensive monitoring or during long periods. High volumes of data are pretty much impossible to analyze manually and even some automated systems do not do a great job of proper processing and categorization of the sounds. Such issues include background noise, overlap between vocalizations for different species, and even the complication of differentiating species-specific sounds from multiple environmental noises (Schmidt and Balakrishnan 2015).

6.1.2 Interference of Environmental Noise

The purity of the biological signal is obscured by environmental noise from wind, rain, traffic, industry, and other human-related activities. Finding faint vocalizations or other animal calls is significantly impacted by noise pollution, particularly in densely populated regions. With the most advanced methods of noise reduction, removal of background noise from biological sound is hard, especially for dense habitats or in general urban settings.

6.1.3 Species Identification and Acoustic Variation

The same species could produce similar vocalizations, and it could also be different among individuals in the same species. Other species may change calls as changes in environmental features or human activity over time. This acoustical variability constitutes the greatest difficulty in achieving identification of a species (Schmidt and Balakrishnan 2015). The call of any given species may vary based on time of day, season, and even weather; hence it becomes difficult to detect and monitor populations of certain species.

6.1.4 Limited Scope and Reach

Bioacoustic monitoring can be conducted on large areas, but it is limited in terms of geography, especially in inaccessible and remote locations. Installation of monitoring equipment is extremely expensive and logistically cumbersome in such locations. Battery life and recording device strength are also major concerns that have to be overcome for effective long-term monitoring, particularly in extreme weather conditions.

6.2 The Prospects for Bioacoustic Monitoring

6.2.1 Advanced Machine Learning and Signal Processing

Numerous obstacles associated with the interpretation of bioacoustic data could be resolved by future developments in machine learning and signal processing techniques. For instance, even at high background noise levels, the application of deep learning algorithms is probably going to significantly enhance species recognition (Gibb et al. 2019). Additionally, machine learning structures can be designed to more accurately recognize calls specific to certain species even in complex acoustic environments, thus reducing the requirement for human verification of the data.

6.2.2 Real-Time Monitoring and Edge Computing

Real-time monitoring systems that are based on edge computing are a giant leap forward. Real-time analysis of bioacoustic data allows for prompt responses to many threats, including illegal poaching and environmental degradation (Gibb et al. 2019). Edge computing facilitates the local processing of data on devices, thus reducing the need to send large volumes of raw data over long distances, which allows for quicker decision-making and more efficient resource use.

6.2.3 Increased Data Storage and Cloud Computing Solution

The growing volume of bioacoustic data requires better storage solutions. Cloud-based platforms provide scalable storage and computation capabilities that facilitate greater accessibility and collaboration between researchers. In addition, the design of cloud-based analytical tools that are easy to use will streamline the process and engage a greater number of researchers, citizen scientists, and conservationists worldwide.

6.2.4 Integration with Other Monitoring Technologies

Future research will probably integrate bioacoustic monitoring with other monitoring methods, such as camera traps, satellite imagery, and GPS tracking. This multi-disciplinary approach will probably offer more comprehensive insights into species behavior, movement patterns, and habitat health (Gibb et al. 2019). Bioacoustic data will probably be integrated with visual or environmental information (e.g., temperature and humidity) to build more accurate models of ecosystem functioning and improve the effectiveness of conservation efforts.

6.2.5 Citizen Science and Public Involvement

Citizen science is envisioned to continue to play an essential role in the further evolution of bioacoustic monitoring. The expansion of public contributions to these monitoring efforts holds promise for gathering data over a wider geographic extent, especially in areas inaccessible to researchers. Growth in citizen science of application use and hand-held recorders is spreading to embrace a wider participation in conservation efforts for wildlife. Increasing public awareness regarding the importance of bioacoustic monitoring is likely to increase public support for conservation activities.

7 Conclusion: New Bioacoustic Monitoring Application for Innovative Wildlife Conservation

Bioacoustic monitoring is the most effective technique for conserving animals and greatly benefits understanding and conserving biodiversity. One can be able to see species, habitat conditions, and ecological change without having to use any invasive or resource-intensive methods because the technique only requires recording of sounds from the ecosystem. Its use can be found from tropical rainforests into cities, and it would give some key insights toward the health of our ecosystems.

This study explicitly shows that bioacoustic monitoring has greatly enhanced our ability to collect extensive, real-time information about animal populations and their environments. Scientists can now study species that are elusive or nocturnal, study animal behavior, and even detect early signs of environmental degradation through the use of sound as a measure for evaluating biodiversity. It is also a non-invasive alternative to traditional approaches, which are costly, labor-intensive, and limited in scope.

Nevertheless, like any technological innovation, some problems persist. Major problems of bioacoustic monitoring are data management and analysis, interference by ambient noise, and species identification. Such an enormous amount of data generated by acoustic sensors drowns researchers, with difficult challenges in processing and interpreting information in time. Background noise, both natural and anthropogenic, complicates the clarity of biological signals, making difficult accurate identification and monitoring of species. The acoustical variability between individuals, species, and environments also makes the analysis more complicated.

Despite these challenges, new developments in signal processing, machine learning, and real-time monitoring are improving the precision and efficiency of bioacoustic methods. With the implementation of ML algorithms, it becomes easy to automate species call classification, reduced hours, and ensured the integrity of data. More than that, edge computing accompanied by real-time data analysis has allowed conducting prompt actions on the matters of conservational activity-such as detection of poaching or habitat destruction, so forth. Besides that, the growing importance of citizen science has democratized wildlife monitoring; it has made it easier for people of different walks of life to contribute to conservation efforts and, therefore, increased the reach and effectiveness of bioacoustic data collection.

The future is bright for bioacoustic monitoring; with advancing technology, tools for bioacoustic monitoring are to become much more complex and sophisticated, capable of resolving data at a high spatial level and facilitating analysis. This technique, if used in conjunction with the extra assessment methods for ecology, such as satellite imaging and camera traps, would give perfect understanding of ecosystems and the interactions of animals. The increased adoption of cloud computing and processing of real-time data will have cooperative efforts in easing the process of sharing information across global conservation networks.

Bioacoustic monitoring is one of the most important fields in the efforts to protect Earth's biodiversity. Better knowledge of wildlife populations, their behavior, and their habitats leads to better-informed conservation decisions and timely interventions that prevent species extinctions and habitat degradation. As we move forward and improve these methods, bioacoustics is going to play an increasingly crucial role in preserving the fragile balance of the environment for future generations.

References

Clink DJ, Klinck H (2019) GIBBONFINDR: an R package for the detection and classification of acoustic signals. arXiv preprint arXiv:1906.02572

Gibb R, Browning E, Glover-Kapfer P, Jones KE (2019) Emerging opportunities and challenges for passive acoustics in ecological assessment and monitoring. Methods Ecol Evol 10(2):169–185

Hayes B, Shier J, Fazekas G, McPherson A, Saitis C (2024) A review of differentiable digital signal processing for music and speech synthesis. Front Sig Process 3:1284100

Kadish D, Stoy K (2022) BioAcoustic index tool: long-term biodiversity monitoring using on-sensor acoustic index calculations. Bioacoustics 31(3):348–378

Kahrs M, Brandenburg K (eds) (1998) Applications of digital signal processing to audio and acoustics. Springer Science & Business Media

Kerry RG, Montalbo FJP, Das R, Patra S, Mahapatra GP, Maurya GK, Rout JR (2022) An overview of remote monitoring methods in biodiversity conservation. Environ Sci Pollut Res 29(53):80179–80221

Kvsn RR, Montgomery J, Garg S, Charleston M (2020) Bioacoustics data analysis—a taxonomy, survey and open challenges. IEEE Access 8:57684–57708

MacCurdy RB, Gabrielson RM, Cortopassi KA (2011) Automated wildlife radio tracking. In: Handbook of position location: theory, practice, and advances. pp 1129–1167

Marcot BG (2022) Sound anomalies of Cornell Swift recorders affect ecoacoustic studies, and a workaround solution. Wildl Soc Bull 46(5):e1363

Nieto-Mora DA, Rodríguez-Buritica S, Rodríguez-Marín P, Martínez-Vargaz JD, Isaza-Narváez C (2023) Systematic review of machine learning methods applied to ecoacoustics and soundscape monitoring. Heliyon

Ramli DA, Jaafar H (2016) Peak finding algorithm to improve syllable segmentation for noisy bioacoustic sound signal. Procedia Comput Sci 96:100–109

Schmidt AK, Balakrishnan R (2015) Ecology of acoustic signalling and the problem of masking interference in insects. J Comp Physiol A 201:133–142

Stahlschmidt P, Brühl CA (2012) Bats as bioindicators—the need of a standardized method for acoustic bat activity surveys. Methods Ecol Evol 3(3):503–508

Teixeira D, Maron M, van Rensburg BJ (2019) Bioacoustic monitoring of animal vocal behavior for conservation. Conserv Sci Pract 1(8):e72

Ulusar UD, Turk E, Oztas AS, Savli AE, Ogunc G, Canpolat M (2019) IoT and edge computing as a tool for bowel activity monitoring. In: Edge computing: from hype to reality. pp 133–144

Xie J, Hu K, Zhu M, Guo Y (2020) Data-driven analysis of global research trends in bioacoustics and ecoacoustics from 1991 to 2018. Eco Inform 57:101068

Ziegenhorn MA (2022) Some like it cool: a study of odontocete ecology in the Hawaiian islands using passive acoustic monitoring. University of California, San Diego

Applying Advanced Deep Learning for Improved Bioacoustic Classification

V. Kumar, T. Sivaprakasam, K. Shanmugapriya, Vanitha Krishnan, and Feslin Anish Mon

Abstract

Acoustic monitoring has gained extensive interest as an essential environmental tool for wildlife population calculation, biodiversity assessment, and conservation efforts. Several species uses regular vocalizations signals that are species specific, which enables monitoring via acoustic recognition even in isolated or difficult-to-access areas. Improvements in computerized sound recorders have empowered data collection at larger temporal and has given rise to enormous datasets. On the other hand, like the development of many data-driven methods, including camera traps and environmental DNA (eDNA), methods for acoustic data collection are progressing more rapidly than those for effective exploration and clarification. But, traditional Machine Learning (ML) methodologies frequently face difficulties with key challenges such as inadequate labeled datasets, the acoustic match between species and background noise which leads to inaccuracy in species classifications. This chapter discovers the usage of cutting-edge deep learning techniques to address these problems and improve the effectiveness of bioacoustic classification systems. This chapter begins with exploring the drawbacks of traditional bioacoustic classification techniques, such as k-Nearest Neighbor, Support Vector Machines (SVMs), Random Forests, Hidden Markov Models, and Gaussian Mixture Models which are frequently depend on hand-crafted features and be unsuccessful to classify efficiently with the rising complexity and large volume of bioacoustic datasets. Deep learning techniques by contrast gives a more effective back ground for bioacoustic analysis by automatically learning features directly from or marginally processed data sets. This chapter highlights certain highly effective deep learning techniques used for bioacoustics classification, that includes Convolutional Neural Networks (CNNs), Recurrent Neural Networks (RNNs), CNN-RNN Hybrids, Long Short-Term Memory Networks (LSTMs)-based models. This chapter ends with exploring the key benefits of deep learning techniques for improved bioacoustics classification, such as automated feature extraction, flexibility, scalability, and accuracy. These techniques have unlocked new way to natural world protection, biodiversity tracking, invasive species finding and environmental research.

Keywords

Bioacoustic · Machine learning · Deep learning · Classification · Biodiversity

V. Kumar (✉)
Department of ECE, Hindusthan College of Engineering, Ingur, Perundurai, Erode, India
e-mail: kumaravelmeae@gmail.com

T. Sivaprakasam
Department of ECE, VSB College of Engineering Technical Campus, Coimbatore, India

K. Shanmugapriya
Department of CSE, Nandha Engineering College, Erode, India
e-mail: shanmugapriya.k@nandhaengg.org

V. Krishnan
Department of Medical Electronics, Velalar College of Engineering and Technology, Erode, India
e-mail: vanitha@velalarengg.ac.in

F. A. Mon
Department of Information Technology, University of Technology and Applied Sciences, Ibri, Oman
e-mail: feslin.anishmon@utas.edu.om

K. Subbaraj et al. (eds.), *Computational Bioacoustic Artificial Intelligence*, Sustainable Artificial Intelligence-Powered Applications, https://doi.org/10.1007/978-3-032-06088-4_2

1 Introduction

1.1 Context of Bioacoustics

Bioacoustics is the process of producing, transmitting, and receiving sounds by animals, often focusing on the analysis of sounds that emanate from a living organism. Sounds provide rich information on behavior, communication, population dynamics, and interaction with the environment. Bioacoustics is thus turning out to be a good monitoring tool in biodiversity, more especially with elusive or nocturnal species that are difficult to monitor with visual observation. Birds, bats, frogs, and sea mammals are just some. From the songs of chirping birds to the roar of the whales, biological acoustics signal transmits information vital to various research studies, conservation explorations, and even public awareness campaigns.

Its application in bioacoustics is pretty wide. It can monitor endangered species, track migration patterns, or even just monitor the changes happening in the soundscapes over time to see how well the health of an ecosystem is performing. These signals also form a part in terms of responses of biodiversity monitoring to habitat destruction or climatic changes. The acoustic data will usually be gathered by using some specially designed microphones-for underwater sounds, hydrophones will do-but for others, some automated recorders spread across different habitats or by drones.

1.2 Traditional Challenges in Bioacoustic Classification

It would consume much time and labor in the analysis of bioacoustic data. In fact, the researchers themselves had to go through individual analyses of spectrograms or traditional methods of signal processing applied manually to obtain features such as pitch, duration, and amplitude. This was generally done by experts, but with a large dataset or very minor changes in the signal, errors begin to creep in.

Bioacoustic classification also presents some difficulties:

Noisy environments: Recordings are made in the natural habitats. Thus, even the clean signal might get affected by some background noise from winds, rains, or human activities.

Overlapping sounds: In areas full of biodiversity, it becomes hard to tell different sounds as various species could make simultaneous sounds.

Absence of data: It is extremely challenging to generate enough labeled data to train the machine learning algorithms on most of the rare or endangered species.

Species variation: The same species may vary with geographic, environmental, or individual differences in their calls, which makes classification even tougher.

1.3 The Role of Deep Learning in Bioacoustics

Deep learning has emerged in many fields as a great transformative tool, including for bioacoustics applications. It provides an answer to the challenges that traditional methodologies face by automatically learning to represent features from datasets, thus significantly reducing manually extracted features. Deep neural networks (Jeantet and Dufourq 2023), which include CNN, RNN, and LSTM networks, would learn complex patterns of interest that can be missed through traditional machine learning models.

Bioacoustic data often would be converted to a spectrogram (Manriquez et al. 2024), which presents audio frequencies graphically over time and so could be processed like images. For example, CNNs can learn and extract frequency modulations or patterns specific to certain species in such spectrograms. On the other hand, RNNs and LSTMs would handle sequential data much better, thus they are good weapons to model temporal dependencies in audio signals. For example, bird song's rhythmic structure will differ; vocalization of whales is periodic, too, and could be very well captured by using LSTMs.

1.4 Importance of Advanced Techniques in Bioacoustic Classification

This indeed requires more advanced techniques to handle larger volumes of data, especially as the field of bioacoustics is growing. Automated bioacoustic signal classification (Potamitis et al. 2014) holds great promise in

Increased accuracy: Deep learning can look for patterns within audio signals that traditional methods cannot, as with manual feature engineering.

Scaling up data processing: With automated tools, researchers can process large datasets in a much shorter time than the same thing done manually, which allows for more comprehensive monitoring over wider geographic areas.

Offer real-time analysis: For example, in applications including wildlife conservation or prevention of poaching, real-time classification may be required, and deep learning systems are possible by providing real-time analysis.

It adapts to other environments: Applying transfer learning techniques, the deep models can perform well even on other bioacoustic datasets with minimal amounts of labeled data while maintaining high levels of performance.

Deep learning can go well through variability and noise within an audio recording; it is therefore useful for research into bioacoustic environments. As opposed to dependency on clean, manually labeled data, it learns from raw, real-world recordings. Additionally, because of recent attentions mechanisms and transformers, attention can be applied to specific regions of the audio, whereas other background noise gets filtered through, and the model is able to perform classification more efficaciously.

1.5 Current Research Trends

The hybrid models—CNNs with RNNs and attention-based architectures—widened the quality of bioacoustic classification. Current research is challenging, including multi-species (Zhong et al. 2020; Gupta et al. 2021) classification, detection of rare or cryptic species, and using behavioral cues for acoustic signals. Most importantly, the transformer models are now seen to be more prominent in the bioacoustics domain since they perform well on sequences involving long-range dependencies.

Another area of rapidly emerging interest is unsupervised or self-supervised learning, where a model might learn useful representations using unlabeled bioacoustic data. This is highly desirable when labeled data is scarce or expensive to obtain. For instance, models can be pre-trained on large, diverse sets and fine-tuned with smaller, species-specific ones, in a process known as transfer learning.

Advanced deep learning techniques in bioacoustic classification will potentially revolutionize the field with far more accurate, scalable, and automated tools for analyzing animal vocalizations. Not only are these achievements scientifically interesting but also important in efforts to preserve biodiversity and threatened ecosystems.

2 Fundamentals of Deep Learning in Bioacoustics

2.1 Overview of Deep Learning

Deep learning is part of machine learning wherein a kind of artificial neural network draws inspiration from the human brain to automatically learn representations from raw data, especially these kinds of neural networks with more than one layer—otherwise known as deep networks. The deep learning models learn to bring out and transform features on their own, whereas the conventional methods in machine learning most rely on manual extraction.

Deep learning emerged as the revolutionary tool in bioacoustics. It enables the analysis of acoustic data far more accurately and scalably than was possible before the development of deep learning. Those models that predated bioacoustic classification, mainly were based on handcrafted features, which were based on domain knowledge and difficult to generalize over various species or environments. Deep learning (Sanchez et al. 2021) can be trained to learn hierarchical representations that would make it able to identify patterns in the data very complex, hence becoming a game-changer in bioacoustic classification.

Deep learning is based in the first place on an artificial neural network, that is, it consists of

Input layers: Here data is fed into the network.

Hidden layers: These are multiple layers where the network performs intermediate calculations and feature extraction.

Output layers: Where the predictions or classifications happen.

The deep learning performance in bioacoustics will thus rely on its ability to capture automatically patterns of bioacoustic signals related to a particular species, behavior, or environmental factor.

2.2 Bioacoustic Data Types

Bioacoustic data are essentially audio recordings of animal vocalizations (Darras et al. 2019; Baldo and Mennill 2011). Generally, records are usually made by sensors such as microphones in terrestrial habitats, hydrophones in underwater habitats, and ultrasonic detectors in other habitats. Those are different depending upon environmental conditions, species being studied, and the technology applied for collecting the data. These major types of bioacoustic data that are applied to deep learning include

(i) **Raw Audio Signals:**

The raw, unfiltered sound waves captured from the environment. These signals are highly dimensional, and such signals in principle are challenging to process directly using neural networks because they convey relevant information about time. Generic audio data, mostly converted into some closer, more tractable forms, spectrograms, or MFCCs, capturing information about frequency over time.

(ii) **Spectrograms:**

Spectrograms are images that display how frequencies change over time across a spectrum for an audio signal. Converts into format resembling 2D images that make the audio extremely suitable to deep learning models, especially to CNN which suits applying on images. Sound is represented by spectrograms: time is along the *x*-axis, and frequency is along the *y*-axis; color or grayscale shading indicates the intensity of sound at each frequency. Quite often, a mel-scale frequency axis version is used in the variants of bioacoustic tasks and serves to provide the information in a meaningful representation of the bioacoustic data-not aligned with the human perception of the sound.

(iii) **Mel-Frequency Cepstral Coefficients:**

Offers one of the widely used feature representations in audio signal processing. This represents a compression in dimension of audio data whereby all the negligible characteristics of sound are retained. These coefficients describe the short-term power spectrum of a sound and are very useful in catching the timbre and pitch of animal vocalizations. MFCCs have conventionally been used in the approaches of traditional machine learning but can be applied on the deep learning pipelines for an efficient representation of the sound.

2.3 Quality of Bioacoustic Data

Bioacoustic data also holds of equal promise for applications, which exhibits some challenges to deep learning techniques:

(i) **High Variability:**

The noises from animals may span quite a wide range considering location, climatic condition, or variations between the species themselves. Birds of the same species might have different calls due to their area of operation. The same would also raise complexity in the classification task. Moreover, the calls may depend on the context where they are voiced (e.g., mating call vs. alarm call).

(ii) **Noise:**

Bioacoustic data often originate in a very heterogeneous, uncontrollable natural scenario with multiple potential noise sources masking calls from the animal of interest. Noise reduction techniques up to the state of the art and deep learning approaches able to extract robust features seem to be particularly useful at this step also.

(iii) **Data Skewness:**

It seems that the class imbalance in the datasets just means that some species—or vocalizations—are less represented. This will result in the majority classes getting overfit, and the models detecting rare species or events will fail. This can be optimized using data augmentation techniques, which include artificially increasing samples for underrepresented classes.

(iv) **Big Data Volume:**

There is a lot of data especially from bioacoustic recordings in almost all the monitoring projects in the long term. Huge such datasets require scalable models that have to be non-computationally expensive for large datasets.

2.4 Bioacoustic Deep Learning Pipeline

Deep learning in bioacoustic signals classification would entail a typical flowchart, as illustrated below Fig. 1.

(i) **Data Collection:**

All these are collected as bioacoustic data and later put into databases for processing. Field recordings differ from forest, ocean, or any urban area since different researchers specialize in different species.

(ii) **Preprocessing:**

Raw audio data usually needs some cleaning and preprocessing before an analysis can be received. Cleaning and preprocessing may include noise removal, cutting off silent parts of the audio, volume normalization, and transformation; for example, transformation of the audio signal into a spectrogram or into an MFCC representation. Sometimes the use of time-domain transformations is involved, like the Short-Time Fourier Transform can be obtained for frequency component, and helpful for the subsequent classification.

(iii) **Data Augmentation:**

This would apply some augmentation techniques (Abayomi-Alli et al. 2022), of which the former is time-shifting with a slight modification in position sounds and pitch-shifting, among others. These are just examples of how audio data may exhibit natural variations due to additions of background noises, among others. Meaning learning invariant features pretty well under that kind of transformation, generalizing well.

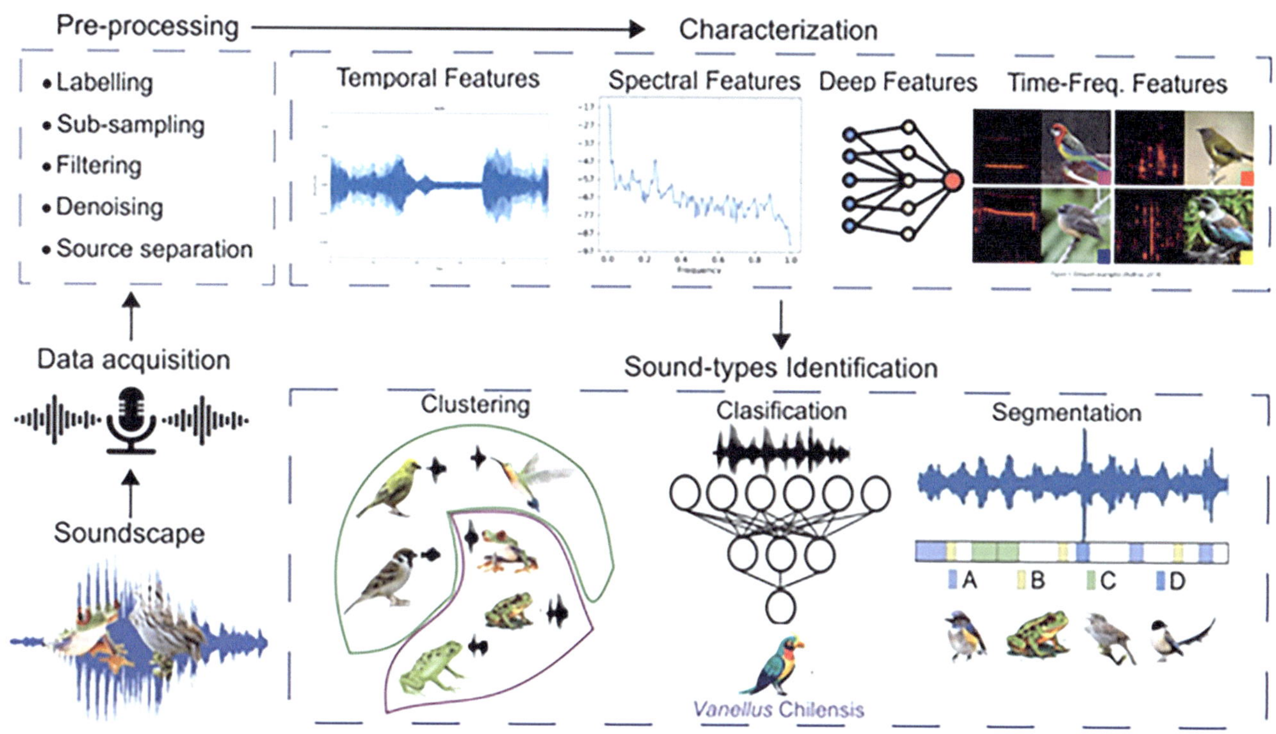

Fig. 1 Bioacoustic signals classification using deep learning

(iv) **Feature Extraction:**

Finally, classical methods of machine learning apply feature extraction. Pitch and spectral bandwidth are calculated in order to represent bioacoustic data in the form of human-made features. However, deep learning has automated (Swiston and Mennill 2009) the process of feature extraction. For example, in CNNs and RNNs, models learn to detect relevant features while training; hence, human intervention is reduced highly.

(v) **Model Selection:**

The nature of the task of bioacoustics dictates what type of deep learning model should be used. In case of spatial data, for instance, in the spectrograms, CNNs can be adopted (Sun et al. 2022). In case there is sequencing of the audio signals, like frequency change with time then RNN's or LSTMs can be employed. These are more complex models which involve hybrid models to combine both spatial and temporal feature extractions like the CNN-LSTM model.

(vi) **Training:**

A labeled dataset trains the deep model, after which the model finds patterns within those bioacoustic signals. Here, "training" refers to tuning of parameters on the model in order to minimize errors in classification. This step is very computationally expensive in terms of size and complexity of data. Transfer learning bypasses this by using pre-trained models as a starting point.

(vii) **Assessment:**

Testing is performed on test dataset in order to get measure for accuracy, precision, recall, etc., after training. Metrics provide information on how well a model might generalize to unseen data and how well it would perform with the challenges introduced by noise or class imbalance.

(viii) **Deployment:**

The trained model can be used directly in real-world applications for automatic classification of bioacoustic recordings. It could be deployed over field recordings, say, over real-time detection of specific calls of a certain species or over long durations analyzing large volumes of bioacoustic data collected from sensor networks.

2.5 Deep Learning Versus Traditional Machine Learning in Bioacoustics

The most significant advantage of deep learning over traditional machine learning methods is that it automatically extracts meaningful features from raw data. In the traditional methods, researchers had to handcraft features such

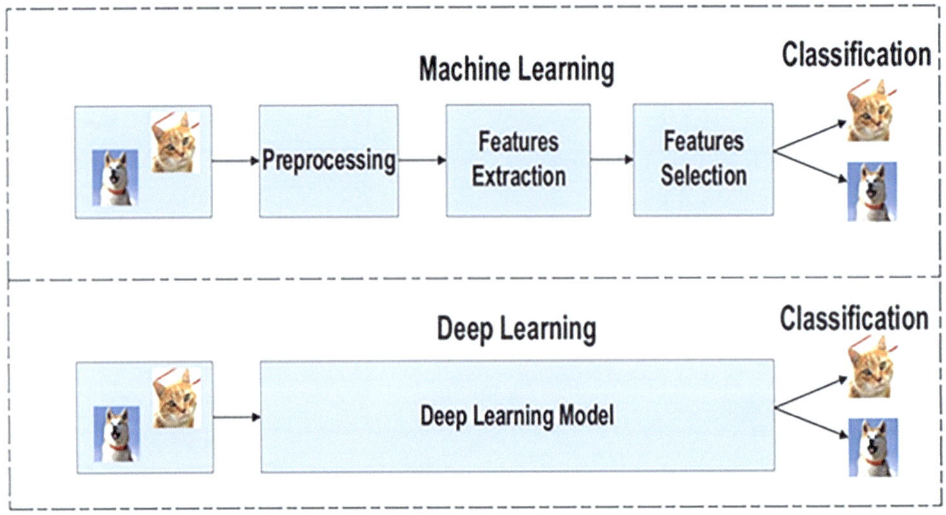

Fig. 2 Comparison between machine and deep learning

as formants, syllable duration, or spectral bandwidth. Such handcrafted features were based on a deep understanding of the species being studied and the acoustic properties of their vocalizations. These handcrafted features often failed to generalize across different environments or species as shown in below Fig. 2.

It does not handcraft feature engineering of inputs instead, because the algorithm learns to use the data for its decision in hierarchical representations. For example, one would likely be looking at deeper features which are being represented in some earlier layer while capturing such specific, higher-level features that would include species-specific call signatures.

Scaling large, complex data also favors deep learning models (Priyadarshani et al. 2018). Within the traditional regime, however, scalability of the model was constrained by the amount of time and expertise to extract meaningful features. Deep learning models can be trained on massive amounts of data and continue to improve their performance as more data becomes available.

3 Convolutional Neural Networks (CNNs) for Feature Extraction

3.1 Overview of Convolutional Neural Networks (CNNs)

Convolutional Neural Networks (CNNs) are a family of deep learning models particularly well-suited for spatial data, including images and audio spectrograms. Convolutional Neural Networks allow for the effective extraction of local patterns within input data, which is valuable for bioacoustic classification because the goal is to analyze the time–frequency representations, or spectra, of animal vocalizations. Hence, CNNs are increasingly being put to use in most audio and image-related processing operations primarily based on automatic detection (Xie et al. 2023; Mutanu et al. 2022) of key features from the spectrogram.

The core innovation of CNNs refers to the use of the convolutional layer: applying a filter or kernel across the input data to detect patterns based on edges, shapes, or frequency modulation, which should be crucial for assessing differences between various animal calls or behaviors in bioacoustic recordings. Unlike traditional machine learning techniques where feature extraction is manual and highly dependent on domain knowledge, the CNN learns these patterns automatically. Therefore, it is well-suited for large-scale and complex bioacoustic datasets.

They consist of the following parts:

Convolutional Layers: They transmit filters to recognize local patterns in input data.

Pooling Layers: This downsampled data by shrinking its dimension but preserving the information; this network is computationally efficient.

Fully Connected Layers: These layers connect all the features extracted by the convolutional layers for the final predictions.

Activation Functions: It needs activation functions like ReLU imposes nonlinearity on the network, which is capable of rich representation.

3.2 CNN Architecture in Bioacoustics

The other main bioacoustic application of CNNs is to spectrograms-a visualization of audio data for which time runs along the horizontal axis and frequency up the vertical axis and intensity by color or shading. Again, for these 2D representations, CNNs can work essentially as they would on images, picking up on patterns in the form of modulations, harmonic structures, and rhythms that are characteristic of certain species or behaviors.

A typical CNN for the classification of bioacoustic classification follows multiple layers of convolutions and pooling followed by layers of fully connected for class. The architecture can go all the way depending upon the task complexity and size but normally, for a good old CNN on bioacoustics, it follows some such:

(i) Input Layer: In most cases, the spectrogram or its equivalent time–frequency representation would be a mel-spectrogram, in which the audio is entered and whose input size would vary due to spectrogram resolution along with duration.
(ii) Convolutional layers apply a few filters or kernels over the input data, learning useful local patterns that might help it distinguish the type of calls. Often early layers focused on basic things like edges and simple frequency modulations; the deeper layers tune to specific harmonics or rhythms.
(iii) Pooling Layers: The pooling layers decrease the dimensionality of the data by extracting the most important features of each local region of the spectrogram. It reduces the computational cost of the network while retaining important information.
(iv) Fully Connected Layers: The flattened high-level features fed into fully connected layers where several convolution and pooling layers might have learned features combined to make a final decision for classification.
(v) Output Layer. This will hold the classification or detection result in this layer. In multi-class bioacoustic classification tasks, the output will be the actual probability distribution over all possible classes, like species or behaviors.

3.3 Feature Extraction with CNNs

This is where the actual strength of CNNs in bioacoustic classification happens: they automatically extract meaningful features from spectrograms. Which, of course, comes in very handy for identification in cases of

Frequency Modulations: Modulations in frequency as a function of time characterize the majority of animal vocalizations. Indeed, the upsweeps or down sweeps themselves often provide discriminative species cues in bird songs as well as in whale calls.

Harmonics: Most animal calls have harmonic structures and are therefore composed of frequencies that are multiples of some fundamental frequency. This renders harmonics very species-specific to be easily identified using CNNs.

Temporal Patterns: The temporal structure—repetitions or rhythm in vocalizations—is highly relevant in differentiating between forms of calls. For example, certain pulses of sound in predetermined time intervals of certain frog species could be extracted by CNN through the analysis of the spectrogram.

Background Noise: Although CNNs are essentially able to pay their attention to where it matters most in the input data, they can also train themselves to become a blur or to filter out the noise made of wind, rain and other environmental sounds.

Hierarchical learning enables CNNs to attain feature extraction. They pick simple features such as edges or low-frequency modulations at early layers, but more complex and abstract features, like harmonic structures or species-specific patterns, when the network is deeper. So, CNNs can easily build a robust representation of the bioacoustic data for purposes of classification.

For instance, in avian vocalizations classification, a CNN can be trained to classify into common elements of songs, including frequency upsweeps, syllables as well as pauses between calls. A CNN applied to marine mammal vocalizations would likely identify and model tonal and frequency modulation patterns of several whale and dolphin species.

3.4 Advantages of CNNs in Bioacoustic Classification

(i) **Automatic Feature Learning:**

While most other approaches require feature extraction in most cases, CNNs learn the features from the data itself. That is a plus in the case of bioacoustics because the data isn't only complex but also has variability in vocalizations in forms that may prove difficult to capture by good designs manually. For example, CNNs can even learn the task of learning classes that classify song from birds across different species with very tiny differences and classify different types of whales by their specific frequency range and pattern.

(ii) **Noise Robustness:**

The bioacoustic recording will always be noisy about wind noise, rainfall, and others in wild settings. CNNs learns patterns in a noise-robust fashion, therefore focus on local patterns and hence such a model will recognize patterns as relevant vocalization within background noises, and data augmentation that is adding of synthetic noises during training helps realize optimal use of such models and tune out as much as possible irrelevant background noises. That makes CNNs particularly effective for real-world bioacoustic applications, especially when recordings are noisy and imperfect.

(iii) **Scalability and Efficiency:**

CNNs are much cheaper computationally than other deep models of learning like RNNs; they can only do local operations since they cannot process the entire input sequence all at once. This makes them quite capable of processing large bioacoustic datasets that contain thousands, or even millions, of recordings. It contributes an extra reduction in dimensionality, and so the process runs fast without compromise in training or inference performance.

(iv) **Feature Transferability**

Features learned by CNNs in one bioacoustic task often generalize to other tasks. For example, if a CNN is trained to classify bird calls, it learns general patterns within acoustic data that will help classifying frog calls or whale songs. This is particularly valuable in bioacoustics because for many species there will be very little training data, and the hope is that transfer learning can partly mitigate performance degradations under these low-data regimes.

3.5 Case Studies and Applications

(i) **Bird Song Classification**

Probably the first obvious application area of CNNs to bioacoustics that can be named would be one on the classification of bird songs (Kotera and Phillott 2022). Given the fact that different species are voiced slightly differently, a CNN can be applied to identifying patterns in spectrograms of birds' calls for different species. Recently, researchers used CNNs on large-scale monitoring projects with the purpose to identify many different species from spectrogram recordings taken from forest areas. The network came to be aware of the diversity of species from how the songs were in pitches, syllable structures, and repetition. Researchers have been able to demonstrate that CNN may produce tremendous accuracy in comparing bird song over old methods especially in dense ecosystems, calls may often be noisy or overlapping.

(ii) **Marine Mammal Vocalization Analysis:**

Marine mammals such as dolphins and whales create different sounds, from clicks that work as echolocation tools to huge complex songs. A deep exploration was performed utilizing CNNs in underwater recordings classifying the marine mammals with each of these types of voices. In one study, CNNs were trained on spectrograms of whale calls and managed to distinguish between different whale species with high accuracy, even in noisy underwater environments. Besides species classification, CNNs are used in behavioral research and in the analysis of the temporal structure of calls to make inferences about behaviors from feeding and mating to navigation.

(iii) **Frog Call Identification:**

Most species of frogs possess distinct calls and are useful in population monitoring and estimating the biodiversity in that region. CNNs were used for classification of frog calls from their spectrograms where by time and frequency signatures may differ with the different species. It classifies frog calls automatically to allow the monitoring of frogs efficiently if their habitats cannot be reached.

3.6 Challenges and Limitations of CNNs

Although CNNs are very powerful tools in extracting bioacoustic features, they have drawbacks.

Challenges:

The inherent noisiness and complexity of data often arise from uncontrolled recording conditions accompanied by background noises of wind, water, or interference from other species sounds. The scarcity of data and imbalance between positive and negative samples occur quite frequently because of the effort that goes into recording and labeling bioacoustic data, and some species are available in very few quantities. Critical temporal dependencies in the bioacoustic signals, due to characteristic patterns, are really difficult for CNNs because they take spectrograms as static images. In addition to this, features in bioacoustic signals like harmonic structures or frequency modulations can be learned by a CNN without expert preprocessing only if relevant to some application. Recording equipment, environmental conditions, and species-specific variability make it difficult to generalize across variously different environments and species.

Limitations:

Key weaknesses in CNN for bioacoustic classification involve the need for more computing power: this usually makes the training as well as deployment more demanding especially in less resourceful systems. These models are usually more complicated to interpret and mostly become black boxes with decisions greatly reliant on lesser ecologically meaningful features but much more so on irrelevant patterns of input changes. They readily give in to noises and adversarial alterations in inputs, hence posing less real-world robustness. They also require large, annotated datasets to work well; hence, they are not as efficient in the much more common situation of small or imbalanced bioacoustic datasets. Finally, CNNs learned in one domain or region rarely transfer well to other environments and, therefore, have limited applicability to a broader scope of biodiversity monitoring tasks.

4 Recurrent Neural Networks (RNNs) for Feature Extraction

The last few years have seen the field of bioacoustics revolutionized by machine learning, primarily through deep learning techniques. Among the very promising models for bioacoustic classification is that of Recurrent Neural Networks (RNNs) (Bahmei et al. 2022). Neural networks are part of a category of networks specifically designed for modeling sequential data, hence putting it in a good position to handle time-series data such as animal vocalizations.

4.1 Overview and Working of Recurrent Neural Networks

RNNs are constructed such that it processes a sequence of data with feedback loops. That is, input from earlier time steps may feedback to inform the model of later time step predictions. This makes RNN suitable for applications with dependent data over time, like in bioacoustic signals.

An RNN works through the means of an internal state or memory that updates for every new input in the sequence. What determines the output at some time step is the current input and previous state. Thus, it's possible for RNNs to capture long-term dependencies in sequential data, highly useful in modeling time-varying signals like animal calls, songs, and other vocalizations.

However, there are also some limitations in a regular RNN regarding learning of long-term dependencies. During training, the gradients either vanish or explode thereby making it impossible for the network to learn long-range temporal patterns. This problem can be solved by introducing specific forms of RNNs; LSTM networks as well as GRUs. Actually, it is the controls of the flow of information which pass through the network to make it possible to obtain longer term dependencies.

4.2 Advantages of Applying RNNs in Bioacoustic Classification

With RNNs, superiority over traditional methods occurs in the following ways:

(i) Time Context: RNN is quite effective for modeling sequential data-a big feature in bioacoustic. Animal calls are typically temporal in nature, while any pattern of sound is invariably time-varying in their nature. The RNN can thus capture the dependence and hence can be better on classes of different calls or even vocalizations.
(ii) End-to-End Learning: RNNs do not need typical feature engineering that usually happens in normal approaches of machine learning. Instead, they can learn from the raw audio data by training themselves on the entire raw waveform or spectrogram of the sound. That makes it possible to automatically learn features that are needed for the purpose of classification and thereby requires very minimal human interference in a huge way.
(iii) Flexibility: RNNs, and specially LSTMs and GRUs, accommodate variable length input sequences reasonably well. This would prove highly beneficial in the bioacoustic application. In reality, there is a huge variation of the length of animal calls. The model does not expect fixed-length inputs; rather, it adapts with different forms of vocalization.
(iv) General Accuracy: In general, unless in those models that exploit time dependencies from a long period ahead, RNN generally achieves higher accuracy than classification which ignores sequential information. Some of the complex features that animal vocalizations would present are small and spread out over long periods of time.

4.3 Applications of RNNs in Bioacoustics

RNNs, especially LSTMs and GRUs, are applied in many bioacoustic classification tasks with really promising results. Some examples are

(i) **Bird Call Recognition**

Classifying songs can form the most common application of RNNs in bioacoustics. Birds are highly vociferous (Sanchez et al. 2021), and their vocalization has been so varied in nature that it required their songs to be classified in examples, for instance, of bird species by the sort of vocalization using an RNN, have sprung out of the time-scale patterns from their spectrograms. It feeds raw audio or spectrogram data to an RNN to learn the temporal structure of each species' song and differ it from other songs.

For example, LSTM-based model in which the classification has been done by a big dataset of bird songs. It has obtained high accuracy due to the huge outperformance compared to the traditional models of machine learning. This approach is also utilized even for the noisy classification of bird songs, a very common situation that occurs in real-world bioacoustic monitoring.

(ii) **Identify Whale Song**

Whales' complex songs aid their means of communication and navigation. Finding and categorizing the whales' songs are very imperative in understanding whale behavior or getting the population sizes. These temporal patterns in underwater recording of sound have been investigated using RNNs; with such an investigation, various specific types of whale song can be identified from some recordings of underwater sounds. The recordings of underwater sound normally contain a combination of different environmental noise and whale songs thus separating multiple sources of the sounds is very challenging.

(iii) **Bat Call Identification**

Bats use echolocation calls to see and locate food, and their characteristics vary among different species. Identification of bat calls is highly necessary to be done for the study of the population and behavior of bats. Applications of RNNs or specifically LSTMs were proved very useful for the classification of bat calls from audio clips.

(iv) **Completeness of Application of RNNs for Bioacoustic Classification**

Although the RNNs have many advantages, their application in bioacoustics also has some disadvantages:

Quality of data: Quality audio data must be available to train good quality RNN models. Environmental noise, poor signal-to-noise ratios, and limitations in field equipment would have not allowed the RNNs to learn good quality representation of the vocalizations.

Overfitting: RNNs, especially more complex variants, such as LSTMs, have such a vast number of parameters that overfitting becomes problematic if there is limited training data. Dropout and data augmentation will help reduce this overfitting, but unfortunately, those techniques need also to be carefully tuned themselves.

Computational Intensity: RNNs, in particular LSTMs, are computationally intensive when it comes to processing large amounts of data. They are quite demanding on the processing capacities and memory needed to train models, which may be a very challenging problem in field-based settings or by researchers who have minimal computing capacities.

Interpretability of the model: One of the criticisms of deep learning models (Karpakam et al. 2022; Malini et al. 2022), including RNN is lack of interpretability. Meaning, it is difficult to tell how an RNN ends up with its decision or which features were more central in the classification. Therefore, this is problematic in terms of finding ecological meaning for predictions done by the model.

5 LSTM Networks in Bioacoustic Classification

5.1 Introduction to LSTMs in Bioacoustics

Bioacoustic classification is not possible without Long Short-Term Memory (LSTM) networks. Advanced networks that can analyze acoustic signals from biological organisms classify them more efficiently. It is what bioacoustics is about: the study of soundscapes in animals and its environment. Some applications comprise species identification, behavior monitoring, and ecosystem health assessment. Since acoustic signals are sequential, traditional methods were put to a test because LSTMs (Kumar et al. 2018; Petmezas et al. 2022) can learn long-term dependencies and temporal patterns, revolutionizing this domain. An LSTM is very effective in discovering patterns over extended audio recordings because of the processing of the temporal data through memory cells and gating mechanisms. Whether it is identification of specific calls for a certain species of birds or analyzing echolocation signals for bats, LSTMs represent the best way to model such sequences.

5.2 Preprocessing for Bioacoustic Signals

Classification with bioacoustic data start from raw audio signal by applying LSTMs usually, bioacoustic data is noisy (Baker and Logue 2007; Mundry and Sommer 2004), there could be background noise due to the wind or some human activity; therefore, noise reduction techniques are applied. Converting the audio signal to meaningful features such as MFCCs or spectrogram hence extracting both time and frequency information. These are fed into LSTM layers where the network learns dependencies between consecutive audio frames. The variants now used more often than not bidirectional LSTMs, processing both sequences forward and backward, with a final layer to compute species labels, behavioral states, or patterns of an ecosystem, followed by some kind of softmax classifiers in a densely configured layer.

5.3 Key Advantages of LSTMs in Bioacoustics

One of the best benefits in applying LSTMs in bioacoustics is control of sequential, variable length character of bioacoustic signals. Signals are able to be variable but also of varying lengths as well as complexity-for example: whale calls, bird songs, or insect stridulations. Traditional machine learning models are challenging with such variability; LSTMs dynamically adapt to varying sequence lengths so they are very well-suited to real-world datasets. LSTMs also have many noise resilience properties because the mechanisms in them filter out noise and focus more on the distinctive patterns associated with the audio (Ventura et al. 2015). Thus, they are perfect for analyzing bioacoustic recordings captured in challenging environments, like those in a rainforest or underwater ecosystems.

5.4 Applications of LSTMs in Bioacoustic Classification

Bioacoustic classification application of LSTMs covers a very wide territory that has impactful coverage. Species identification comes under species differentiation between the bird and its song or between marine mammals and their underwater calls, among others. Behavioral analyses (Aide et al. 2013) have also been found in use because LSTMs are capable of predicting specific activities, such as foraging or mating, and even detect distress calls from the pattern of vocals. They also find applications in biodiversity monitoring through large-scale analysis of soundscapes with the purpose of inferring species richness or detecting alien species in an ecosystem. The filtering of noise as well as identifying anomalies gives them special merit for usage in conservation, including observing endangered species in real-time down to the illegal poaching activity through acoustic surveillance.

5.5 Challenges in Using LSTMs for Bioacoustics

Some of the disadvantages of LSTMs in bioacoustic classification are: the amount of labeled data on bioacoustic scenes is quite limited, and even more so for endangered or rare species. It generally causes overfitting where the model memorizes the training examples rather than generalize to new ones. Dropout is one kind of regularization addressing such an issue. Again, computational complexity can be a bottleneck in case of long audio sequences. Lightweight architectures and truncating sequences along with some optimization techniques solve such issues. Environmental noise and overlapping calls of various species further complicate the analysis and require sophisticated preprocessing and robust feature extraction methods.

5.6 Enhancements and Variants of LSTMs

A few architectures have been proposed recently to serve as a step beyond LSTMs for the task of classification of bioacoustics. Convolutional LSTMs are particularly robust when complex soundscapes have to be handled in processing spectrograms because strengths are incorporated from convolutional layers into LSTMs. The other important augmentation in these contexts is attention, where attention mechanisms help a standard LSTM to focus on appropriate segments of an audio input, and thereby improve accuracy besides interpretation. In this context, pre-trained LSTM models dependent on transfer learning have progressively been used for fine-tuning on small bioacoustic datasets. Data scarcity is being approached from different angles, making LSTMs an even more potent tool for processing bioacoustic data.

5.7 Real-World Use Cases of LSTMs in Bioacoustics

Real-world applications of LSTMs in bioacoustics are full of promise to revolutionize the field. For example, LSTMs analyze continuous forest soundscapes to automatically identify bird species in real-time monitoring projects. LSTMs classify whale and dolphin calls in marine environments, which assists in population tracking and migration studies. Behavioral patterns and tracking species diversity are revealed from the analysis of bat echolocation signals using LSTMs. LSTMs can analyze a soundscape in tropical rainforests to infer not only the presence of species but also environmental change over time. Applications like these open up flexibility and ability for LSTMs to solve complex bioacoustic challenges.

5.8 Future Directions for LSTMs in Bioacoustics

There are three futures of LSTMs in bioacoustics: real-time systems, multi-modal learning, and explainable AI.

It could eventually be deployed on edge devices such as acoustic sensors that monitor species and habitats in remote places without frequent human monitoring. The rich insights provided by acoustic data combined with other sensory inputs, such as temperature or camera data, will be further enhanced. Furthermore, interpretable LSTM models will help researchers and conservationists understand why certain patterns are classified in a certain way, thus establishing trust and aiding them in their decision-making within ecology. With all that is to come, LSTMs will play an even more important role in the comprehension and preservation of biodiversity.

5.9 Case Studies in Bioacoustic Classification Using LSTM

(i) **Bird Species Identification in Tropical Forests**

One such case study that can be cited is bird classification in tropical forests by an LSTM network, where the biological acoustic data is particularly challenging and noisy due to high environmental levels and overlapping calls. For analysis, the researchers used LSTM-based processing models for continuous audio recordings at these forest monitoring stations. Audio data in this case was pre-processed into spectrograms to capture as much temporal and frequency aspects as possible. LSTMs proved to be very accurate in distinguishing between hundreds of species of birds using the analysis of the temporal structure of their songs, outperforming traditional machine learning techniques, such as SVMs. Even greater improvements in classification performance were obtained by employing bidirectional LSTMs to account for the dependencies in forward and backward directions. This application especially helped the conservation efforts as the population of species could be monitored real time and even identify endangered or migratory birds in vast ecosystems with the least intervention of humans.

(ii) **Marine Mammal Detection and Classification**

This area has also been applied in classifying and monitoring marine mammals like whales and dolphins. Another related work made use of LSTMs for the processing of underwater acoustic records taken by the ocean hydrophones. They pre-filtered the MFCCs of the marine mammal vocalizations, which comprised echolocation clicks, whistles, and songs, and then sent it to the LSTM model. This is able to differentiate between blue whales, humpback whales, and bottlenose dolphins species by observation of the temporal patterns exhibited within the calls. This model, too, separated different species' vocalizations if it overlapped with others-a very hard task in traditional methods. The app is very useful in tracing migration patterns, monitoring health-related population data, and observing human impact-related issues of shipping and sonar on the marine ecosystem.

(iii) **Bat Echolocation Signal Analysis**

Another important case study is that of bats, which use echolocation signals for both navigation and hunting. Scientists used LSTMs to analyze a large number of high-frequency calls from various species of bats, so this model can classify several species and even the types of behavioral states, such as feeding or foraging. Preprocessed spectrograms were fed into LSTM models to absorb the complex and variable timing patterns of bat echolocation signals. It is successful to identify the species, even in noise-rich habitats such as urban or thick forests. It showed that LSTMs can indeed work with overlapping calls and produce meaningful patterns in order to correctly classify the species. These applications are essential for observing biodiversity, in tracking population changes, and ecological roles particularly as pollinators and pest controllers of bats.

(iv) **Soundscape Analysis for Ecosystem Monitoring**

LSTMs were also used in massive soundscapes analysis. As an example, it helped in monitoring the health of ecosystems and their biodiversity. Its use in analyzing continuous recordings such as animal vocalizations can help in the detection of species, measuring species richness, and detecting invasive species through learning with LSTM models about the relations between sounds over time in continuous recordings of environmental sounds such as wind and animal vocalizations. This improved the outcome's interpretability since attention mechanisms were indeed implemented into the LSTM models to focus on the most important parts of the audio sequences. That approach has provided invaluable insights for ecosystem dynamics and helped conserve biodiversity changes due to issues such as deforestation, climate changes, or human activities.

(v) **Illegal Wildlife Monitoring and Poaching Detection**

Another very important application of LSTMs in bioacoustics is anti-poaching. Researchers in Africa created LSTM-based systems that could analyze audio recordings from wildlife reserves with the aim of detecting gunshots or other illegal activities, such as chainsaw noise from logging. LSTM models were tested to classify the natural sounds of the animals against man-made poaching sounds. LSTM models were used in real-time, and there were detections such that rangers could intervene on illegal activities at the rates. In this case study, the applicability of LSTMs is reflected by bioacoustic data, where it not only helps in monitoring species but also saves endangered wildlife from humankind.

(vi) **Insect Bioacoustics for Agricultural Monitoring**

It has been applied to analyze insect bioacoustic signals in the agriculture domain as a means of monitoring pest populations and considering their effects on crops. According to reports, it has been reported that scientists have recorded grasshopper and other agricultural pest stridulations and processed the obtained data using LSTM models in which the LSTM network will classify different species and judge population densities based on the examination of the temporal patterns in calls of the insects. This application, developing early pest detection through timely interventions, minimizes pesticide use and increases crop yield. Applications of LSTMs in this field indicate that they are multi-functional, besides conservation applications.

(vii) **Monitoring Endangered Amphibians Through Acoustic Signals**

They have also been applied in the tracking of amphibians, mainly frogs, in which vocalizations are used as a means of conveying a mate. The latest application is the use of acoustic sensors in wetlands for acquiring frog call data for months. Audio data were noisy, obtained through the combination of overlapped calls and background noises which LSTMs were used in species classification and population dynamic changes modeling. This was key information for conservationists in ascertaining the impact of habitat loss and climate change on the populations of amphibians. The LSTMs in this study thus showed the utility in handling noisy and complex datasets, thus making it a vital resource in monitoring amphibians.

(viii) **Biodiversity Monitoring in Urban Areas**

These LSTM networks are used in monitoring biodiversity in classifying urban wildlife sounds where noise pollution complicates bioacoustic analysis. In a metropolitan area study, recordings were analyzed using LSTMs in urban parks with the influence of strong background noises due to traffic and human activity, but still identified bird and insect species. It made it a very effective solution for assessing the biodiversity of urban landscapes due to its ability to filter out irrelevant noise while focusing on the temporal structure of wildlife vocalizations. Therefore, this case study draws attention to the adaptability of LSTMs to challenging environments and their potential to inform strategies of urban conservation.

6 Conclusion

In this chapter, we extend the application of more advanced deep learning techniques, including Recurrent Neural Networks, Convolutional Neural Networks, and Long Short-Term Memory networks, toward improving bioacoustic classification. Applications like these have been taken from scientific understanding of biological sounds such as animal calls and naturalistic environmental soundscapes toward new approaches for species identification, behavioral monitoring, and assessing health of ecosystems. Even though the traditional techniques were experiencing problems as the acoustic data is generally messy and not ordered in nature, introduction of RNNs, CNNs, and LSTMs helped avoid such obstacles and provide strong, scalable, and accurate systems for bioacoustic classification.

Recurrent Neural Networks happen to be the fundamental method of sequential information, wherein bioacoustic classification has principally focused. Animal calls and environmental sounds tend to be inherently sequential. The contextual input in one part of the signal can affect the meaning within another. RNNs utilize such sequences by maintaining hidden states that evolve over the input. This enables prediction, taking into account relevant prior information.

RNNs capture the temporal dependencies in such acoustic sequences, especially bioacoustic applications, such as to identify a particular species of bird from its song or classifying bat echolocation signals. However, one of the prime disadvantages of traditional RNNs is that they are incapable of capturing long-term dependencies because of vanishing gradients. Despite such a limitation, RNNs led to more specific architectures eliminating such weaknesses. Advanced LSTMs were actually propounded precisely to overcome that very weakness of the basic RNNs. In an LSTM memory cells and gating mechanisms are presented, which enable to better control the flow of information over longer sequences of long-term dependencies than with simple RNNs. Because of this, LSTMs are especially appropriate to bioacoustic tasks concerning extended temporal patterns.

For instance, LSTMs may recall intricate patterns that crop up in species identification tasks by analyzing long recordings, such as the songs of whales or the frog calls, which a basic RNN cannot remember. Another type of bidirectional LSTMs, looking into both forward and reverse directions in a sequence, helped considerably in the bioacoustic classification application since these LSTMs can perceive the context from both previous and later audio frames, hence resulting in a highly increased accuracy for classification. Instead, the

CNNs focus on the spatial information hence perfect to be used while processing a spectrogram visual representation of sound. Actually, in spectrograms, frequency and the temporal aspect appear in both axes. For distinguishing various acoustic events, feature-rich representation features from sounds are necessary in such systems. They could then be applied to more complex patterns such as frequency bands or harmonic structures that might represent pitches or rhythmic changes in the call. They would, thus be highly useful in classifying many species of animals based on calls-for instance, bird species by learning features of different species of frequencies and rhythm of songs. Feature extraction CNNs can identify features not readily apparent in raw waveform data, thereby increasing their accuracy of classification.

A major breakthrough in the bioacoustic classification came from the merging of RNNs, CNNs, and LSTMs into hybrid models that effectively capture both spatial and temporal features of acoustic signals. The CNNs are most commonly used as feature extractors, where they look at spectrograms to recognize the relevant spatial patterns within the sound. Then, these features are forwarded to RNNs or LSTMs for processing the temporal aspects of an audio sequence. This fusion helps in understanding the data in a much more comprehensive way, where the system is able to recognize not only specific patterns of sounds but also how those patterns evolve over time. This allows CNN-LSTM to effectively perform call separation as well as classify individual calls belonging to separate species via the frequency characteristic and even through the signal temporal dynamics even when there might be different species calling within the acoustic environment.

Integration in a system of CNNs with LSTMs may often be used to benefit many real-world applications for bioacoustic classification. Species identification is probably the most important application. They have been used to analyze and classify bird calls, marine mammal sounds, and insect vocalizations. Hybrid CNN-LSTM models have proven very useful in identifying subtle differences in species' calls that might be hard for humans to distinguish in noisy, natural environments. The LSTM-based models have been used for the classification of specific activities, such as mating, foraging, or distress calls, thus giving insight into the behavior of species without direct observation. In large-scale biodiversity monitoring, deep learning models are being used to classify species from environmental soundscapes, thus providing valuable data regarding ecosystem health, species richness, and the detection of invasive species. These advanced deep learning techniques also profoundly affect the impact on conservation. Traditional species monitoring is often impossible or extremely expensive in most of the remote regions or in the protected habitats. Now with deep learning models being placed on edge devices like acoustic sensors or drones, scientists are able to remotely monitor wildlife and ecosystems in real time. This would be of utmost importance for detecting anti-poaching activities, monitoring status in endangered species, or observing migratory patterns with no interference by human beings. Here, power tools include LSTM that deals with handling long periods of recordings and CNN to extract spatial features.

However, despite the progress made, several challenges remain in applying deep learning techniques like RNNs, CNNs, and LSTMs to bioacoustic classification. The primary challenge is that labeled data is scarce, particularly for rare or endangered species. It takes time and is expensive to gather large amounts of labeled bioacoustic data, so it becomes challenging to train robust models. Some methods being pursued in overcoming this hurdle include data augmentation, transfer learning, and self-supervised learning. Furthermore, the calls of other species overlap, and the environmental noise creates complicated analysis requiring advanced preprocessing techniques and noise reduction methods for proper classification. Bioacoustic classification seems to have an emerging future in developing new improved deep learning techniques. Advancements in hybrid models, involving combinations of RNNs, CNNs, and LSTMs, that also apply attention mechanisms, will facilitate further classifications, including fine and accurate analyses.

For example, such models can attend to relevant parts of an audio sequence because the attention mechanisms let a model achieve better performance along with interpretability. Now as computing power increases, the deployment of such a model in real time into the edge devices is pretty more viable so that this type of remote species or ecosystem can be continuously monitored. Multi-modal learning where acoustic data assimilate other environmental data such as temperatures and even video footages further makes the field of bioacoustics a rich area, adding comprehensive views of these ecosystems. Conclusion The combination of RNNs, CNNs, and LSTMs represents a powerful and versatile approach to bioacoustic classification with significant improvements in species identification, behavioral monitoring, and ecosystem assessments. These advanced deep learning models revolutionize the field to make monitoring biodiversity more accurate, scalable, and in real time. Techniques developed will play a crucial role as the field develops with the rising challenges in conservation and ecosystem management to eventually conserve biodiversity for future generations.

References

Abayomi-Alli OO, Damaševičius R, Qazi A, Adedoyin-Olowe M, Misra S (2022) Data augmentation and deep learning methods in sound classification: a systematic review. Electronics 11(22):3795

Aide TM, Corrada-Bravo C, Campos-Cerqueira M, Milan C, Vega G, Alvarez R (2013) Real-time bioacoustics monitoring and automated species identification. PeerJ 1:e103

Bahmei B, Birmingham E, Arzanpour S (2022) CNN-RNN and data augmentation using deep convolutional generative adversarial network for environmental sound classification. IEEE Signal Process Lett 29:682–686

Baker MC, Logue DM (2007) A comparison of three noise reduction procedures applied to bird vocal signals. J Fieldornithol 78:240–253

Baldo S, Mennill DJ (2011) Vocal behavior of great curas-sows, a vulnerable neo tropical bird. J Field Ornithol 82:249–258

Darras K, Batáry P, Furnas BJ, Grass I, Mulyani YA, Tscharntke T (2019) Autonomous sound recording outperforms human observation for sampling birds: a systematic map and user guide. Ecol Appl 29(6):e01954

Gupta G, Kshirsagar M, Zhong M, Gholami S, Ferres JL (2021) Comparing recurrent convolutional neural networks for large scale bird species classification. Sci Rep 11(1):17085

Jeantet L, Dufourq E (2023) Improving deep learning acoustic classifiers with contextual information for wildlife monitoring. Eco Inform 77:102256

Karpakam S, Malini P, Kumar S, Vijayakumari G, Thirukkumaran R, Hanne LS (2022) Deep learning and fuzzy decision support system for visually impaired persons. IEEE xplore 7

Kotera MM, Phillott AD (2022) Calls for conservation: a review of bioacoustics monitoring with case studies from India. Asian J Environ Ecol 19(4):142–150

Kumar V, Gokulpriya D, Subharatha R, Dineash V (2018) Banana tall plant disease detection and classification using image processing and artificial neural network. Int J Adv Sci Eng Res 3(1):452–459

Malini P, Vijayakumari G, Ganesh C, Femina V, Ramani U, Basha SI (2022, October) A deep learning based framework implementation and artificial intelligence using optical communication. In: 3rd international conference on smart electronics and communication (ICOSEC). IEEE, pp 1077–1080

Manriquez PR, Kotz SA, Ravignani A, De Boer B (2024) Bioacoustic classification of a small dataset of mammalian vocalisations using deep learning. Bioacoustics 33(4):354–371

Mundry R, Sommer C (2004) Tonal vocalizations in a noisy environment: an approach to their semi-automatic analysis and examples of its application. An Acad Bras Ciênc 76:284–288

Mutanu L, Gohil J, Gupta K, Wagio P, Kotonya G (2022) A review of automated bioacoustics and general acoustics classification research. Sensors 22(21):8361

Petmezas G, Cheimariotis GA, Stefanopoulos L, Rocha B, Paiva RP, Katsaggelos AK, Maglaveras N (2022) Automated lung sound classification using a hybrid CNN-LSTM network and focal loss function. Sensors 22(3):1232

Potamitis I, Ntalampiras S, Jahn O, Riede K (2014) Automatic bird sound detection in long real-field recordings: applications and tools. Appl Acoust 80:1–9

Priyadarshani N, Marsland S, Castro I (2018) Automated birdsong recognition in complex acoustic environments: a review. J Avian Biol 49(5):jav-01447

Sanchez FJB, Hossain MR, English NB, Moore ST (2021) Bioacoustic classification of avian calls from raw sound waveforms with an open-source deep learning architecture. Sci Rep 11

Sun Y, Maeda TM, Solís-Lemus C, Pimentel-Alarcón D, Buřivalová Z (2022) Classification of animal sounds in a hyperdiverse rainforest using convolutional neural networks with data augmentation. Ecol Ind 145:109621

Swiston KA, Mennill DJ (2009) Comparison of manual and automated methods for identifying target sounds in audio recordings of pileated, pale-billed, and putative ivory-billed woodpeckers. J Field Ornithol 80:42–50

Ventura TM, de Oliveira AG, Ganchev TD, de Figueiredo JM, Jahn O, Marques MI, Schuchmann K-L (2015) Audio parameterization with robust frame selection for improved bird identification. Expert Syst Appl 42:8463–8471

Xie J, Zhong Y, Zhang J, Liu S, Ding C, Triantafyllopoulos A (2023) A review of automatic recognition technology for bird vocalizations in the deep learning era. Eco Inform 73:101927

Zhong M, LeBien J, Campos-Cerqueira M, Dodhia R, Ferres JL, Velev JP, Aide TM (2020) Multispecies bioacoustic classification using transfer learning of deep convolutional neural networks with pseudo-labeling. Appl Acoust 166:107375

Deep Neural Networks and Wavelet Transforms for Bioacoustic Classification

Wasswa Shafik

Abstract

This chapter proposes the utilization of expert systems such as deep neural networks for bioacoustics classification and detection with continuous wavelet transform features. Acoustic monitoring through bioacoustics and machine learning research has gained considerable attention from the scientific community since both tools may provide great help in constructing automatic systems that are able to identify and catalog acoustic recordings. These means help us reach faster analysis tools that can filter and identify target signals. This is an exciting task in bird bioacoustics, where biologists are given the difficult task of listening to the signal of interest and identifying bird species, for instance, but in other contexts, different acoustic signals might be of interest. There is an increasing number of electronically based devices to capture audio data, equipped with multiple microphones and capable of operating in several kinds of environments. Monitoring and processing these recordings have become more complex, especially in the acoustic domain. Acoustic alarms against wildfires or intrusive species, bioindicator extraction, monitoring of vocal animals, including whales and dolphins, undersea analysis, observation and conservation of birds of interest over cities, estimating populations using audio data, insect diversity and biodiversity estimation, and selective capture of wildlife calls are a few examples of ecological and applied science problems that are highly time-consuming. In addition, acoustic detection has an important role in urban and rural areas to prevent possible natural disasters. The chapter protocol aims to contribute to a quick, accurate, and reliable automated bioacoustic event classifier, primarily focusing on the identification of birds that are present in a given captured audio query.

Keywords

Bioacoustics · Artificial intelligence · Deep learning · Sustainable development · Feature engineering

1 Introduction

Bioacoustics is the study of sound production, dispersion, and reception by living organisms. Understanding biological sounds has applications in many fields, such as wildlife conservation, environmental assessment, monitoring, and husbandry. The main goal when training a model for bioacoustic classification is to have a precise model to aid in the pertinent classification for each scenario. In this context, bioacoustic data can be heavy, necessitating demanding analysis steps (Nolasco et al. 2023). This is where deep learning technologies excel, avoiding manual feature extraction. Bioacoustics deals with the study of sounds produced by living organisms such as birds, humans, and other mammals. The mechanisms used to produce them, their dissemination, and even their perception are studied, addressing scientific issues related to physiology, behavior, and ecology. Biological sounds can be utilized in study areas specific to ecologies, such as biodiversity, population dynamics, or energetics, supporting sustainable development targets of ensuring a sustainable future (Dufourq et al. 2022).

Their analysis can be utilized in species classification, detection of individuals, and study of the activity rhythms of these animals, to name a few, being today an essential tool in natural research, serving as a non-invasive method of sampling. With the uptick in data produced through acoustic surveys, the contestability among the community to provide

W. Shafik (✉)
School of Digital Science, Universiti Brunei Darussalam, Gadong, Brunei Darussalam
e-mail: wasswashafik@ieee.org

Dig Connectivity Research Laboratory (DCRLab), Kampala, Uganda

K. Subbaraj et al. (eds.), *Computational Bioacoustic Artificial Intelligence*, Sustainable Artificial Intelligence-Powered Applications, https://doi.org/10.1007/978-3-032-06088-4_3

the most precise automatic approaches to the domain has erupted (Dufourq et al. 2022). Investigating deep learning methods on acoustic data proved to be a successful avenue. Sound data has been utilized in tasks such as speech synthesis, music generation, cancerous tumor localization, or even during pandemic surveillance. In bioacoustics, they have been collaboratively used in several circumstances. Research questions addressed include the raw sound deep learning studies on biological sounds, the preprocessing applied, the type of architectures that were explored, utilized error metrics, and the software used in the proposed methods (Kirsebom et al. 2021; Shafik 2024a).

Deep learning for bioacoustic classification poses several clustering methodologies, including unsupervised, semi-supervised, or even transfer learning, as Fig. 1 presents different forms of expert systems. This chapter primarily focuses on the supervised methodology for the classification of audio samples. The term automatic speech recognition is also a study of acoustics. Deep learning is an approach applied for feature representation from audio data in these studies (Escobar-Amado et al. 2024). These methods can utilize deep learning either to reduce the data through unsupervised learning to classify samples of audio data or a hybrid of both. The following presents the existing methods proposed in audio classification in bioacoustics in the last five years. The selected fields are bird songs or animal calls, mainly birds. The obtained results from the deep learning strategies for bioacoustics are establishing a state-of-the-art in the domain (Ferrari et al. 2020; Shafik 2024b).

1.1 Definition and Scope of Bioacoustics

Given that there are no unique definitions for "bioacoustics," it can be considered to be the study of acoustic phenomena relating to the biology of living organisms. Thus, when bioacoustics focuses on signal production, the acoustic signal produced by the living organism is most frequently referred

Fig. 1 Expert systems

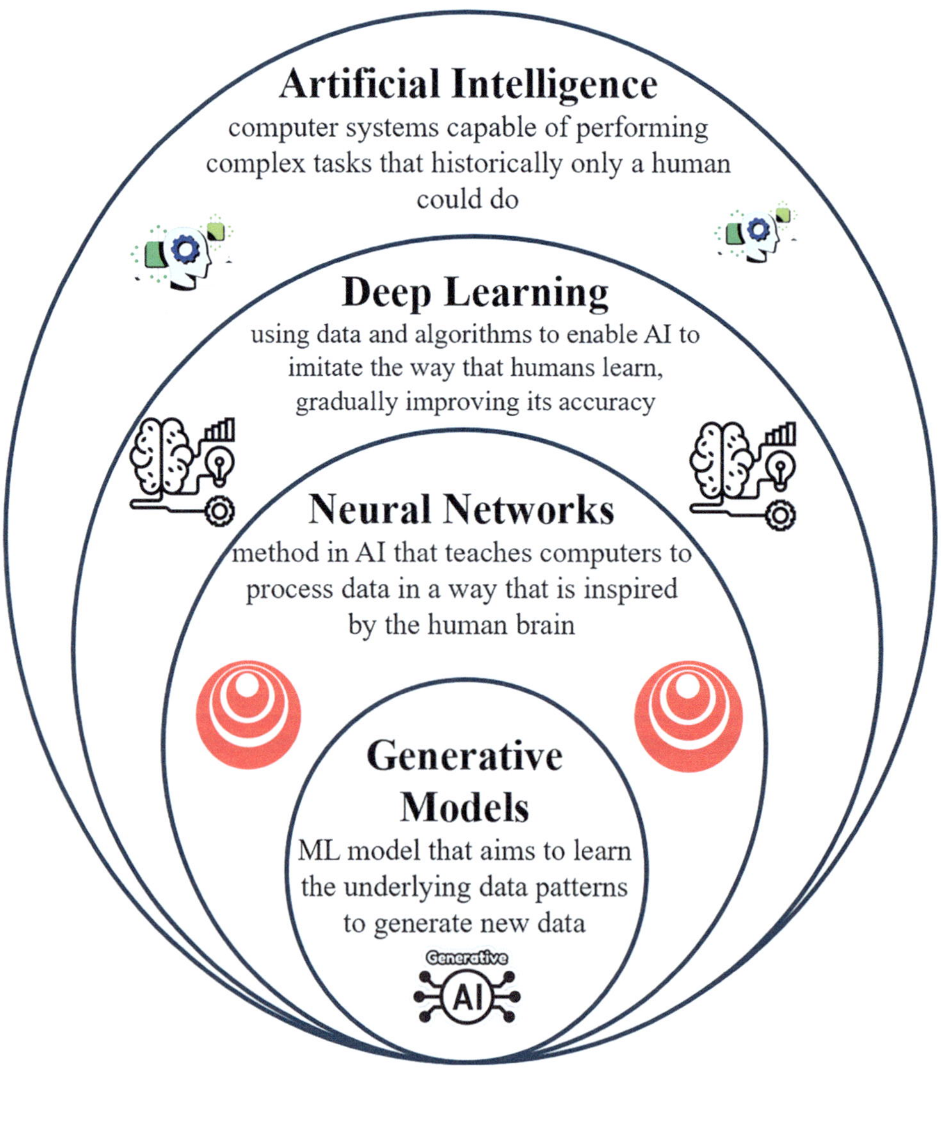

to as the call, although note that organisms considered to produce song are birds. When bioacoustics deals primarily with signal transmission, this is often referred to as sound propagation. When bioacoustics focuses on signal reception, such as the capture, recording, and playback of signals, the field is frequently referred to as bioacoustic engineering (Kiskin et al. 2020). In the context of the present text, bioacoustics is defined as the study of the acoustic signaling or communication processes that take place within and between living organisms, particularly but not only those processes observed in animals. Although deep learning applies to a host of different signal classification tasks across many scientific and engineering fields, this chapter tackles only those tasks that are essentially, or strongly, bioacoustics problems (Thomas 2019). The decision to offer a particular scientific focus on deep learning in bioacoustics was based on the number and increasing availability of large-scale labeled datasets that are currently available.

1.2 Importance of Bioacoustics in Ecology and Biodiversity Conservation

Bioacoustics is the science that studies sound production, modification, and transmission in animals, and its application is essential for many ecological studies and biodiversity conservation. This field of science is very interdisciplinary, using knowledge from several areas of physics, mathematics, and biology. It is widely used, among other tasks, for the delimitation of habitats, population monitoring, studying species interactions, and even for animal identification. Research in bioacoustics can be divided mainly into three themes: bioacoustics for basic research, bioacoustics for wildlife research, and bioacoustics for animal research (Goëau et al. 2016). The development of technology capable of acquiring long sound sequences can accelerate field research and improve scientific studies since it allows data acquisition for long periods without the need for the physical presence of researchers. However, the large amount of data restricts the use of traditional tools for classification, requiring the development and use of new techniques capable of processing extensive musical (sound) data. It is in this context that deep learning techniques appear as a valid alternative. Within deep learning, convolutional neural networks are the most widely used in practice, mainly due to their robustness for high-dimensional inputs and the possibility of training based on limited annotated data (Dufourq et al. 2021).

2 Fundamentals of Deep Learning

In this section, we briefly review current methods in training deep neural networks, which are the foundation for state-of-the-art performance in many machine learning tasks, including natural language processing, computer vision, and our domain of interest, bioacoustics classification. The goal of training a neural network model, as in the supervised setting for classification tasks in bioacoustics, is to minimize a given misclassification criterion or loss function on the training dataset (Escobar-Amado et al. 2024). One popular and currently the most effective method to train deep learning models is stochastic gradient descent-based optimization, updating the neural network parameters of the cost function gradient computed on a single training example or a small mini-batch of examples. The optimizer then updates the model parameters, typically using a decaying factor of the computed gradient based on the learning rate, moving the model parameters in the opposite direction of the gradient and minimizing the empirical risk on the training dataset (Kirsebom et al. 2021).

Increasing the training dataset size usually improves the generalization of the trained model, and deep neural network models are known to scale with large amounts of data. However, high computational requirements for dynamic learning of large-scale models present one of the systemic limitations of these architectures. Modern parallel computing resources, such as graphics processing units interfacing with libraries specifically designed to optimize tensor manipulation, make a variety of deep learning models efficiently trained, taking advantage of the massive parallelism offered by this computational infrastructure (Kiskin et al. 2020; Shafik 2024c). More powerful libraries commonly used to define deep learning models and the corresponding training loops are typically Keras, integrated with TensorFlow or the high-level PyTorch framework. Several components help classification, as depicted in Fig. 2.

2.1 Neural Networks and Deep Learning Architectures

Deep learning is a subfield of machine learning algorithms that are able to learn through hierarchical levels of representation. The representation uses the form of the input. Most machine learning algorithms require the features to be identified and separated from the data before they can be used to train the algorithm. However, deep learning is able to discover these features through the model training process. Neural networks are the basic computational models behind deep learning (Thomas 2019). Deep learning architectures

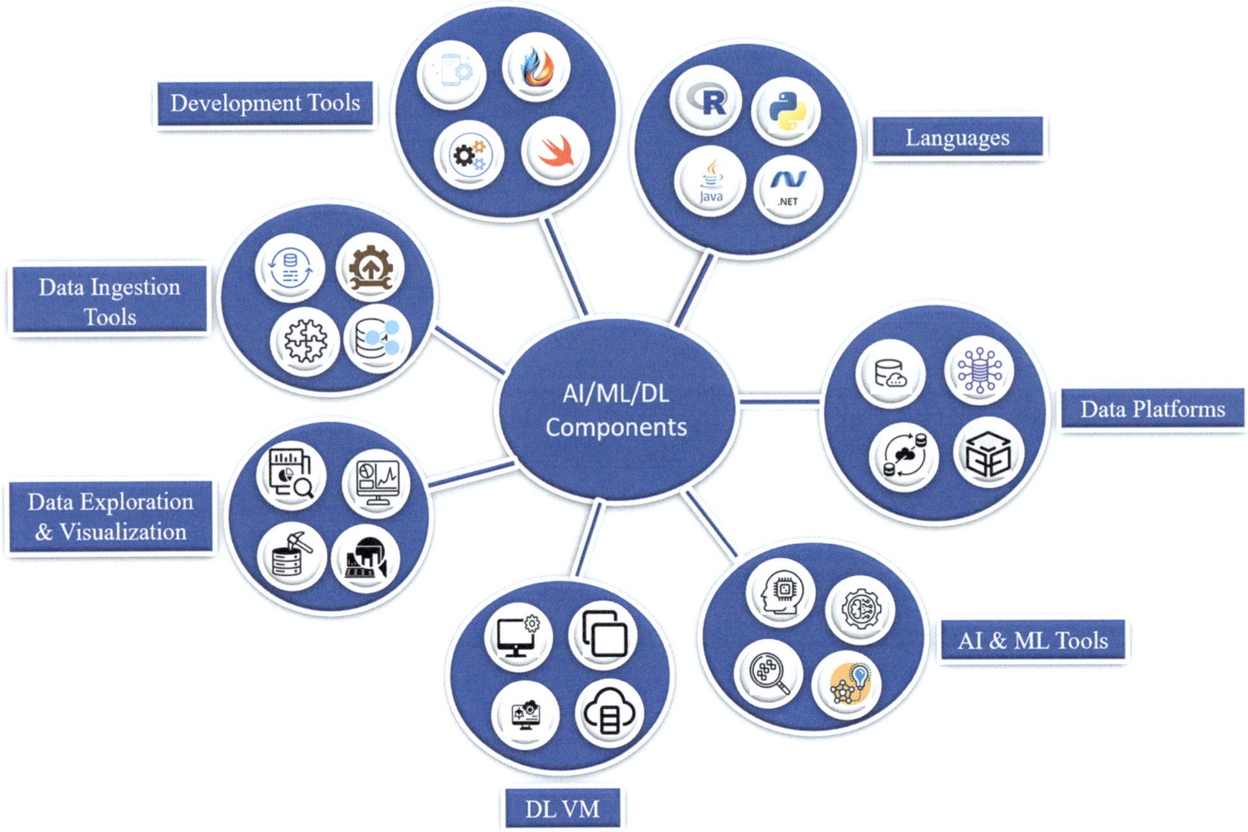

Fig. 2 Artificial intelligence/machine learning/deep learning components

have then evolved into network designs with many layers, called deep networks, where a top-layer representative uses the output or prediction of the model. Many of these new deep architectures are inspired by or mimic neuroscience in some ways, creating a revolution in the fields of speech recognition, object segmentation, and classification, as examples. In the context of bioacoustic classifiers, recent deep network architectures have been shown to improve results based on previous models based on audio significantly (Dufourq et al. 2021).

The fundamental component of deep networks, the neural networks, consists of a succession of many simple, interconnected processing units designed to simulate the way the human brain works. In neural networks, data values enter each processing unit and contribute to a decision or output value. Each unit is often referred to as an artificial neuron, organized in a different number of layers, including an input layer and an output layer. There are also one or more hidden layers; however, these cannot be directly observed. Each artificial neuron is connected to the artificial neuron of the next layer (Zhou et al. 2023). These are essentially connection weights, which indicate the strength of the connection that changes over time. Using the input value, weight, and activation function, a neuron will generate an output value. Thus, a network will classify a set of input signals in specific areas of the feature space and train different deep learning models with their corresponding regularizations. In deep networks, those that have a high number of layers are referred to as such (Thomas 2019). The training of deep networks is usually done using the backpropagation algorithm, which can take a long time as it requires a large amount of data; an illustration of how classification is done is presented in Fig. 3.

2.2 Training and Optimization Techniques

To achieve the best performance on these datasets using deep learning approaches, we will describe some of the most common training and optimization techniques. In general, the most common models employed in bioacoustic tasks are based on convolutional neural networks or recurrent neural networks, such as the long short-term memory model, GRU, and variants. Each of these models has particular characteristics; because of this, constructing an ensemble of these architectures is a common practice to enhance final

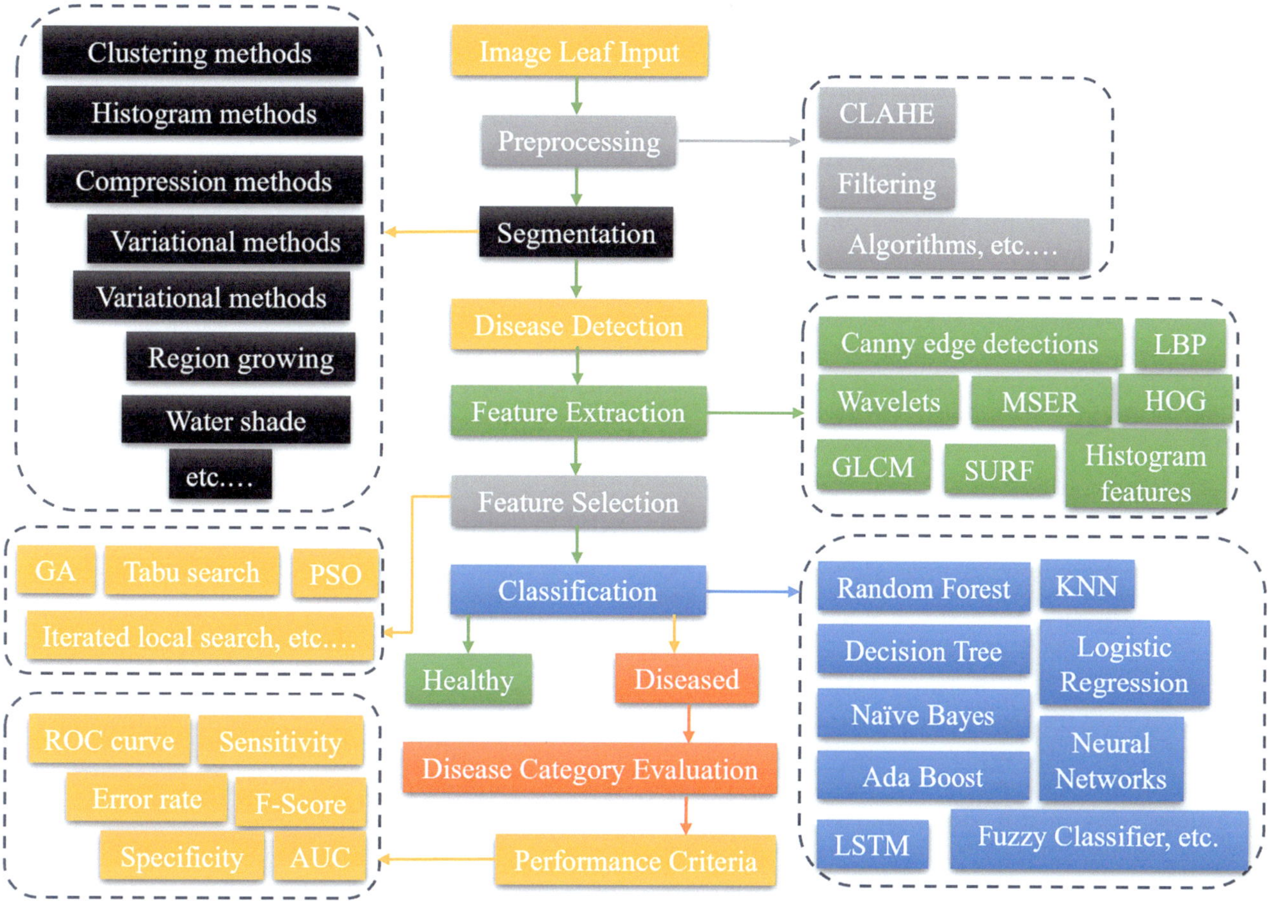

Fig. 3 Systematic diagram of deep learning classification

performance. Furthermore, the ensembled models produce different types of errors when working from the end-to-end approach, which can be beneficial to semantic segmentation or sequence-to-sequence tasks (Zhou et al. 2023). From the optimization standpoint, there is a need to regularize the weights of the deep model artificially, explicitly enforcing some inductive bias as constraints to allow better generalization for unseen examples at test time. In CNN models, weight decay over weight units and dropout mechanisms tend to be a strategy that stands out. Constrained optimization can be transformed into unconstrained optimization with the use of a parameter to which the function to be optimized is limited (Khalighifar et al. 2022). We have popular methods for the optimization of neural network architectures characterized by which refers to the feasibility envelope.

3 Preprocessing and Feature Extraction in Bioacoustics Data

In the case of bioacoustic classification tasks, it is particularly challenging to preprocess the data as well as extract meaningful features to train machine learning models to recognize specific types of sound. There is no well-established method in the preprocessing and feature extraction for bioacoustic data. Standard audio features usually include mel-frequency cepstral coefficients, power spectral density, mel-frequency cepstral coefficients (MFCCs) statistics, power spectral density statistics, perceptually weighted power spectral density, pitch, intensity, fundamental frequency, zero-crossing rate, wavelet transform, audio waveform, spectrogram, and others (Escobar-Amado et al. 2024). Furthermore, it has been found that a combination of different features can enhance the performance of a model given the appropriate combination. Data augmentation can be widely applicable in training sound classification models to increase the size and diversity of a training dataset. Note that it only artificially increases the size of the training dataset and does not correct for an imbalanced dataset or lack of diversity in the training data, which would make the model prone to overfitting and perform worse on an unseen test set (Kiskin et al. 2020).

3.1 Audio Signal Processing Techniques

A brief discussion of audio processing techniques that have had remarkable results in environmental sound classification, for example, urban noise, birdsongs, and voice sounds. Relevant concepts and audio features essential to the design and implementation of bioacoustics classification are also discussed. The audio signal manipulation and statistical description of the acoustic feature framework are the main paths of our analysis. Signal manipulations such as transformation, short-time Fourier transform, constant-Q transform, and wavelets, as well as acoustic analysis for time and frequency-domain features, are also considered (Thomas 2019). In audio processing, features allow capturing certain aspects of the audio, resulting in good features for the classification of such signals, especially in a supervised learning context. Yet, the calculation is computationally expensive, especially for real-time applications, although it has demonstrated significantly good classification performance for audio datasets larger than 5 GB. Uses filter banks and calculators to sum energy values in certain frequency ranges with an exponentially spaced mel-frequency scale and map those sums in a logarithmic space to denote the time-varying frequency pattern of an audio signal (Dufourq et al. 2021). Are kernel-based transfer functions in time and frequency domains that allow the separation of audio into time-scale decomposition basis?

3.2 Feature Extraction

Most recent models for bioacoustic audio analysis based on deep learning in feature extraction, we extract the most important information from input acoustic data. As audio signals carry minimal information, we have to extract the most meaningful representation of sound samples. If raw audio data is used directly as input for deep learning models, it may not enhance the performance of the model. However, by converting raw audio signals into useful features, the model may learn the most discriminating characteristics of the given problem and potentially improve performance. Depending on the requirement and problem, we can use a variety of feature extraction methods (Khalighifar et al. 2022). Following are some of the popular and widely used methods for feature extraction: mel-frequency cepstral coefficients—spectrogram—power spectral density—waveform. Feature extraction methods that we describe in this section also include both time and frequency characteristics, which can aid in distinguishing differences between sounds. Although audio files contain a wide range of information such as frequency, channel, time, room reverberation, amplitude, and so on, researchers use a combination of top *N* features because using all features may not always enhance performance, especially for real-time applications (Kirsebom et al. 2021; Thomas 2019). We require an in-depth investigation of the time complexity, features, and dimensionality of the method selected for feature extraction. An example of feature extraction has been illustrated, which shows audio clip features and makes an easy distinction between the voice of a frog, nightingale, cock, and cow. Frequently, researchers face the problem of an insufficient number or irrelevant choice of audio data features due to poor feature selection, which can lead to an unsatisfactory classification model. Several implementations and open-source libraries are available for extracting input features using various programming languages (Aradi and Karoly Varga 2023).

3.3 Spectrogram Generation and Feature Engineering

Sound (audio) is a sequence of waves recorded from various sources in the environment. A sound recording is transformed into a spectrogram as features using tools from digital signal processing. The spectrograms are generally based on the short-term Fourier transform. The generated images are a representation of each time frame of samples. The frequency is computed, and they are generally spaced with a Hamming window. As a result, we obtained information about different frequencies present in the audio recording and plotted for amplitude at different time steps. Mel-frequency cepstral coefficients are other commonly used features to represent audio recordings (Lauha et al. 2022). The log-scaled short-time Fourier transform has been shown to work well in practice. It is an intensity representation rather than the phase data. High-quality audio is more vulnerable to noise and sensitive to high dimensionality and computational complexity and requires more specific types of audio/microphone equipment, which makes using STFT of little relevance. The frequencies are converted into the mel-scale using the mel-frequency cepstral coefficients, and thus MFCCs are derived. Teager energy operators from the spectrographic region were selected from the spectrogram (Salamon et al. 2017). With the proposed FE-temporal, FE-related artifacts, FE-related calculation, and power spectral estimate, the images are then optimized using time series techniques.

3.4 Data Augmentation Depending to the Dataset

To tackle limited datasets, bioacousticians carrying out deep learning research frequently rely on data augmentation to increase the number of training samples and the diversity of the data. Various methods exist for creating augmented versions of samples in acoustic time series data. Time stretching and pitch shifts create variations of the original signals and have been employed with considerable success. Similarly, adding noise and reverberation for tasks with recorded audio data has led to improved model performance for bioacoustic classification. Data augmentation can also be useful to prevent overfitting due to the relatively small amount of training data often available for bioacoustic tasks (Clink et al. 2023). Time-domain augmentation that preserves the original strength of the audio information or adds reasonable variations can lead to consistent results across bioacoustic classification datasets.

An overview of methods and gain percentages through augmentation is presented by tailored test cases based on four different classification tasks, which include airborne and hydrophone classification. Augmented audio snippets were created by conserving the temporal structure and specific characteristics of each input stimulus. Among the five test cases/modalities in each bioacoustic classification task, augmentation demonstrated the highest gain in the correct classification of positive and negative examples in three modalities/tasks, as follows. Preprocessing tools have also been developed to allow end users to specify how they would like their data augmented (Morgan and Braasch 2020). Quantification of the augmentation procedures showed that increasing the number of augmented data increases the percentage gain, and different concerns influence the classification process, as sampled in Fig. 4.

Fig. 4 Data quality

3.5 Normalization

Audio data deposited in bioacoustics repositories often come from different recorders, which may have different sampling rates. Sample rates are usually greater than the minimum requirement of the domain experts. Thus, in applications, audio data needs to be down-sampled first. After this, audio data need to be normalized to the same singer for training and validation. Generally, this procedure aims to reduce the effects of confounding factors, such as temporal and spectral features, that help in classifying audio signals, making the classification method more robust and able to transfer between experiments. Nevertheless, the confounding factors are beyond the scope of this paper (Bergler et al. 2019). Many optimization algorithms, such as gradient descent, work better with standardized or normalized input data. This is because unstandardized or unnormalized data mean that the optimization algorithm cannot use faster convergence rates in whichever dimension. Speeding up convergence is particularly important in neural network training, where computational resources such as time and memory are usually in short supply.

Based on this, it is usually part of best practice to normalize features as this speeds up learning, helps avoid the issue of vanishing gradients, avoids numerical difficulties, and improves the shape of the optimization cost function of the network. Thus, this also improves the convergence property of the weights in the network (Lauha et al. 2022). Normalization typically ranges from simple rescaling to zero mean and unit variance. Min–max scaling is the simplest method where the raw data is scaled to a fixed range. Other scaling and normalization techniques include Z-score normalization with zero mean and unit variance. Normalization is essential when the output data is close to 0 or when the data does not have a Gaussian distribution. Analysis under different noisy environments shows that the classifier convergence speed in feature learning is the quickest under normalization. A lack of normalization can lead to the network's predictive accuracy and recall decreasing. If low-level feature extraction is poor and the input is not normalized, then the network will not learn effectively. Moreover, when using transfer learning on small training or validation datasets, particularly with data augmentation (Khalighifar et al. 2022). This provides an analysis of preliminary experiment results showing that deep learning in bioacoustics fundamentally requires input audio data to be normalized.

4 Training Deep Learning Models

The dataset preparation step is essential for the training phase. It includes the data cleanup step as well as the split of the data into a training, validation, and test set. The model to be trained can be chosen from many state-of-the-art acoustic models used in environmental sound classification. In bioacoustics, however, it has been shown that time–frequency representations of the data perform well for a variety of different tasks. The loss function plays an important role in guiding the optimization of the model (Aradi and Karoly Varga 2023). It quantifies the degree of dissimilarity between the network's predicted output and the ground truth target values. There are many different types of loss functions available, each of them being used in a specific case for a specific problem. Optimization algorithms are then used to change the parameters of a network such that the loss calculated at the output of the network decreases. The way this is iteratively done works. These algorithms are able to find good values for the parameters of the network to allow learning (Lauha et al. 2022). Some of the most commonly used optimization algorithms for neural networks include Stochastic Gradient Descent, Adagrad, RMSProp, or Adam. Some of the most commonly used regularization techniques include dropout, weight noise, weight decrease, or variations of L2. In practice, dropout is typically used in most neural network architectures due to its simplicity and efficiency. In the context of training deep learning models to classify bioacoustic events, it is important to choose an appropriate batch size, learning rate, and evaluation protocol for the model (Clink et al. 2023). The chosen batch size should be large enough for a significant number of training samples to be passed to the model during each iteration yet small enough to fit into the memory of a graphics processing unit. It is important to continuously monitor the learning process before any model is considered to have been fully trained in order to ensure that the model generalizes its performance. In practice, this involves training a model, evaluating it on a separate validation set, and stopping the training if an improvement is not seen over some specified number of epochs (Morgan and Braasch 2020). For sampling-based protocols, making sure that the test and validation sets are diverse is especially crucial.

4.1 Loss Functions

Good loss functions are essential to training deep neural networks effectively. In selecting a loss function, one needs to consider the characteristics of the classification task: Is the output discrete? If so, how many classes are there? Are the classes mutually exclusive? It is important to make sure that the chosen loss function aligns well with the criterion that is desired to be optimized as well. In most cases, the choice of loss function may, in fact, be influenced by the metric used for evaluation since these two are closely related. In this subsection, we motivate our design choices for the suggested application of our models and give the mathematical foundation and implementation of the loss functions, possibly together with rules on weighting for imbalanced datasets (Khalighifar et al. 2022). Loss functions provide a measure of the error for the predictions made by a model. By backpropagating this error, the model adjusts its parameters in an attempt to minimize the loss of the training data. The choice of loss function plays a crucial role in the convergence speed of the model and its generalization ability. As such, the properties of the dataset at hand should be taken into account when selecting the loss function. For regression tasks, one could use a mean squared error loss if the solution has unbounded ranges or absolute errors if the solution is bounded. For classification tasks, one of the most popular choices is the cross-entropy loss, which is used to train most modern deep learning models for multi-class classification (Aradi and Karoly Varga 2023). Behind cross-entropy loss, the softer classes are given higher weights, thereby increasing the influence of these data points on the total objective. For binary classification problems, the binary cross-entropy loss is commonly used. These options depend on whether or not the classes are mutually exclusive. When the dataset is imbalanced, a more favorable loss would weigh down the influence of the majority class. Outside of these loss functions, other examples include margin-based loss functions for discriminative embeddings or losses involving memory mechanisms. Practical examples of the loss functions follow (Lauha et al. 2022).

4.2 Optimization Algorithms

This play a crucial role in deep learning models to perform model learning more effectively. Optimization refers to searching for the best route from the current position to reach a certain goal. In the context of deep learning models, the goal is to minimize the loss functions to get the best possible performance. During the training process, variations of weights need to be updated in order to guarantee that predictions from the model are getting closer to the true value, and the main role of optimization algorithms is to determine the size, direction, and frequency of how the weights should be updated. In turn, the performance of the deep learning model is impacted by the selection of optimization algorithms (Salamon et al. 2017). One of the well-known and most frequently used optimization algorithms in the deep learning field is the Stochastic Gradient Descent. Unfortunately, the SGD algorithm has several challenges despite its simplicity; for example, the choice of learning rate is a challenging task and is sensitive to the selection of the learning rate for convergence. Currently, the algorithm's performance can be enhanced by determining the specific learning rates for the weights as implemented in the Adam optimizer. The Adam optimizer updates the average size of gradients for training separately as per the training steps. As a result, it minimizes the components that have faster convergence rates in comparison with the components that have slower convergence rates in individual training steps (Clink et al. 2023). Overall, the Adam optimizer has been found to have better performance in terms of both computational efficiency and convergence rate, depending on the dataset as well as network complexity over different models, learning rates, and batch sizes.

4.3 Regularization Techniques

Regularization techniques, together with appropriate network architecture, are crucial to ensuring that deep learning models generalize well to perform classification of audio data in the bioacoustic domain. Regularization aims to mitigate the effects of overfitting by imposing penalties on the learning model's complexity. This discourages the model from learning the noise present in the training data and ensures the learned parameters of the model can classify unseen examples correctly. Common regularization techniques include L1 and L2 regularization, dropout layers, and early stopping (Ruff et al. 2020). L1 and L2 regularization add a term to the objective function of the network that penalizes the weight magnitudes. Both L1 and L2 regularization enforce sparsity and prohibit very large weights. Dropout layers temporarily drop a random subset of the network units during training, thus forcing the network to learn more robust features. Dropout layers can also be seen as an ensemble method given that different subsets of the network are trained on the same input data. Early stopping terminates training after the accuracy on a validation dataset ceases to improve (Prince et al. 2019). A typical scenario where regularization techniques are crucial is when the collected bioacoustics training data is highly specific to the environment where it was collected. When developing models for real-world applications, it is vital that the models are able to tolerate input from a wide range of recording environments, each with their characteristic sounds. Data collected from multiple recording sites around a study area are often used to build such models (Padovese et al. 2023). Models trained using this multi-site data are more exposed to environmental noise and contain a level of noise that is more representative of field conditions than models trained on data collected from a single site. However, interpreting the multi-site data from different biological points of view is more difficult than working with data collected from a single site.

Classifying unseen data correctly using single-site acoustic monitoring is challenging given that classification accuracy can be degraded by overall low numbers of training examples and low SNR (Dias et al. 2022). To balance our discussion, we mention that some properties of regularization techniques can lead to underfitting on the unseen dataset. Most of the regularizations require selection of appropriate parameters and can be computationally intensive. The values of these parameters can directly affect the training process, and choosing the right values may not be simple and generally requires further data-dependent parameter assessment. Different parameters of the regularization techniques affect the training in different ways (Bravo Sanchez et al. 2021). For example, higher dropout rates make the learning process noisier and may slow down the convergence but may in some other cases help the optimizer to escape from local minima or avoid overfitting. Using large values for regularization parameters like weight decay can lead to underfitting, a very important aspect to bear in mind in bioacoustics, as it directly affects the classifier's performance when the data is limited by its nature in some field applications, and its associated uncertainty can vary greatly from one site to another (Michaud et al. 2023). There are also computational aspects to take into account depending on the application.

5 Deep Learning Models for Bioacoustics Classification

Deep learning models for classification, such as convolutional neural networks and recurrent neural networks, are applied to bioacoustics signal classification. We also present best practices that can be added to an existing deep learning model, and this will be beneficial to the community. Even though a recurrent neural network is capable of capturing long-term information, the output of the RNN model is a mean pool generated from the frames of the input signal to reduce the computational load, which results in neglecting some significant regions of the signal (Khalighifar et al. 2022). Unlike a convolutional neural network, a recurrent neural network is best applied to sequence classification tasks such as emotion recognition, autoregressive audio signal generation, and automatic lyrics transcription. To improve the results of our bioacoustics classification study, instead of proposing complex solutions from scratch, we add some best practices or improvements to a baseline CNN or RNN. We found that spectrogram representation outperforms other representations. For this reason, it is primarily used in the audio analysis domain (Ferrari et al. 2020). However, for the duration of the signal at time–frequency, this representation is restricted by a fixed size of the analysis window. Additionally, a CNN or RNN needs to have a fixed input size that does not include frameshift operations. A consecutive batch of training samples receives a batch size with a fixed sequence length similar to RNNs. These are unaligned and change the energy separation of context windows. The output of the classifier is used to provide the decision function of the method. The recurrent neural network and convolutional neural network models often assume a fixed-sized frame input, meaning the lengths of all input sequences are assumed to be the same (Thomas 2019).

5.1 Convolutional Neural Networks (CNNs) for Audio Data

The building block of CNN networks is convolutional filters. A wide range of settings of the neural network, such as the number of layers, the specific parameters of the convolutional layer, and others should be well chosen through a validation process for a given dataset. The challenge of using CNNs to classify bioacoustic data is the expensive resources that the wide range of potential settings may consume. Nevertheless, because of their simplicity and efficiency, CNNs have been widely used for bioacoustic classification tasks, leading to a variety of works, especially during a specific period (Escobar-Amado et al. 2024). Another interesting work involved designing a series of networks, such as relatively small networks focused on the aforementioned bioacoustics taxonomy classification challenge and other more sophisticated networks aimed at distinguishing cross-species with a wide frequency range between them, evaluating the different results based on the ROC curve and the area under the curve. Further innovations on the architectural aspects of CNNs include the use of tree-shaped filter design that greatly and flexibly alleviates the fact that convolutions lose resolution and frequency and time decoupling. Other innovative advanced scenarios aim to build a detection network guided by known audio events to detect mixtures, birds, insects, etc. (Dufourq et al. 2021). Another interesting work used a specific dataset for the detection of various types of audio events for a technological merger proposal.

5.2 Recurrent Neural Networks (RNNs) and Long Short-Term Memory (LSTM) Networks

These are neural networks that have connections feeding the output of a layer back to the input of the same layer, either directly to itself or to its previous or later layer. There are single sequence-to-label RNNs and windowed sequence-to-label RNNs. Their architecture includes simple cells that, hierarchically, transform signals that arrive in the form of time series. Their cells can be implemented using several strategies. For event-based bioacoustic classification, recurrent neural networks with long short-term memory (LSTM) cells are recommended (Lauha et al. 2022). LSTMs have been applied in many machine learning problems, especially those aiming to model sequential data. They have some very desirable characteristics, such as avoiding the vanishing gradient problem, and are designed to detect patterns in sequences when they are clean and when various elements, including time steps, are missing. A large part of the RNN instability comes from the sequential nature of SGD, which makes it hard to clean up the potentially big issues before feeding them throughout the RNN. The application of LSTMs deals with that issue, which is quite relevant for both single sequence-to-label RNNs and windowed sequence-to-label RNNs (Clink et al. 2023).

One of the advantages of RNNs is their ability to handle time sequences of varying lengths by processing them one at a time. However, by design, sequence processing RNNs, including single sequence-to-label RNNs, do not have access to other sequences in the batch when the entire time sequence is used as input. One potential solution to this issue is to use windowed sequence-to-label RNNs. Since the computation is done on sub-sequences of a fixed length, sequences collected in a batch can be of variable length. As a result, the entire batch can be processed in one go in GPU memory (Bergler et al. 2019). Moreover, the state or signals of the representations for previous sequences in the batch can be used to inform the representation of sequences to come, as long as the updates for all sequences in all inferences are indeed applied. One of the biggest problems with RNNs is that for time series with a window size, their convergence is extremely slow. At least, this is a known issue for the vanilla RNN. Other cell types may not have that issue. As far as we know, for time series scheduling during training, "scratch" training is needed to reach desired levels of accuracy. For this, the RNN certainly takes longer to make predictions because the temporally local processing rules are intrinsic. So, constrained global and maybe cross-global communications will only grow with time, slowing global processing and transient response (Prince et al. 2019). The source of this extra burden may lie in the nature of the vanishing gradient and the necessity to pass the state of all cells with time—much like with a scanner of a generator. This problem is believed to explain why training takes so long for longer window sizes using sequential SGD.

6 Transfer Learning and Domain Adaptation in Bioacoustics

Deep learning has seen a strong surge in popular use cases from image, video, audio, and natural language processing. These models seem to be amazingly good at capturing features that predict relevant outputs related to the trained domain or task. When a pre-trained model on a different domain or task is used to extract features, the process is called transfer learning. Deep learning models usually have a huge number of latent features that are good at capturing multiple levels of representation. Some of these features have been shown to be generalizable across domains within an application (Dias et al. 2022; Koops et al. 2015). Using strong pre-trained models to extract these generalizable features strongly improves performance in a related domain. The nature of pre-trained models is what makes it possible to use them to train new kinds of datasets. The hierarchy in CNNs can learn representations that are independent of the dataset and can be used in datasets that are different from the one the model was initially trained on. Continued advancement in computer vision fields has made available the pre-trained models on more and larger datasets. This has led to the models being able to capture more generic representations. The models also vary greatly in architectures and are usually used in four ways for transfer learning: as feature extractors, fine-tuning the last layers, transfer learning with a different task, and as pre-trained model initialization in unsupervised feature learning (Michaud et al. 2023). Domain adaptation occurs when the target domain is shifted but relevant for the task and also not the same as the original domain where the pre-trained model was trained. It is different from transfer learning because it does not use the pre-trained weights of the last layers on the new set of images. Domain adaptation requires the inner representations of the model to be specific to the task and the target domain of interest. Deploying a pre-trained model requires many layers for the optimization of the target domain or dataset. When the new dataset used for a biological classification domain contains the category of interest, domain adaptation is considered, and when it is worked on without including categories, it is a type of transfer learning (Morgan and Braasch 2021). A successful adaptation requires the dataset to be of the same format and size.

6.1 Concepts and Techniques for Transfer Learning

This method of adapting pre-trained models for problems other than the ones they were originally built for allows them to be trained with less data and computational effort than training from scratch, while still getting highly accurate results. Pre-training, in the domain of deep learning, means training the model with a large dataset made up of authentic and varied samples from within its task domain so that it can learn in a relatively unsupervised manner about a wide variety of feature patterns present in that domain that probably contribute to solution tasks, especially those of supervised learning, without knowledge of their labels (Bermant et al. 2019). This allows for knowledge and experience of the model to be stored in its inner layers, referred to as lower or deep layers, in the form of mathematical weights that can be transferred to new layers that are built to accommodate purposeful new tasks, which are referred to as upper or task-specific layers. In practice, many researchers will flatten or significantly restrict the learnable aspects of a model, replacing the upper layers with new ones, or fixing task-specific layers at a stationary state by setting the learnable weights to non-trainable values, or altered values of fixed and processed trained embedding to deactivate learnable weights, labels, or some layers, also by freezing kernel layers (Lauha et al. 2022). Then, novel experimental designs are set up to make use of initially locked low-level information at the start of a procedure and to gradually unlock it step by step as typically progresses by means of exploration linking and task-specific learning or fine-tuning procedures, whose patterns can be specialized for several applied sciences applications. Thus, the models can be optimized for the problem's demands while still retaining much of the knowledge extracted from diverse sources in their first layers (Escobar-Amado et al. 2024; Kirsebom et al. 2021).

6.2 Adapting Pre-trained Models to Bioacoustics Datasets

Transfer learning is a deep learning approach widely used when the amount of annotated training data is scarce. By transferring the learned features from a dataset for a large, representative, and general domain, the retraining time for the target domain is significantly accelerated. Currently, it is highly recommended to take a pre-trained model and adjust the network layers to match the characteristics of the target dataset and fine-tune the model to converge to better performance with new data. For acoustic classification, transfer learning can involve datasets for a similar biological domain, human speech, sounds, images, and data that are easy to collect, such as using websites for image search and including the image labels as source-level information (Kiskin et al. 2020). In recent years, there has been an increase in the adaptation of computer vision techniques to bioacoustic analysis. The starting point for this approach is to consider our input sound spectrogram as an image, which can guide the reuse of popular CNN architectures trained on large image datasets by simply defining an analogous frequency spectrogram to the red, green, and blue color channels. These universally recognized architectures can be seen as powerful feature extractors that can distinguish patterns from different image datasets, which led to the development of transfer learning: we should use only the prelearned feature extractor and add at the end a small neural network that learns how to classify the specific task of our data (Thomas 2019). Moreover, as the learned features of datasets with complex contents are frequently conducive to the extraction of transferable features in distinct domains, the method allows for an increase in model generalization power by the transfer of learning through domain adaptation.

7 Evaluation Metrics for Bioacoustics Classification

When presenting bioacoustics classification models, it is common to provide the sensitivity, precision, and F1 score. These metrics are also measured by the researchers of the most related works. However, to our knowledge, no previous work has shown the use of all these metrics, or AUC and the loss function of the model, to indicate how the model behaves in various degrees of confidence. We believe that these metrics are important because an AI for use in the real world should not predict only with high confidence. We must know what the possible errors are and their degree of confidence. Researchers should report multiple evaluation metrics regarding classification tasks (Dufourq et al. 2021). It is useful that the metrics evolve to a perfect score according to the proportion of classes predicted with a given level of confidence. Morphoacoustic classification is an example: in the morphoacoustic classification tasks, we are usually very confident that a recording will be the class we expect. That is because the characteristic sound associated with these species depends on the class. The resulting trend is that we have an imbalanced dataset, where the majority class has a large number of recordings with a specific call annotation and the other classes have a small number of recordings with a specific call annotation (Khalighifar et al. 2022).

7.1 Accuracy, Precision, Recall, and F1 Score

The accuracy of the output of a deep neural network measures the proportion of inputs in a test set on which the network agrees with the ground truth. In bioacoustic classification problems, accuracy is the usual baseline metric. If considering a problem as binary classification, create a confusion matrix by calculating the true label probability from the training set. Use the entire validation sample to predict the output, and after rounding it, also create a confusion matrix. However, accuracy can be misleading if class distributions in the test data are skewed. Networks with billions of parameters learn weights to generate high confidence near one-hot predictions for small or biased sets of the training partition (Aradi and Karoly Varga 2023). In such solutions, one parameter often represents more than one category of the inputs. Thus, precision, recall, and F1 score are also calculated.

7.2 Receiver Operating Characteristic (ROC) Curve and Area Under the Curve (AUC)

Receiver operating characteristic (ROC) curves are a very useful diagnostic tool for the illustration and interpretation of binary classifier predictions. Moreover, they provide important insights into the behavior of the classifier. ROC curves are presented in a way that more numbers would have a good trade-off to a higher number of other classes among the target class, making it difficult to decide which one to pick (Salamon et al. 2017). ROC curves also provide visual evaluations of such trade-offs between false positives, the number of wrong alarms, and true positive rates rather than prediction values, which are part of visual metrics with their default prediction thresholds between 0 and 1. Area under the curve (AUC) is the most famous scalar performance metric that summarizes the receiver operating characteristic curve and provides scalar performance for comparison (Prince et al. 2019). AUC presents the ability of the algorithm to separate the target class from the other class. The ROC curve visualizes predictions or probability values resulting from a classifier made for a random distribution of threshold values between 0 and 1. AUC clearly provides language-independent scalar performance. AUC can also provide advice on choosing a threshold because the trade-off between true positive rate and false positive rate is significant information for decision-makers. The result is that better classifiers give larger AUC values and, as a result, enable a user to accurately find problems when choosing a classifier for a task. The diagonal line represents an algorithm with completely random predictions, while the $y = 1$ and $x = 1$ line represents ideal classification with perfect trade-off and best performance (Koops et al. 2015). The target class, with probability scores, can be combined for multiple models and drawings that represent trade-offs. In some cases, the problem is that incorrect labeling of some true/false flags in the dataset creates such a scenario and leads to increasing thresholds to minimize this error.

8 Challenges and Future Directions in Deep Learning for Bioacoustics

Bioacoustic data typically show strong, locally unstable patterns and backgrounds. We envision that models developed to address bioacoustic tasks using simple supervised models will primarily operate in "sweet spots" on various trade-off curves. Although 2D CNN can be stacked for 3D sound learning, no studies currently demonstrate the superior performance of layer-stacked 2D CNN compared to 1D CNN in the field of sound discrimination. Simply put, current applications of deep learning deploy the multi-physical network to combine different inputs (Escobar-Amado et al. 2024; Kirsebom et al. 2021). The combination is an alternative in the last phase of the network. If we want to have sound recognition, the signal input is not involved in the training module. In order to realize comprehensive bioacoustics, we argue that deep learning should capture more naturalistic feature sets of sound than current benchmarks and understand these features from a perception-endorsed point of view. We argue that creating 3D sound processing capable of automatic learning with a local feature hierarchy may lay the foundations for sensimetrics, extending the pareidolia-like detection capabilities of humans and other highly auditory organisms in rapid, dynamic sound environments. With assistance from human expert knowledge, deep learning can support bioacoustic monitoring for a diverse array of scientific and conservation efforts (Ferrari et al. 2020). Last but not least, with resultant validated networks and metrics, an automatic identification system can make extraordinary human-like capabilities mainstream.

8.1 Data Imbalance and Label Noise

It is not an unusual problem for bioacoustics that the training data are given for a small number of classes, but the implemented species library in a monitoring system could be applied to a larger set of classes. In contrast to the existing common solution of using a mismatch error or the background class by training a model that aims to reject other classes, the work improved the learning-to-reject procedure for training those models. The proposed exponential weight rejection method partially mitigates the problem of less data during training by penalizing misclassification of the background class, which should be more challenging (Goëau et al. 2016). Although the proposed procedure has shown robust performance in experiments, an important future work direction assumes further study of the influence of model uncertainty on output classification and decision threshold rules of vote-based ensemble methods, implied in the practical implementation of bioacoustic monitoring systems. This work addressed the task of bird species classification in bioacoustic monitoring scenarios by exploiting deep learning-based systems that support uncertainty measures of model predictions (Dufourq et al. 2021). To mitigate the challenge of dealing with a highly imbalanced multi-label dataset in bioacoustic classification, a rejection-based procedure was introduced which improved model learning at the training stage using the existing abundance of labeled data. The proposed exponential weight rejection method minimizes the loss contributions from supportive training data via a weighting function, which is inversely proportional to the prediction scores. Experimental results on two different bioacoustic monitoring datasets showed the robustness of the rejection-based procedure and encouraged further experimentation on the practical application for the model ensemble, which tackled the problem above in experiments (Escobar-Amado et al. 2024; Kirsebom et al. 2021).

8.2 Interpretable Deep Learning Models for Ecological Studies

This chapter discusses the usage of deep learning algorithms and models for the classification of bioacoustics data, which is regarded as a fundamental component of ecoacoustics research. It is shown that state-of-the-art deep learning algorithms and models such as convolutional neural networks, recurrent neural networks, attention models, dense neural networks, and deep metric learning models can be employed for the extraction of high-level discriminative features from the bioacoustics data (Dufourq et al. 2021). Architectural configurations, data preprocessing steps, and training strategies are explained comprehensively. In addition, discussions on the interpretability and fairness of the models alongside the open challenges and future research directions within the context of bioacoustics classification are also provided. A benchmark of model performances based on bird audio datasets is implemented and the results are presented. This aims to provide a comparative reference for the meta parameters and the performance of the deep learning models (Kirsebom et al. 2021).

Furthermore, the interpretability and fairness of the learned features and the model make a key contribution in relation to long-term bioacoustic classification in the context of ecoacoustics research. Existing deep learning models and domain knowledge can be regarded as the correct mirror for the disruptive understanding of the fine-grained ecological dynamics from the mass species-oriented bioacoustics data. Through interpretations of the features such as species functional groups, bioacoustic signals, and geographical diversity, high-level reasoning concerning the population dynamics within ecosystems can be made. Optimized conservation strategies can be suggested for wildlife and habitats (Goëau et al. 2016). Various user-driven applications and systems such as interactive species recognition and unsupervised clustering of ecoacoustic data can also be developed with the evolution of interpretability.

9 Case Studies and Applications

Automated classification of bioacoustic signals is an active area of research with applications in bioacoustics indexing and monitoring species presence in a certain location. In this paper, we explore the performance of deep learning on five different datasets. We evaluated the performance of deep learning for bioacoustic signal classification on six different base architectures for an open set of multi-label classification tasks, with as many as 42 training classes. We show that considerable performance exists between different deep learning models, and a deep and wide classifier is able to outperform competition-winning approaches by combining inception, residual, and DenseNet architectures and adding some convolutional layers to modify the network input shape (Zhou et al. 2023). Automated classification of bioacoustic signals is of great importance to help scientists and the general public analyze large amounts of recordings for bioacoustic indexing and monitoring of species presence in a certain location. The implementation of an automated system is hampered by the high dimensionality of audio, which makes manually extracting features for a fractional number of samples, a significant challenge. Additionally, a high proportion of audio recordings are annotated and labeled, so a method capable of performing the tasks of training only with known annotations and being able to adapt to new annotations would be of significant value (Khalighifar et al. 2022).

9.1 Bioacoustic Monitoring of Bird Species

The monitoring of bioacoustic data for species identification has been a very active research area in the past decade. Indeed, bird vocalizations, being species-specific, reliable, and relatively easy to detect and record, are among the most commonly used sounds for research purposes. The recording process is often performed during field studies of bird behavior, which leads to the accumulation of vast datasets. This has raised the need to automate the individual species detection tasks for all descriptions of the life cycle of bird species over time and even in specific conditions. Recently, deep learning techniques have been proposed to automatically perform birdsong detection tasks from raw audio signals (Salamon et al. 2017). This has paved the way for the creation of more extensive controlled databases of birdsong for both training and validation of machine learning models. The research community leverages these techniques to analyze larger regions. Among the other applications, the use of bioacoustic transmission to estimate species numbers or even predict species extinction in certain areas has been investigated. More generally, a complete bioacoustic system consists of three logical parts: the bioacoustic acquisition of the signal, the treatment and event detection after the decomposition of the audio signal into elements, and other signal properties, and, the assignment of descriptive characteristics to each concentration area (Morgan and Braasch 2020).

9.2 Marine Mammal Vocalization Classification

The goal of our project is to create a classifier to help environmental scientists classify animal sounds. We trained our classifier on a subset of a large ecological dataset, which contains over 10,000 h of audio with recordings of a variety of Arctic marine animals. Our classifier tackles the problem of

Table 1 Application of deep learning in bioacoustic classification

Application domain	Data type	The model objective	Deep learning techniques	Challenges	Real-world use cases	Reference
Aquatic insect detection	Aquatic insect larvae movement and sounds	Identify aquatic insect species for water quality and habitat monitoring	RNNs for detecting time-sequential patterns	Distinguishing between movement and environmental noise	Monitoring water bodies for ecological changes through insect population studies	Kirsebom et al. (2021)
Human-wildlife conflict mitigation	Animal distress calls	Detect signs of stress or conflict between humans and wildlife (like the elephants)	Hybrid CNN-RNN architectures for pattern recognition	Low-quality data from remote regions, real-time detection	Monitoring elephant herds to prevent crop-raiding or conflicts with humans in rural areas	Escobar-Amado et al. (2024)
Endangered species detection	Rare animal calls or endangered species' vocalizations	Automate the identification of rare or endangered species	CNNs with data augmentation to handle small datasets	Limited training data, noisy environments	Detecting endangered species like the Javan Rhino by identifying their vocalizations in remote areas	Ferrari et al. (2020)
Bat echolocation monitoring	High-frequency bat calls	Identify bat species from their echolocation calls for conservation efforts	High-resolution CNNs and wavelet transforms	High-frequency noise interference, short-duration calls	Monitoring bat populations for changes in biodiversity and disease surveillance (like, White-Nose Syndrome)	Kiskin et al. (2020)
Invasive species detection	Vocalizations or movement-related sounds of invasive species	Detect the presence of invasive species before they cause ecological damage	Spectrogram classification using CNNs	Detecting rare or subtle sounds, low data availability	Early detection of invasive species like cane toads through their unique calls in regions they are encroaching	Thomas (2019)
Terrestrial soundscapes	Environmental soundscapes (composite of many sounds)	Monitor ecosystem health by analyzing patterns in biodiversity soundscapes	CNNs and autoencoders for unsupervised anomaly detection	Differentiating between species in complex soundscapes	Using soundscapes as ecological indicators in forests, grasslands, or wetlands to detect biodiversity shifts	Goëau et al. (2016)
Amphibian call detection	Frog and amphibian vocalizations	Detect specific species' calls to assess habitat health and biodiversity	CNNs for spectrograms, transfer learning for small datasets	Overlapping calls, seasonal variability	Monitoring amphibian populations in sensitive ecosystems for conservation purposes	Dufourq et al. (2021)
Insect sound recognition	Insect buzzes and calls	Detect and classify insect species for ecological and agricultural monitoring	LSTMs and RNNs for time series analysis	Complex background noise, short sound events	Detecting pollinators in agricultural fields or identifying pests in crops	Zhou et al. (2023)

(continued)

Table 1 (continued)

Application domain	Data type	The model objective	Deep learning techniques	Challenges	Real-world use cases	Reference
Marine bioacoustics	Underwater sounds (for example, whale, dolphin)	Classify marine animal calls to monitor species and track migration	CNNs for underwater acoustic spectrograms	Sound distortion in underwater environments, data scarcity	Tracking endangered whale populations and monitoring the effects of ship traffic on marine mammals	Khalighifar et al. (2022)
Wildlife monitoring	Animal vocalizations (birds, mammals, etc.)	Automatically identify species from vocalizations to track biodiversity	CNNs for spectrogram analysis, RNNs for time series classification	Noisy data, overlapping sounds, environmental noise	Bird call classification to monitor population trends and migration patterns	Aradi and Karoly Varga (2023)

multi-label bioacoustic classification by building upon unsupervised deep learning models. We report our classifier's performance against rule-of-thumb classifiers on a three-species vocal classification task. For this narrowly focused task, our classifier measured a maximum F1 classification score of 0.941 at the 0.005 FPR level (Chen et al. 2020; Lauha et al. 2022). Overall, our work contributes evidence that deep learning unsupervised feature learning can be a prevalent tool for applied bioacoustic tasks. Two main acoustic features distinguish marine mammal vocalizations from ambient biologic noise in Arctic Sea ice environment audio: structure at non-trivial temporal frequencies and harmonic peaks. Previous work introducing a hand-crafted classifier leverages these characteristics. The hand-crafted classifier achieved excellent performance by exploiting these two distinct manual features but does not harness recent advances in signal processing or deep representation learning. For this project, we explore the performance of an unsupervised deep representational learning model, find difficulties in proving the model, and propose other similar models. If successful, we will build a bioacoustic classification system for field deployment (Prince et al. 2019). Table 1 summarizes a broader application of deep learning in bioacoustic classification.

Key findings and insights in bioacoustics classification can be closely related to the evolution of directly applicable deep learning techniques. One of the largest blocks to progress in bioacoustics classification is the current size and distribution of collected datasets, and it is imperative to focus on methods for bioacoustic data collection and verification. The continual evolution of deep learning models and algorithms will be a key driver for uncovering multi-modality patterns and improving bioacoustic-based time series classification. In addition, new issues for large-scale hybrid models and techniques for combining these competing models will need to be developed (Lauha et al. 2022). Further extensions dealing with effectively combining multiple weakly correlated base models will also need to be developed. In order to make efficient decisions, we will not just have a few different models that give a fair performance for different species in different regions with varying sound environments; we will have knowledge of the collective, overall performance of algorithms that have the potential to be applied globally in areas such as conservation and automated field guides (Kirsebom et al. 2021).

10 Conclusion

Many challenges exist in the use of deep learning techniques to assist in bioacoustics classification. The size and representation of current datasets are the first place for research to improve. A larger, more evenly distributed dataset that covers a wide range of species across a global range of sound environments would help combat the challenges introduced by the variability of bioacoustic data. Algorithms will need to be improved, and new bioacoustic-based algorithms will need to be developed to deal with this variability, including environmental and species-specific transformations. It is important to remember when discussing bioacoustics research within certain industries and applications, for example, species distribution and classification, that humans have worked for hundreds of years putting together metadata on individual species. Without this metadata, typical of large data sources, many areas of automated classification and interpretation could not take place, as individual species is not the final goal in bioacoustics classification. Therefore, more complex algorithms capable of recognizing cross-species attributes that we might not yet know about are required. Continued collaborations between taxonomists and data scientists will be necessary to make advances in these areas. Algorithms will need to become more computationally efficient, and further research is required in the areas of few-shot learning and transfer learning to deal with the scarcity of labeled data across large numbers of species. By developing better algorithms for transfer learning models, we will not just have a few good models, but we will have knowledge that can be applied to many species in areas such as conservation and species monitoring. Finally, the ethics surrounding machine learning-based bioacoustics classifications as part of decision-support tools for arguments such as hunting, culling, protection, or release need to be discussed.

References

Aradi A, Karoly Varga A (2023) Artificial intelligence-based machine learning data classification for the analysis of underwater noise of ship traffic and bottlenose dolphin sounds. In: IEEE international workshop on metrology for the sea; learning to measure sea health parameters, MetroSea 2023—proceedings. https://doi.org/10.1109/MetroSea58055.2023.10317350

Bergler C, Schmitt M, Cheng RX, Schröter H, Maier A, Barth V, Weber M, Nöth E (2019) Deep representation learning for orca call type classification. Lecture notes in computer science (Including subseries lecture notes in artificial intelligence and lecture notes in bioinformatics), 11697 LNAI. https://doi.org/10.1007/978-3-030-27947-9_23

Bermant PC, Bronstein MM, Wood RJ, Gero S, Gruber DF (2019) Publisher correction: deep machine learning techniques for the detection and classification of sperm whale bioacoustics. Sci Rep 9(1). https://doi.org/10.1038/s41598-019-53408-7

Bravo Sanchez FJ, Hossain MR, English NB, Moore ST (2021) Bioacoustic classification of avian calls from raw sound waveforms with an open-source deep learning architecture. Sci Rep 11(1). https://doi.org/10.1038/s41598-021-95076-6

Chen X, Zhao J, Chen Y, Zhou W, Hughes AC (2020) Automatic standardized processing and identification of tropical bat calls using deep learning approaches. Biol Conserv 241. https://doi.org/10.1016/j.biocon.2019.108269

Clink DJ, Kier I, Ahmad AH, Klinck H (2023). A workflow for the automated detection and classification of female gibbon calls from long-term acoustic recordings. Front Ecol Evol 11.https://doi.org/10.3389/fevo.2023.1071640

Dias FF, Ponti MA, Minghim R (2022) A classification and quantification approach to generate features in soundscape ecology using neural networks. Neural Comput Appl 34(3). https://doi.org/10.1007/s00521-021-06501-w

Dufourq E, Durbach I, Hansford JP, Hoepfner A, Ma H, Bryant JV, Stender CS, Li W, Liu Z, Chen Q, Zhou Z, Turvey ST (2021) Automated detection of Hainan gibbon calls for passive acoustic monitoring. Remote Sens Ecol Conserv 7(3). https://doi.org/10.1002/rse2.201

Dufourq E, Batist C, Foquet R, Durbach I (2022) Passive acoustic monitoring of animal populations with transfer learning. Ecol Inform 70.https://doi.org/10.1016/j.ecoinf.2022.101688

Escobar-Amado C, Badiey M, Wan L (2024) Computer vision for bioacoustics: detection of bearded seal vocalizations in the Chukchi shelf using YOLOV5. IEEE J Oceanic Eng 49(1). https://doi.org/10.1109/JOE.2023.3307175

Ferrari M, Glotin H, Marxer R, Asch M (2020) DOCC10: open access dataset of marine mammal transient studies and end-to-end CNN classification. In: Proceedings of the international joint conference on neural networks.https://doi.org/10.1109/IJCNN48605.2020.9207085

Goëau H, Glotin H, Vellinga W-P, Planqué R, Joly A, Lifeclef Bird AJ (2016) LifeCLEF bird identification task 2016: the arrival of deep learning. In: CLEF: conference and labs of the evaluation forum

Khalighifar A, Gotthold BS, Adams E, Barnett J, Beard LO, Britzke ER, Burger PA, Chase K, Cordes Z, Cryan PM, Ferrall E, Fill CT, Gibson SE, Haulton GS, Irvine KM, Katz LS, Kendall WL, Long CA, Mac Aodha O et al (2022) NABat ML: utilizing deep learning to enable crowdsourced development of automated, scalable solutions for documenting North American bat populations. J Appl Ecol 59(11). https://doi.org/10.1111/1365-2664.14280

Kirsebom OS, Frazao F, Padovese B, Sakib S, Matwin S (2021) Ketos—a deep learning package for creating acoustic detectors and classifiers. J Acoust Soc Am 150(4_Supplement). https://doi.org/10.1121/10.0007998

Kiskin I, Cobb AD, Wang L, Roberts S (2020) Humbug Zooniverse: a crowd-sourced acoustic mosquito dataset. In: ICASSP, IEEE international conference on acoustics, speech and signal processing—proceedings, 2020-May. https://doi.org/10.1109/ICASSP40776.2020.9053141

Koops HV, Van Balen J, Wiering F (2015) Automatic segmentation and deep learning of bird sounds. In; Lecture notes in computer science (Including subseries lecture notes in artificial intelligence and lecture notes in bioinformatics), 9283.https://doi.org/10.1007/978-3-319-24027-5_26

Lauha P, Somervuo P, Lehikoinen P, Geres L, Richter T, Seibold S, Ovaskainen O (2022) Domain-specific neural networks improve automated bird sound recognition already with small amount of local data. Methods Ecol Evol 13(12). https://doi.org/10.1111/2041-210X.14003

Michaud F, Sueur J, Le Cesne M, Haupert S (2023) Unsupervised classification to improve the quality of a bird song recording dataset. Ecol Inform 74.https://doi.org/10.1016/j.ecoinf.2022.101952

Morgan M, Braasch J (2020) Holistic, long-term soundscape monitoring in Upstate New York using convolutional long short-term memory deep neural networks. J Acoust Soc Am 148(4_Supplement). https://doi.org/10.1121/1.5147610

Morgan M, Braasch J (2021) Improving data labeling efficiency for deep learning-facilitated bioacoustics monitoring. J Acoust Soc Am 149(4_Supplement). https://doi.org/10.1121/10.0004505

Nolasco I, Singh S, Morfi V, Lostanlen V, Strandburg-Peshkin A, Vidaña-Vila E, Gill L, Pamuła H, Whitehead H, Kiskin I, Jensen FH, Morford J, Emmerson MG, Versace E, Grout E, Liu H, Ghani B, Stowell D (2023) Learning to detect an animal sound from five examples. Ecol Inform 77.https://doi.org/10.1016/j.ecoinf.2023.102258

Padovese B, Kirsebom OS, Frazao F, Evers CHM, Beslin WAM, Theriault J, Matwin S (2023) Adapting deep learning models to new acoustic environments—a case study on the North Atlantic right whale upcall. Ecol Inform 77.https://doi.org/10.1016/j.ecoinf.2023.102169

Prince P, Hill A, Piña Covarrubias E, Doncaster P, Snaddon JL, Rogers A (2019) Deploying acoustic detection algorithms on low-cost, open-source acoustic sensors for environmental monitoring. Sensors (Switzerland) 19(3). https://doi.org/10.3390/s19030553

Ruff ZJ, Lesmeister DB, Duchac LS, Padmaraju BK, Sullivan CM (2020) Automated identification of avian vocalizations with deep convolutional neural networks. Remote Sens Ecol Conserv 6(1). https://doi.org/10.1002/rse2.125

Salamon J, Bello JP, Farnsworth A, Kelling S (2017) Fusing shallow and deep learning for bioacoustic bird species classification. In: ICASSP, IEEE international conference on acoustics, speech and signal processing—proceedings. https://doi.org/10.1109/ICASSP.2017.7952134

Shafik W (2024a). Environmental conservation and sustainable development. In: Developing digital inclusion through globalization and digitalization. IGI Global, pp 142–176. https://doi.org/10.4018/979-8-3693-4111-7.ch008
Shafik W (2024b) Industry 4.0 technologies' opportunities and challenges for realising net-zero economy. In: Singh R, Khan S, Kumar A, Luthra S, Chokshi H (eds) Net zero economy, corporate social responsibility and sustainable value creation. CSR, sustainability, ethics and governance. Springer, Cham. https://doi.org/10.1007/978-3-031-55779-8_2
Shafik W (2024c) Sustainable development goal 14: explainable AI (XAI) for ocean health. In: De D, Sengupta D, Tran TA (eds) Artificial intelligence and edge computing for sustainable ocean health. The springer series in applied machine learning. Springer, Cham. https://doi.org/10.1007/978-3-031-64642-3_8
Thomas M (2019) Towards a novel data representation for classifying acoustic signals. In: Lecture notes in computer science (Including subseries lecture notes in artificial intelligence and lecture notes in bioinformatics), 11489 LNAI. https://doi.org/10.1007/978-3-030-18305-9_67
Zhou X, Hu K, Guan Z, Yu C, Wang S, Fan M, Sun Y, Cao Y, Wang Y, Miao G (2023) Methods for processing and analyzing passive acoustic monitoring data: an example of song recognition in western black-crested gibbons. Ecol Ind 155.https://doi.org/10.1016/j.ecolind.2023.110908

Optimizing Bioacoustic Signal Classification Using AI-Enhanced Traditional Algorithms

G. Vijayakumari, N. Gomathi, P. Kokila, J. Amudha, and Chithirai Pon Selvan M.

Abstract

Computational bioacoustics is using the source of artificial intelligence to detect and classify the animal behavior, interactions, and vocalization. Signal processing and analysis play an important role in study of bioacoustics. Various machine learning (ML) and deep learning (DL) classification algorithms are explained in this chapter. At the beginning of the chapter, it aims to provide an overview of the traditional ML and DL algorithms for computational bioacoustics and clarification of concepts. To analyze audio signals for species detection, conventional classification algorithms like support vector machines (SVM), Hidden Markov Models (HMM), k-nearest neighbors (KNN), decision trees, and random forest are used. To improve the accuracy of the large dataset, the features are extracted from raw audio data using deep learning algorithms like convolutional neural networks (CNNs), recurrent neural networks (RNNs) and CNN-RNN Hybrids, long short-term memory networks (LSTMs) with different architectures ResNet, Inception. Pre-trained audio neural networks (PANNs) are trained on a large-scale dataset called AudioSet can be used as an architecture for animal sound detection. The classification of bioacoustics sounds using artificial intelligence follows several key steps. Preprocessing is a crucial stage, where raw audio data is prepared for analysis and classification. Noise reduction techniques are applied to remove common background sounds like wind, rain, and human activity from the recordings. Data augmentation is used to artificially increase the size of the dataset by the software packages of python. Acoustics analysis made by mel-frequency cepstral coefficients (MFCCs) is used to compressing the spectral information into a small number of standardized measurements. Nowadays CNN architectures like WaveNet replaced the MFCCs to detect the animal sounds accurately. This chapter ends with comparing all the traditional algorithms for the computational bioacoustics and provide the best classification algorithm based on the accuracy of the extracted features. The key benefits of this chapter include traditional ML and DL techniques for classification of bioacoustics, biodiversity protection, different animal species identification.

Keywords

Machine learning · Deep learning · Biodiversity · Data augmentation · Mel-frequency cepstral coefficients (MFCCs)

1 Introduction to Computational Bioacoustics

This interdisciplinary field called computational bioacoustics encompasses the use of computation, including signal processing, machine learning, and data analytics, toward analytical and interpretive skills to determine what there is

G. Vijayakumari (✉)
Department of ECE, Kangeyam Institute of Technology, Tirupur, India
e-mail: viji.ece05@gmail.com

N. Gomathi
Department of ECE, JKK Munirajah College of Technology, Erode, India
e-mail: gomathin@jkkmct.edu.in

P. Kokila
Department of ECE, Nandha Engineering College, Erode, India
e-mail: kokila.paramasivam@nandhaengg.org

J. Amudha
Department of ECE, Sri Shanmugha College of Engineering and Technology, Salem, India
e-mail: amudhaj@shanmugha.edu.in

M. Chithirai Pon Selvan
Curtin University Dubai, Dubai, United Arab Emirates
e-mail: pon.selvan@curtindubai.ac.ae

K. Subbaraj et al. (eds.), *Computational Bioacoustic Artificial Intelligence*, Sustainable Artificial Intelligence-Powered Applications, https://doi.org/10.1007/978-3-032-06088-4_4

to say about birds, mammals, and insects that inhabit an environment. Biological sounds could be thought of as bearing key information about a broad diversity of species, and, of course, environmental change would be discernible by listening. Traditionally, bioacoustics has made heavy use of the recognition of these sounds through human expertise. The field of computational bioacoustics uses modern algorithms to deal with much larger volumes of audio signals accurately and with greater efficiency.

1.1 *Role of Artificial Intelligence in Bioacoustics*

The introduction of artificial intelligence remarkably changes the discipline of bioacoustics by permitting thorough analysis, interpretation, and classification of complex acoustic signals originating in natural environments. Traditional techniques face many challenges from extremely vast and complex characteristics in bioacoustic data from background noise and minor difference in animal calls. The correct approaches through advanced machine learning and deep learning algorithms in AI, toward better outlooks of ecosystems through identification of species and population monitoring as well as interpretation of animal behavior with greater effectiveness and precision in scalable modes, can be found helpful in making it possible for streamlined procedure like signal detection, noise reduction, and behavioral pattern analysis (Stowell 2022). New opportunities for real-time surveillance and conservation initiatives could then become available. With bioacoustics, evolution comes with AI technology that translates raw sound data to proven valuable insights for conservation and protection of biodiversity and the environment.

1.2 Animal Behavior Sound Analysis Using AI

ML and DL have greatly enhanced our understanding of animal sound behaviors and their interactions because they offer very powerful methodologies for untangling complex acoustic patterns. The artificial intelligence methodologies of analyzing large datasets of animal vocalizations identify subtle characteristics that may indicate communication, mating calls, territorial signals, or distress alerts. It trains the model on labeled datasets so that it can identify species-specific sounds, which means which animal, thus allowing very intricate studies of social structures and behavioral dynamics. Deep learning, even in CNNs and RNNs, is able to identify very fine-grained temporal patterns and relationships within the vocalizations that will then feed into knowing how animals relate and respond to their environment. These technologies allow researchers to interpret the meaning beneath different calls and are also helpful in understanding behavioral changes induced due to environmental stressors, thereby providing important information for conservation actions and enhancing our knowledge of animal communication systems.

2 Signal Processing for Audio Analysis Using Traditional Algorithms

The core of audio analysis, particularly in bioacoustics is signal processing. There are established algorithms in this field that can extract the needed insights from complex acoustic signals. Algorithms applied decompose raw audio data into components that are easily interpretable for sound identification and classification. The steps are usually preprocessing, feature extraction, and classification. Time-domain signals are transformed into frequency domain representation in techniques like Fourier transform and Wavelet transform, which is useful to analyze the patterns found within audio data (Alves et al. 2021). Techniques like filtering remove noise, whereas statistical techniques help extract relevant parameters like pitch, duration, and amplitude. Traditional signal processing leads toward very advanced analyses involving bioacoustic data through artificial intelligence. Figure 1 shows the flowchart of processing steps for bioacoustic signal process.

2.1 Key Steps and Challenges in Bioacoustic Signal Processing

Bioacoustic signal processing is a carefully structured process designed to obtain meaningful information from animal vocalizations. Yet, it involves several novel challenges. Data acquisition of this process is the recording of different sounds using microphones or sensors within different settings. Then, during preprocessing, the extracted recordings are filtered to remove useless noise and audio signals standardized; thus, ensuring uniform sound quality. It is later split into portions of data, and feature extraction is done based on some properties like pitch, amplitude, or duration in the time-domain, frequency-domain, or even time–frequency. The extracted features classify based on applying algorithms such as HMM or machine learning approaches classifying the species or behavior. That excluding, it has its difficulties, much more so when there are noises or overlapping sounds by natural backgrounds that add up to the obscurity of the signal. Furthermore, a large amount of data requires high-computational power for processing and storage. Variability in animal vocalizations and poor annotated datasets are critical challenges for accurate classification as well as effective model training.

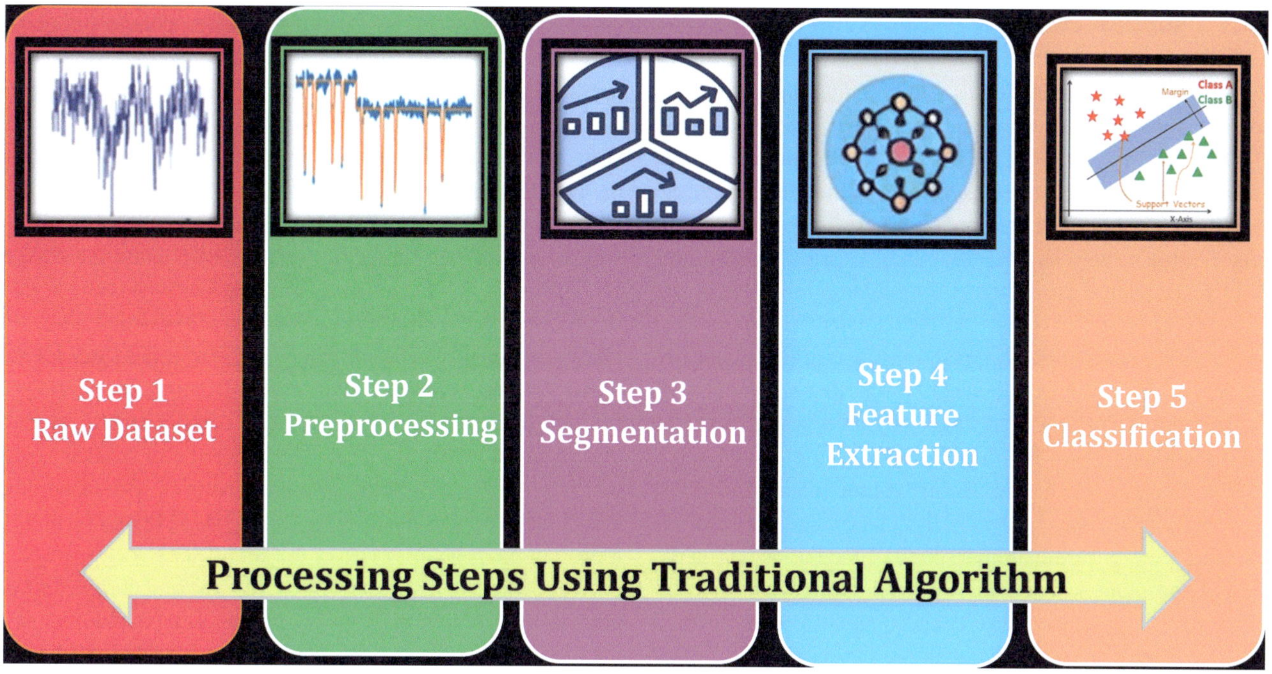

Fig. 1 Flow chart for audio signal processing steps

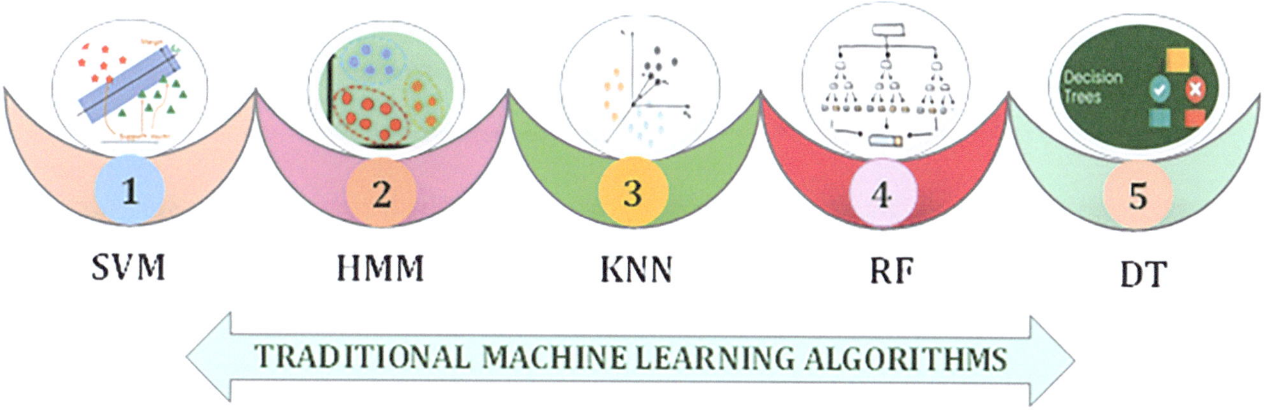

Fig. 2 Traditional ML algorithms in bioacoustics

So, efforts should be made to resolve these problems to enhance accuracy and reliability in bioacoustic analysis especially in ecological monitoring and conservation applications (Nieto-Mora et al. 2023).

3 Traditional Machine Learning (ML) Algorithms

3.1 Introduction to ML Algorithms in Bioacoustics

Conventional machine learning is very significant in bioacoustics, since it represents the basis of fundamental methodology for the study and analysis of animal vocalizations. Algorithms like SVM, HMM, and GMM are used to classify and identify acoustic signals through the identification of patterns in features extracted, such as pitch, duration, and frequency. Unlike deep learning techniques, most conventional machine learning techniques rely on engineered features that require relatively small datasets and are hence applicable to many bioacoustic tasks where labeled data might not be abundant. They are superior in species identification, call segmentation, behavior recognition, etc., with models that are naturally interpretable and can be used for special ecological studies. Although traditional machine learning algorithms are not as user-friendly with complex or noisy data as modern deep learning methods, they are still basic tools for the field of bioacoustics and a basis for more advanced AI-based analyses (Fig. 2).

3.1.1 Support Vector Machines (SVM)

Reliability and accuracy distinguish the SVM, which makes it very popular in using cases for classification of animal vocalizations in bioacoustic research. The process starts with preprocessing of audio to eliminate unwanted noise that might mislead the algorithm and split the data of audio into more manageable segments. Then, the most prominent features such as mel-frequency cepstral coefficients (MFCCs), spectral characteristics, and temporal patterns are extracted so that the sound can be depicted in a numerical format well compatible for machine learning applications (Zhong et al. 2020a). The above properties are input into the support vector machine which applies a kernel function often the radial basis function to enhance the class's separability through transformation in feature space. The SVM uses this labeled data; it learns the classification of sounds resulting from different animal species based upon best separating hyperplane learned. Having been trained, SVMs classify novel, previously unheard audio samples at very high accuracy levels. Its high dimensional capacity and subtlety of acoustic variation make SVM particularly useful in bioacoustics applications such as species identification and monitoring ecosystem health through sound. SVM aims to find the optimal hyperplane that maximizes the margin between two classes. The decision function is given by

$$f(x) = w^T x + b$$

where:

w is the weight vector
b is the bias vector

The margin can be calculated as:

$$\text{Margin} = \frac{1}{\|w\|}$$

3.1.2 Hidden Markov Models (HMMs)

HMMs are very widely used in bioacoustics for improving the classification of animal vocalizations due to their robust capability to capture sequential data and time-dependent patterns. The performance of an HMM is outstanding because it can capture the temporal dynamics that are critical for differentiation among various animal calls, which often have distinctive temporal features. The first step involves preprocessing audio data by removing the background noises and further subdividing the audio into frames. Every frame will have features, consisting of MFCCs and spectral attributes describing the frequency content and timbral properties of the sounds.

For any class of animal sound under consideration, in HMM-based classification, a different HMM is used. Now each specific acoustic pattern or phonemes in the sound can be represented by using HMM. The transitions between such HMM states represent temporal variations occurring in the signal. There have been computations of the three elements-transition probabilities, emission probabilities as well as initial state for training HMM with using the Baum-Welch algorithm. This stage of training facilitates the model's acquisition of the customary sequence of acoustic characteristics associated with each category. New audio samples are classified and processed then compared to trained HMMs. Such comparison is done with the help of the Viterbi algorithm that produces the most probable sequence of hidden states that might have caused the observed features. Therefore, this algorithm produces the most probable HMM to be able to best represent the probable class of the sound. This method highly efficiently addresses variability in the time and complexity of animal calls, which may, in fact, make it an excellent candidate for species classification, monitoring behavior, or even tracking changes in ecological environments. It provides a rich model for the classification of animal sounds even in noisier and more complex natural habitats by including spectral and temporal characteristics of bioacoustic signals. For classification, the probability of a sequence of observations $O = (O_1, O_2, \ldots, O_T)$ given a particular model is calculated and it is represented as:

$$P(O|\lambda) = \sum_{\text{all states}} P(O, S|\lambda)$$

where λ represents the HMM parameters.

3.1.3 K-Nearest Neighbors (KNN)

It is a simple but robust classification technique for bioacoustic signals like animal calls by the k-nearest neighbors algorithm, which employs non-parametric and instance-based learning. Bioacoustic analysis typically includes preprocessing of audio signals with the removal of noise followed by framing into segments to obtain significant features, including MFCCs, spectral, and temporal. These features represent the sounds with numbers, so with them, KNN would be able to measure the similarities between audio samples. KNN compares a new, unlabeled audio sample with a set of labeled training samples by using a distance metric and is commonly the Euclidean distance. The distance between two points $x_i = (x_{i1}, x_{i2}, \ldots x_{id})$ and $x_{\text{test}} = (x_{\text{test}1}, x_{\text{test}2}, \ldots x_{\text{test}d})$ in a dimensional feature space is calculated as:

$$D(x_i, x_{\text{test}}) = \sqrt{\sum_{j=1}^{d} (x_{ij} - x_{\text{test}j})2}$$

where d is the number of features.

It has then determined the "k" nearest neighbor—those training samples like the new data points—and assigned a class by majority vote from the nearest neighbors. Such makes KNN quite useful for the classification of animal vocalizations in which unique sounds may have a subtle difference picked up by similarity of features. KNN is particularly suitable for bioacoustic applications, especially when the dataset is relatively small, or the classification boundaries are not linearly separable. However, the selection of the parameter "k" and feature scaling should be done with care to avoid misclassification. Although it is computationally expensive for large datasets, KNN gives robust performance for species identification and behavior analysis, capturing the variations in animal vocalizations and adapting to diverse acoustic environments.

3.1.4 Decision Trees and Random Forests

Decision trees and random forest are used very frequently for the classification of bioacoustic signals, like animal voice because they can be used in complex, non-linear interaction still with interpretability. The decision tree works based on a procedure that is applied in the data partition step in each of its nodes using iterated steps, the most discriminative feature, i.e., the MFCC, spectral centroid, zero crossing rate among others metrics for Gini impurity or information gain. The criteria of evaluation for every node in the tree are the branching points that lead the input data to a series of conditions to reach a terminal node, which can represent a specific type of an animal or call type. It is intuitive and works perfectly on categorical and numerical data. Therefore, it comes as helpful in detecting patterns of bioacoustics.

Random forest is an ensemble approach, with improved decision trees in the overall accuracy of classification and generalized performance. Each individual decision tree in the forest was trained on a bootstrap sample extracted from the training set; in each partitioning step of each tree, a subset of features is randomly selected for application. This minimizes the effect of overfitting but increases robustness. Classifications are made by combining votes from all the decision trees, and the voting procedure determines the class labels to be assigned. This collective methodology enables random forests to detect subtle differences in animal vocalizations, handle noisy data, and reduce the likelihood of bias.

Both have an efficiency in the classification of bioacoustic signals to handle high-dimensional feature spaces and missing data more efficiently. The random forest algorithms are more effective for complicated datasets that involve diversified acoustic features, making it very suitable for species identification, behavioral monitoring, and ecological surveillance. Their rank feature importance also gives some insight into which of the acoustic features are the most important in distinguishing among various types of animal vocalizations.

4 Deep Learning (DL) Algorithms in Bioacoustics

4.1 Overview of DL and Its Role in Processing Large Audio Datasets

Deep learning methods altogether changed the way of thinking about biosound analysis and classification, especially when it comes to dealing with big data. Applications, including CNN and RNN, are proven highly successful in recognizing complex relationships or patterns in raw audio, independent of the need for extensive preprocessing data for feature extraction by hand. The general process begins by translating the audio signals into spectrograms or mel-spectrograms: frequency composition over varying instances of time. The spectral, along with the temporal, properties of the vocalization by the animal are thus inputs of visual representation to deep architectures inside the model. CNNs are very efficient at spatial pattern recognition in spectrograms and thus very well suited to recognizing distinctive species calls or vocalization types. Filters can be applied to feature-detecting functions like frequency contours or harmonics to recognize even noisy audio patterns. RNNs and LSTM networks are good for analyzing temporal dependencies in sequences and thus suited to analysis of the development and structure of longer animal vocalizations or song sequences.

Deep learning architectures require large amounts of data to learn to generalize well, but their ability to learn large amounts of data makes classification accuracy much better. Data augmentation and transfer learning further improve performance by mitigating issues of insufficient labeled data or uneven class distributions. The models are very robust to the large variability that bioacoustic signals possess and can identify subtle differences in species and behavioral patterns. Thus, deep learning has emerged as an effective tool in the field of bioacoustics and aided applications like species monitoring, behavior evaluation, and ecological assessments with great accuracy and scalability.

4.1.1 Convolutional Neural Networks (CNNs)

As part of their native competency, they automatically get out the spatial pattern in-directly from complex data into audio for achieving extremely efficient enhancement of the classifying capacity for bioacoustic signals. In bioacoustics, the first move taken after preprocessing these signals transformed them into spectrogram and mel-spectogram depicted visually with all such characteristics that involve frequency and time.

These spectrograms are then fed into the CNN, which applies a few convolutional filters to spot features such as frequency contours and harmonics and temporal patterns. At each convolutional layer the representation of the sound has become increasingly abstract, yet more meaningful, from very basic acoustic features to complex structures applicable for identification of species or behavioral analyses (Manriquez et al. 2024). This gives CNNs some form of robustness over noisy inputs to distinguish between very close animal sound images by learning minute differences over the spectrogram. Layering pooling in the neural networks also helps reduce their dimensionality to make more efficient models while retaining more information.

Finally, the fully connected layers classify categories like different species, call types out of the features extracted here. Feature extraction and hierarchical learning automatically remove the requirement for manually engineering features. It improves accuracy and scalability and makes CNNs highly applicable in bioacoustics in species monitoring, habitat assessment, and changes in the ecological system by detecting sound with high performance even in complex and variable acoustic environments.

4.1.2 Recurrent Neural Networks (RNNs)

Improvement of bioacoustic signals classification is made especially strong by RNNs since they are really high performing on sequential data and spotting temporal dependencies. Bioacoustic signals often contain animal calls in them that represent the changing patterns over time. That makes RNN especially apt for analyzing such kinds of sequences. The protocol usually transforms audio signals into the feature sequences, such as MFCCs or spectrogram frames, that serve the inputs for the RNN. Unlike traditional neural networks, RNNs have hidden states or "memory" that contains information about previous time steps and can therefore model sound progression and identify temporal relations (Tosato et al. 2023).

4.1.3 Long Short-Term Memory (LSTM) Networks

Architectures like LSTM and GRU architectures further improve the performance of RNN by overcoming problems such as vanishing gradients, which enables the network to learn long-term dependencies in the audio signal. For example, for classifying complex animal sounds, as in the case of bird song or whale call, often with intricate and extended temporal structure. It learns to identify patterns in sequence data and predict class labels that go with them, for example, species or call types. Therefore, RNNs can well capture the dynamics of bioacoustic signals by exploiting their capabilities for temporal modeling and are powerful tools for species identification, behavioral analysis, and monitoring ecological changes over time.

4.1.4 CNN-RNN Hybrids

Recently, Hybrid CNN-RNN has emerged as a powerful tool that aims to leverage the strengths of both architectures in optimizing classification bioacoustic signals. In this hybrid setting, CNN is used to extract spatial features from an audio signal; typically, such signals are presented as either mel-spectrograms or spectra. The convolutional layers identify patterns of interest from this spectogram: the frequency contour, harmonics, as well as local temporal structures. CNNs capture low-level acoustic features and reduce the size of data with filtering and pooling operations while retaining important data. That CNN's output will then be fed into RNN in most cases by constituting LSTM or GRU cells because that is where processing a natural sequential nature that's come along has been extracted, providing room for the second phase to create learning not only through how dependent times are but also a change from the sound pattern across a given time. This difference in the structure is important while distinguishing between the two animal voices with the same spectral structure but with a different timeline. For instance, this would include distinguishing between the simple call of a monotonous bird and the complex song of a whale in identifying spatial and temporal patterns.

At the time of training, this architecture can absorb spatial and temporal properties very well into a meaningful representation to enhance the accuracy of classification. This approach is very helpful in managing noisy or noisy data conditions quite common in natural scenarios. This method can capture subtleties in bioacoustic signals such that even minute changes in pitch or rhythm, so important to distinguish species with precision or to estimate behavioral parameters, can be well captured. These hybrid models form a robust framework for advanced bioacoustic applications, including wildlife monitoring, ecosystem assessment, and early detection of environmental changes, by capitalizing on the complementary strengths of CNNs and RNNs.

4.2 Popular Architectures: ResNet, Inception

Transfer learning has been applied with pre-trained architectures such as Residual Networks (ResNet) and Inception to classify bioacoustic signals using animal sound datasets. Originally developed for image classification, the models were readily adapted for the classification of bioacoustic signals when the auditory data is represented in an image-like format such as spectrograms or mel-spectrograms. Some of the ways that the pre-trained models have been applied in bioacoustics include:

4.2.1 Audio Signal to Image-Like Representation

The first step of applying the ResNet or Inception architecture for bioacoustic applications is the preprocessing of raw animal sound recordings, usually converted to either spectrograms or mel-spectrograms, which are temporal representations of frequency. That is, audio is translated into 2D images, such that CNN-based architectures could be applied on it-the ResNet and Inception.

4.2.2 Using Pre-trained Models with Transfer Learning

A key methodology of transfer learning is to allow the use of so-called pre-trained models like ResNet or Inception to solve other, but related tasks such as, for example, animal call classification (Tsalera et al. 2021). While these models learn about another domain, such as ImageNet, the knowledge acquired is useful for the actual task and, thus, particularly useful if labeled bioacoustic datasets are scarce or small.

ResNet: It is the class of deep convolutional networks that is distinguished because of a different approach to the residual learning concept, hence defining structural design. Additional "skip connections" or residual links between the layers enhance network efficiency by preventing vanishing gradients within the architecture that occurs specifically in very deep architecture. This therefore enabled ResNet to learn appropriately from long data sets involving many layers. For bioacoustics, a pre-trained ResNet model trained on image data can be fine-tuned to the spectrogram images of animal sounds so that the learned feature extraction capabilities from the former can be transferred to the latter. It is good at extracting hierarchical features from spectrograms to identify both low-level acoustic patterns and higher-order features that help distinguish between different species or call types (Stowell 2022).

Inception: A novel architecture generally known as Google Net gives a new shape to the idea of extraction of feature using multi-scale features. The model comprises the number of filters working on several different scales; one for getting various spatial clues related to differences in scale. Such a method is particularly well-suited for classification of bioacoustic sources because animal calls can sometimes differ significantly in frequency ranges, rhythm, and duration. The architectural framework of Inception is built to incorporate a wide range of features, which turns out to be beneficial in differentiating complex animal vocalizations. Fine-tuning a pre-trained Inception model on bioacoustic data, that is, spectrograms representing animal sounds, the model learns to identify patterns across scales and levels of abstraction, thus improving its ability to classify a variety of animal vocalizations.

4.3 Fine-Tuning the Model

The model then gets fine-tuned over the bioacoustic dataset after choosing either a pre-trained ResNet or Inception model. Fine-tuning, in general, typically adjusts the final layers of the network to fit the task at hand, for instance, to classify between the different species or types of calls. This is done by replacing the final fully connected layers of the pre-trained model with a new set of layers designed specifically for the task of bioacoustic classification. Then, it learns how to map spectrogram features onto the right output labels; that could be species or call type.

Freezing Layers: The first layers of a neural network, which have learned basic low- and mid-level features—like edges or textures in images—can be "frozen" in fine-tuning such that their weights cannot be modified. This technique allows for preserving the important information in the pre-trained model while forcing the attention on adjusting the subsequent layers to better fit the bioacoustic domain.

Training: Train the model on that bioacoustic dataset considering more epochs. Here in that process, use a number of transformed spectrograms as the number of input data. With training, the weights get adjusted in the terminal layers toward efficient classification of animal sounds.

4.4 Classifying Animal Sounds

After training the model, it will then classify unknown animal calls. The new audio file is transformed into a spectrogram and then forwarded to the fine-tuned ResNet or Inception model. The model produces an output, which would generally be the prediction, perhaps even a class label (Jeantet and Dufourq 2023). The potential to extend from wide training on huge image datasets enhances its ability to recognize intricate patterns in the bioacoustic data and, thus, become a useful instrument for applications such as

Species Identification: Identification of the species by classification of animal vocalizations.

Behavioral classification: This refers to the identification and classification of various behaviors or vocalizations. These include mating calls, alarm signals, and feeding sounds.

Monitoring Biodiversity: Monitoring changes over time in a species' presence or abundance.

4.5 Advantages of Using Pre-trained Models

Effective Learning: The above-mentioned pre-trained models significantly reduce large annotated bioacoustic datasets. Thus, the model may learn from a relatively smaller dataset while obtaining the generalized features learned from large images datasets.

Better Performance: The pre-trained models such as ResNet and Inception are already optimized for extracting the high-level features from the images that transfer well to spectrograms. Thus, the model performs better than training one from scratch.

Reduced Training Time: This saves a lot of time when using a pre-trained model, as the base layers already have learned good features, and the only thing needed to be tuned are the top layers.

Very strong approaches to classification of animal vocalizations when using transfer learning for bioacoustic signal classification are pre-trained architectures such as ResNet and Inception. These models, once fine-tuned to analyze spectrograms obtained from bioacoustic datasets, capture all the necessary spatial and temporal characteristics required for accurate classification. Fine-tuning these models on domain-specific datasets allows researchers to tap the power of deep learning even when the labeled data is very limited, which improves accuracy and efficiency in bioacoustic monitoring, species identification, and behavioral analysis.

5 Advanced Pre-trained Models for Bioacoustics

5.1 Introduction to Pre-trained Audio Neural Networks (PANNs)

Among the major classes of deep learning frameworks which are designed to analyze and classify audio signals, including the types used in bioacoustics, by applying principles from transfer learning is known as pre-trained audio neural networks. They train in big heterogeneous audio datasets so bestow within them strong features and these serve rather well on broad sweeps of audio classification workloads. PANNs have been very promising in the classification of animal sounds, including vocalizations, calls, and songs, based on their pattern and structure recognition ability in audio data. In contrast to conventional machine learning methods, which usually require a lot of feature engineering, PANNs can automatically learn hierarchical features from unprocessed audio inputs, which makes them particularly efficient for applications such as species identification, behavioral analysis, and ecological monitoring (Stowell et al. 2018). Using the lessons learnt from large audio sets usually applied to speech or environmental sound classification, the finetuning of pre-trained networks on bioacoustic data sets may reach very high classification accuracy even with comparisonally modest data sets. Especially now models such as VGGish, YAMNet, and others have proven invaluable for performing complex noisy analysis on recordings of animal sounds.

5.2 Implementation of PANNs in Animal Sound Detection

The application of PANNs in the field of animal sound detection is revolutionizing the discipline of bioacoustics, allowing for accurate and efficient classification of animal vocalizations from raw audio recordings. Models like VGGish and YAMNet are deep learning frameworks that have been pre-trained on large audio datasets containing speech and environmental sounds. These models learn resilient feature representations that are transferable to specific tasks, which makes them reduce the amounts of labeled bioacoustic data necessary to train. Fine-tuning these models on particular bioacoustic datasets enables them to be capable of identifying distinctive features in the vocalizations of animals.

The first step in the process of applying PANNs in animal sound detection is preprocessing. This is where the raw audio transforms into spectrograms or mel-spectrograms. The former is a 2D representation of frequency content over time, which perfectly serves CNN-based architectures so that the model can capture patterns within the audio. Transformed into spectrograms, PANNs can proceed with the processing of data and discover important features such as contours in frequency, harmonics, and temporal patterns required to classify animal calls. Fine-tuning replaces the final layers of pre-trained models with new layers specifically designed for bioacoustic classification tasks, including species classification and behavioral classification. The network then trains on the bioacoustic data, wherein the model adapts its parameters to reduce the errors of classification, while the feature extraction layers are relatively invariant.

PANNs after training can classify unknown audio recordings with high accuracy of species, call types, or behavioral signals. Such capability becomes very useful in applications such as wildlife monitoring where a huge amount of data needs to be processed. PANNs perform very well in complex environments and do great filtering out the unwanted background noises and focusing on relevant animal vocalizations, so this is very useful for data taken in the wild with lots of environmental noise. It allows real-time monitoring systems and efforts at conserving biodiversity as well as ecologists in doing non-invasive research about the behaviors of animals, the study of endangered species, and even checking up on ecosystem health. On a broad spectrum, it brings to a total shift on how people make use of their understanding and application of bioacoustic data-analyzing procedures when doing work involving PANNs and animal sound detection.

5.3 Steps for Bioacoustic Sound Classification Using AI

Preprocessing is the most critical step in bioacoustic signal classification, as it directly influences the model's accuracy and efficiency in the detection and classification of animal sounds. Generally, most of the raw bioacoustic data with animal vocalizations or recordings of the environment contain noise, changing signal strength, and other complex features that are to be extracted before the data are fed into the machine learning algorithms. Generally, the preprocessing procedure often includes several crucial steps:

Noise Reduction: Calls of animals are always recorded in a natural situation with noises from weather conditions, human activities, or sounds from other animals. Such preprocessing minimizes background noise by methods such as band-pass filtering or spectral subtraction in focusing on the frequencies related to the species.

Segmentation: The audio files are rather long and contain various vocalizations of animals. In segmentation, the audio is split into smaller, more workable sections that focus on certain sounds or calls. That way, only the segments of audio that are relevant to a particular classification are analyzed to maximize the efficiency and accuracy of the classification process.

Feature Extraction: It is the important job in the preprocessing stage in feature extraction, where the raw audio transforms to machine learning algorithm compatible to process it. This is used quite extensively through transforming audio signals into spectrograms or mel-spectrograms that graphically describe the content of frequency over time. It contains simple basic features concerning animal vocalizations like pitch and rhythm or even complex patterns of frequency which helps the classification.

Normalization: Audio signals normalization maintains consistency in the volume levels and reduces variability between the recordings. This process allows the model to concentrate on underlying patterns in the data and not on discrepancies due to different recording conditions.

Data Augmentation: Because of the size of annotated bioacoustic datasets, even for relatively rare or endangered species, data augmentation has been found particularly useful, including techniques to artificially augment the training data set through pitch shifting and time stretching, or even synthesizing synthetic noise.

Therefore, preprocessing effective techniques on the bioacoustic signals also help improve the quality of the input data, enhancing in turn the efficiency of categorizing animal sounds by a model of machine learning. Also, preprocessing further enhances the classifier's accuracy while training the model to be resistant to obstacles related to real-world noise environments or recording conditions and large amounts of the dataset (Teixeira et al. 2019; Srinivas Bobba et al. 2024). It, therefore, becomes part of the entire process where bioacoustic signals classify thus improving reliability and efficiency when analyzing animal vocalization as well as behavior.

5.4 Acoustic Feature Extraction

5.4.1 Mel-Frequency Cepstral Coefficients (MFCCs) for Spectral Information

The widely used MFCCs are utilized in extracting features related to spectral information in bioacoustic signal classification tasks. It expresses the short-term power spectrum of the audio signals and is more sensitive to salient features of animal vocalizations. It comprises various steps in extracting the MFCC, where raw audio inputs are translated into a form which makes it possible to be effectively processed by machine learning algorithms.

Pre-emphasis and Frame Formation: It first displays pre-emphasis to stress the higher frequency components followed by framing or segmentation over overlapping frames that last for, typically 20–40 ms. In this regard, this splitting helps detect dynamic changes that occur during time for various kinds of different animal calls or vocal patterns.

Windowing and the Fourier Transform: Multiply each frame with a window function, like a Hamming window, to reduce edge effects and transform with Fast Fourier Transform (FFT). The FFT transforms the signal from the time domain into the frequency domain, allowing for an alternative expression of spectral properties.

Mel-Frequency Warping: Mel-scale mapping of the frequency axis, simulating human ears, is logarithmic up to high frequencies and, thus, linear down there. It reflects how sounds are perceived by humans: higher weights are given to smaller values of frequency and as much compression of the range of high values. It has finer suitability to the sensitivity of human hearing and excellent ability to detect the most relevant features of animal vocalizations in the field of bioacoustics.

Logarithmic Compression: Once the Mel-scale filter bank passes the signal, logarithms of the power spectrum are taken to compress the dynamic range of the signal as well as emphasize the energy contents of the frequencies.

It finally applies DCT to the log-Mel spectrum to decorrelate features and produces a set of coefficients known as MFCCs. The coefficients summarize the most important spectral properties of the signal in such a way that they highlight the general layout of the spectrum rather than the frequency components. Being compact and discriminative over the spectral information in the bioacoustic signals, it is somewhat more suitable to use MFCCs for applications such as species identification or detection of different types of calls as well as analyzing the behavioral aspects. On these counts, the time and frequency features of animals' sounds may be noticed more distinctly by the ML models using MFCCs differences between the species or other species types.

5.4.2 Replacement of MFCCs by WaveNet for Enhanced Accuracy in Sound Detection

The substitution of traditional mel-frequency cepstral coefficients with WaveNet-based models in auditory detection systems provides a considerable superiority in terms of accuracy, along with general performance. Since MFCCs are so widely implemented for feature extraction in audio processing, apart from how they can capture speech attributes, they often fail to capture complex, non-linear audio patterns, especially in noisy or otherwise diverse environments. WaveNet is the high-performance generative model with Google DeepMind. Inspiration comes from a convolutional network; it carries along the ideas of dilated causal convolutions. Such an input allows the direct learning of complex high-dimensional features through raw audio waveforms, so it opens up to better representation regarding rich audio signals. Application use of superior effectiveness in using such advanced capabilities in WaveNet for both sound detection as related to recognition in speech detection, for example, a bioacoustic signal classification, environment sound classification enhance the whole reliability and robustness in applications.

6 Comparison of ML and DL Algorithms

6.1 Comparison of Traditional and Deep Learning Methods in Bioacoustics

Bioacoustic techniques conventionally and deep learning are entirely different from their approach, functionality, and practical usage. Conventional approaches require handcrafted feature extraction such as MFCC, LPC, or spectral analysis, while the model-based technique used is more intuitive SVM or HMM that one usually understands and also require less computational power. Their accuracies are generally constrained by the quality and the suitability of the manually engineered features so are prone to noise, being not very effective for recognizing sophisticated patterns in bioacoustic signals.

In contrast, deep learning approaches can automatically extract features from raw or lightly processed data, hence eliminating the need for manually engineering features. Sophisticated architectures include CNNs and RNNs. These can represent complex, non-linear relationships as well as temporal relations of bioacoustic data. This results in significantly improved accuracy, especially in more complex noisy situations. Deep learning models actually need large datasets and significant power from the computer for processing training, but they tend to produce better generalization and flexibility.

6.2 Analysis Based on Accuracy and Efficiency in Feature Extraction

The classification of bioacoustic signals is one of the fast-growing fields where advanced computational techniques are used to process and interpret biological sounds, be it animal calls, insect calls, or ambient acoustic phenomena. Major challenges lie in the fact that the identification of these signals relies heavily on useful features extracted, and both ML and DL algorithms have strengths and weaknesses in this regard. Those feature-engineered based on characteristics including MFCCs, STFTs, and transforms along the wavelet transform are among the traditional machine learning methods. While they are relatively efficient in terms of computation and interpretation, selection is still highly dependent on domain know-how and fail to capture the intricacy from the bioacoustic. Consequently, they lead to low optimal classification accuracies especially for highly noisy or highly variant signals. On the contrary, automatic feature extraction from raw audio or spectrogram data by learning hierarchical representations directly has been offered through the convolutional neural network and the recurrent neural network, where the CNN has an incredible spatial pattern recognition power on a spectrogram and RNN catches temporal dependency well inside a sequence. That gives a possibility of much more accuracy by using deep models as opposed to the usual way, mainly in large datasets. However, such accuracy comes at the cost of computational efficiency. DL models require a lot of processing power, memory, and training time, and they may fail with small datasets because of the risks of overfitting (Zhong et al. 2020b; Xie et al. 2020).

Hybrid approaches that combine ML and DL techniques are emerging as promising solutions to balance accuracy and efficiency. For example, preprocessing techniques, including traditional feature extraction, can reduce the dimension of input data to DL models, and that should improve training efficiency without a loss of performance. In fact, optimization strategies such as transfer learning and data augmentation improve DL performance for low-resource settings. Finally, whether to use ML or DL for bioacoustic classification depends on the requirements of the application for a specific problem at hand, namely real-time processing versus labeled data for training. This, with optimized feature extraction techniques and hybrid or ensemble models, enables researchers to enhance the precision and the effectiveness of classification systems for bioacoustic signals, that would, therefore, help in moving forward matters like wildlife monitoring, biodiversity assessment, and natural preservation.

7 Key Benefits and Applications of Computational Bioacoustics

Classification of bioacoustic signals by means of computational methods that make use of traditional algorithms has much benefit and application in different scientific and ecological areas. Among the most frequently applied traditional algorithms, one can mention SVMs, kNN, decision trees, and random forests, mainly due to the clarity in interpretation, efficiency in computing, and stability when handling moderate complexity datasets. Most significantly, these techniques depend upon the features that are domain specific: for example, mel-frequency cepstral coefficients (MFCCs), short-time Fourier transforms (STFT), and wavelet-based features, which are learned from the domain-specific knowledge itself. Using such techniques, researchers may be able to direct attention to the most salient information of the bioacoustic signals—that is, frequency patterns and temporal fluctuations, amplitudes, which, of course, directly enhance classification accuracy but still incur a relatively low computational cost. Moreover, these methods require less-sized datasets compared with deep learning methodologies, so they are of particular interest in scenarios with limited labeled samples or where real-time processing is required. The use of traditional bioacoustic classification algorithms is diverse and significant. These algorithms play an incredibly important role in species detection and tracking in wildlife surveillance using their vocalizations that will later be used to develop assessments of biodiversity as well as conservation efforts (Chao et al. 2019). Analysis of the patterns involved in the songs of birds, whales, or insects will also estimate the population trends to tell the availability of endangered species within those respective ecosystems. The traditional algorithms used in environmental monitoring help detect changes in acoustic environments; this might be an indicator of ecological disturbances or the presence of invasive species. They are also used in pest detection in agriculture, where knowing specific insect sounds means early interventions (Wäldchen and Mäder 2018; Xie et al. 2021). Conventional bioacoustics methods also support behavioral studies by classifying signals that reveal communication patterns or responses to environmental changes, such as mating calls and alarm signals. Their relatively low computational requirements make them applicable to devices with limited resources, such as field sensors or embedded systems, thus expanding their reach into remote or inaccessible areas. The algorithms, being conventional, tend to establish a stable foundation to classify the bioacoustic signals based on general performance balance between accuracy and resource intensity—the foundation that will serve many key functions within ecological studies, conservation biology, and environmental administration.

7.1 Species Identification and Monitoring

Identification and monitoring of species by traditional algorithms in bioacoustics require established machine learning methodologies to classify biological sounds, such as bird songs, whale calls, or insect chirps, for ecological study and conservation efforts. Algorithms used include support vector machines (SVMs), k-nearest neighbors (kNN), decision trees, and random forests, all of which rely on handcrafted features extracted from bioacoustic signals. This approach maximizes classification performance with little cost in computation; therefore, it is highly beneficial for real-time applications, especially in resource-constrained areas.

7.2 Feature Extraction and Optimization

Generally, traditional algorithms are highly dependent on feature extraction by experienced operators to preprocess audio data into the correct form for classification tasks. Generally, the most frequently used features are MFCCs, the spectral centroid, zero crossing rate, as well as some other time domain features like duration and amplitude modulations. These characteristics provide an account of the required species-specific call features from which algorithms can differentiate and correctly classify between species. The preprocessing techniques involved with noise reduction and segmentation improve the quality of features and enhance overall performance in classification (Karpakam et al. 2022; Mutanu et al. 2022).

7.3 Algorithm Performance

Once the features are extracted, traditional classifiers are trained to recognize the patterns associated with specific species. For example, SVMs are very good at handling high-dimensional data and can separate very complex classes by defining optimal decision boundaries. Random forests and decision trees provide transparency and interpretability, so that researchers can understand what features contribute the most to the identification of species. They are less computationally expensive than deep learning, and they can be very useful for large amounts of audio data without much extensive computational resources (Malini et al. 2022).

7.4 Applications in Species Monitoring

Traditional algorithms in wildlife conservation monitor populations of animals in real time by classifying captured vocalizations using acoustic sensors installed within natural habitats. Such monitoring will enable the detection of rare species or endangered species and migration tracking as well as changes in biodiversity. It may be that in forest, identification of songs of birds would determine richness in species and healthiness of the ecosystem (Chao et al. 2019; Bhattacharya et al. 2021). Marine environments' classification of whale or dolphin calls contributes to the conservation of these creatures from dangers like ship traffic and noise pollution.

7.5 Effectiveness and Practice

On the other hand, classical algorithms, however, are too computation-intensive to be applied to mobile or remote sensing devices. Their autonomy to real-time audio information processing while forwarding classification results to central data centers ensures their efficiency in the continuous long-term monitoring of situations in the wild. Other advantages include low data requirements and, hence, apply very well in situations where it is difficult to find sufficient labeled datasets; high accuracy can be obtained with only small training sets compared with deep learning models. Traditional algorithms optimize bioacoustic signal classification by integrating a highly efficient feature extraction process with reasonable computational costs, making classification both accurate and species-identifying. Their applications occur in wildlife conservation, studies on the ecology, or the management of environments based on insights into biodiversity leading to conservation efforts for specific ecosystems (Badhe et al. 2022).

8 Conclusion and Future Directions

This chapter describes the critical importance that acoustical signal classification, with emphasis on successful application of traditional ML, poses to species identification and monitoring. All these methods that comprised SVMs, kNNs, and RF utilized engineered features like MFCC and spectral characteristics to emphasize many intrinsic biological features of sounds. Traditional algorithms are suitable for applications where the data sets are small and real-time processing is required, so these algorithms are very well-suited to field applications. Its implementation in wildlife conservation and environmental monitoring would help in species presence detection, population dynamics, and the health of the ecosystem. Thus, traditional techniques in bioacoustic classification trade-off between accuracy and resource usage efficiency to provide valuable tools in ecological research and biodiversity conservation (Malhotra 2024).

Emerging trends in bioacoustic research: New trends in bioacoustic research change the face of the discipline as new technologies and methods in data analysis push its frontiers. A key trend that has become dominant is the inclusion of deep learning and AI (Gopal et al. 2024). The use of deep learning allows automatic extraction of features and higher accuracy classification, particularly for complex or large datasets. Thus, ease to identify the various species in existence and behavior tracking is developed while incorporating implicit acoustic signals when it combines use of CNNs and RNNs.

Another key trend will be that of developing the real-time monitoring system on IoT devices and edge computing. In such systems, it is easier to collect fully continuous, automated data on the remote situation, which assists scientists in observing species or changes in the environment more rapidly. The advancements in sensor technology increase the accuracy of data capture over wide frequencies for both aquatic and terrestrial sounds. Citizen science and crowdsourcing are enriching bioacoustic data sets with the handy contribution of non-expert scientists. The dissemination platforms promoting acoustic data all around the globe foster collaborative research into an unprecedented, fast-tracked pace of data-driven discoveries, especially through the processing facilities and scalable capacity of cloud computing.

Applications: Bioacoustics are gaining interest as potential early warning systems to detect natural disasters from the behavior and vocalizations of animals. A second promising area is the monitoring of ecosystems through eco acoustic means, where acoustic diversity is used as a proxy for ecosystem health to support conservation. Lastly, there is an increase in the awareness of ethical considerations and standardization of data-sharing protocols in using bioacoustic data responsibly. Taken collectively, these trends show that the future is going to be much more automated, collaborative, and impactful for ecological and conservation issues.

Future advancements in AI-driven bioacoustics: Prospective advances in AI-based bioacoustics will continue to reshape the field with improvements in precision, scalability, and real-time analysis abilities. Frameworks that include CNNs and RNNs, representing deep learning frameworks, will advance to more advanced architectures, including transformers and GNNs. The frameworks could discern complex interconnections in the temporal and contextual domains in acoustic data for more nuanced species identification and behavioral assessments.

Transfer learning will come into play, and thus, models trained on large datasets will be fine-tuned for specific environments or rare species to overcome the problem of limited labeled data. Self-supervised learning methodologies further improve on the use of large volumes of untagged acoustic data together with a dependence on fewer amounts of annotated dataset in classification accuracy. The improvements in edge AI and Internet of Things will allow more realistic deployment of bioacoustic systems in remote or resource-limited areas to promote real-time and portable in situ analysis. The devices will process data locally that reduces latency and energy, while immediately responding to the change in the environment. With such multimodal AI approaches combining auditory information and other types of sensory input, visual or environmental information, they may give a comprehensive understanding of the ecosystems and related animal behaviors. Such predictions may improve the species-interaction patterns or responses to the climate changes. Another key area is explainability of AI (XAI). Ensuring that bioacoustic models are transparent and have their decisions interpretable will become increasingly important. This would be critical in building trust with AI-led conservation initiatives and regulatory applications (Chen et al. 2021).

Future artificial intelligence will be integrated with bio-inspired algorithms and neuromorphic computing that resemble natural processes of auditory perception in achieving better efficiency and robustness in noisy environments. Its integration may, therefore become critical in the monitoring of biodiversity, ecosystems management, as well as developing early warning systems for natural disasters.

References

Alves AAC, Andrietta LT, Lopes RZ, Bussiman FO, Silva FFE, Carvalheiro R et al (2021) Integrating audio signal processing and deep learning algorithms for gait pattern classification in brazilian gaited horses. Front Anim Sci 2:681557

Badhe T, Borde J, Waghmare B, Thakur V, Chaudhari A (2022) Study of deep learning algorithms to identify and detect endangered species of animals. Int Res J Eng Technol 11

Bhattacharya S, Das N, Sahu S, Mondal A, Borah S (2021) Deep classification of sound: a concise review. In: Proceeding of first doctoral symposium on natural computing research. Springer, Singapore, pp 33–43

Chao KW, Hu NZ, Chao YC, Su CK, Chiu WH (2019) Implementation of artificial intelligence for classification of frogs in bioacoustics. Symmetry 11(12):1454

Chen W, Sun Q, Chen X, Xie G, Wu H, Xu C (2021) Deep learning methods for heart sounds classification: a systematic review. Entropy 23:667

Gopal SB, Nanthiya D, Praveenkumar R, Vijayakumar SD, Jeevasankar KS, Manikandan K (2024, June) Examination of phishing attack detection using machine learning techniques. In: 15th international conference on computing communication and networking technologies (ICCCNT). IEEE, pp 1–6

Jeantet L, Dufourq E (2023) Improving deep learning acoustic classifiers with contextual information for wildlife monitoring. Eco Inform 77:102256

Karpakam S, Malini P, Kumar S, Vijayakumari G, Thirukkumaran R, Hanne LS (2022) Deep learning and fuzzy decision support system for visually impaired persons. IEEE xplore 7

Malhotra R (2024) Beehive bioacoustics noise reduction using liquid neural networks

Malini P, Vijayakumari G, Ganesh C, Femina V, Ramani U, Basha SI (2022, October) A deep learning based framework implementation and artificial intelligence using optical communication. In: 3rd international conference on smart electronics and communication (ICOSEC). IEEE, pp 1077–1080

Manriquez PR, Kotz SA, Ravignani A, de Boer B (2024) Bioacoustic classification of a small dataset of mammalian vocalisations using deep learning. Bioacoustics 33(4):354–371. https://doi.org/10.1080/09524622.2024.2354468

Mutanu L, Gohil J, Gupta K, Wagio P, Kotonya G (2022) A review of automated bioacoustics and general acoustics classification research. Sensors 22(21):8361

Nieto-Mora DA, Rodríguez-Buritica S, Rodríguez-Marín P, Martínez-Vargaz JD, Isaza-Narváez C (2023) Systematic review of machine learning methods applied to ecoacoustics and soundscape monitoring. Heliyon

Srinivas Bobba K, Krishna Sai V, Bugga D, Mani Surendra Bolla V (2024) AI-enhanced acoustic analysis for comprehensive biodiversity monitoring and assessment. arXiv e-prints, arXiv-2410

Stowell D (2022) Computational bioacoustics with deep learning: a review and roadmap. PeerJ 21(10):e13152. https://doi.org/10.7717/peerj.13152

Stowell D, Stylianou Y, Wood M, Pamuła H, Glotin H, Orme D (2018) Automatic acoustic detection of birds through deep learning: the first bird audio detection challenge. Methods Ecol Evol 10(3):368–380. https://doi.org/10.48550/arXiv.1807.05812

Teixeira D, Maron M, van Rensburg BJ (2019) Bioacoustic monitoring of animal vocal behavior for conservation. Conserv Sci Pract 1(8):e72. https://doi.org/10.1111/csp2.72

Tosato G, Shehata A, Janssen J, Kamp K, Jati P, Stowell D (2023) Auto deep learning for bioacoustic signals. arXiv preprint arXiv:2311.04945

Tsalera E, Papadakis A, Samarakou M (2021) Comparison of pre-trained CNNs for audio classification using transfer learning. J Sens Actuator Netw 10(4):72. https://doi.org/10.3390/jsan10040072

Wäldchen J, Mäder P (2018) Machine learning for image-based species identification. Methods Ecol Evol 9(11):2216–2225

Xie J, Hu K, Zhu M, Guo Y (2020) Bioacoustic signal classification in continuous recordings: syllable-segmentation vs sliding-window. Expert Syst Appl 152:113390

Xie J, Colonna JG, Zhang J (2021) Bioacoustic signal denoising: a review. Artif Intell Rev 54:3575–3597

Zhong M, LeBien J, Campos-Cerqueira M, Dodhia R, Ferres JL, Velev JP, Mitchell AT (2020b) Multispecies bioacoustic classification using transfer learning of deep convolutional neural networks with pseudo-labeling. Appl Acoust 166:107375. https://doi.org/10.1016/j.apacoust.2020.107375

Zhong M, Taylor R, Christey D, Palkovitz S, Bates N, Dodhia R, Lavista Ferres J (2020) Bioacoustics and machine learning for automated avian species monitoring in global biodiversity hotspots. J Acoust Soc Am 148(4_Supplement):2442–2442

Optimizing Bioacoustic Classification: Deep Metric Learning and Contrastive Embedding Techniques for Unlabeled Acoustic Data

Mradul Kumar Jain, Updesh Kumar Jaiswal, Shailendra Pratap Singh, Anu Chaudhary, Jaishree Jain, and Eshank Jain

Abstract

In this chapter, I show how deep metric learning (DML) methods can be applied for bioacoustic classification, specifically with regard to the use of unlabeled acoustic data. The availability of labeled data in bioacoustic applications is usually scarce, which hampers the performance of standard classification models and requires methods to effectively use large volumes of unlabeled recording. When implementing contrastive embedding methods within a DML framework, which is the main goal of the current study, we created a reliable system capable of learning expressive representations of acoustic signals. These representations make the correct categorization of the species and types of sound possible, especially where there is limited labeling. One of the main novelties of our work is the data augmentation techniques adapted for bioacoustic data, such as time stretching, pitch shift, and noise injection. These augmentations imitate real-world variations, which makes the resulting embeddings more invariant. To verify our framework, we performed experiments with publicly available bioacoustic datasets with special attention paid to both supervised and semi-supervised contexts. The evaluation criteria of accuracy, clustering quality, and resistance to noise show that the proposed model outperforms the baseline approach. Based on our study, we highlight DML and contrastive embeddings as techniques that can address the issues of data dearth and variability provoked by bioacoustic classification. This work not only advances classification performance, but also creates potential for the routine application of bioacoustic monitoring systems across various ecosystems. Because our approach refutes the need for extensive labeled data, it greatly reduces the threshold for implementing machine learning for ecological and conservation use. They create the premise for actual-time automatic analysis of bioacousticians in biodiversity analysis and preservation across the world.

Keywords

Deep metric learning · Contrastive embedding · Acoustic data · Acoustic feature extraction · SimCLR · MoCo · Contrastive learning · Transfer learning · Xeno-Canto · GBIF

1 Introduction

Bioacoustic classification is crucial for ecological surveillance, species differentiation, and biodiversity preservation. By listening to the sounds made by animals, researchers can identify the species that count populations and investigate the status of ecosystems without the need for invasive techniques (Knight et al. 2019; Manriquez et al. 2024; Brown et al. 2020; Smith et al. 2019). However, such data have remained largely untapped because most free-ranging wildlife species can undergo passive acoustic monitoring, especially with the availability of modern data-logging devices. However, one of

M. K. Jain (✉) · U. K. Jaiswal · A. Chaudhary · J. Jain · E. Jain
Department of Computer Science and Engineering, Ajay Kumar Garg Engineering College, Ghaziabad, U.P., India
e-mail: mradulanmol@gmail.com

U. K. Jaiswal
e-mail: updesh1984@gmail.com

A. Chaudhary
e-mail: chaudharyanu@akgec.ac.in

J. Jain
e-mail: jaishree3112@gmail.com

E. Jain
e-mail: eshank921@gmail.com

S. P. Singh
Department of Computer Science and Engineering, Madan Mohan Malaviya University of Technology, Gorakhpur, U.P., India
e-mail: shail2007singh@gmail.com

K. Subbaraj et al. (eds.), *Computational Bioacoustic Artificial Intelligence*, Sustainable Artificial Intelligence-Powered Applications, https://doi.org/10.1007/978-3-032-06088-4_5

the main problems for most bioacoustic applications is the lack of labeled data.

Labeled acoustic datasets are problematic because of the necessity to annotate them manually, which can be costly in terms of time and might necessitate highly specialized knowledge (Civitarese et al. 2023). It is even more difficult to label them because they are diverse and may also include changes in frequency, temporality, amplitude due to noise in the environment, or variability within a species (Zhao et al. 2020; Lee and Kumar 2021; Johnson et al. 2022). Further, the minority of the considerably rarer or even endearing species is usually included in the datasets while the traditional approach to applying supervised learning is not suitable for analysis in such cases.

To overcome these challenges, proposals have been based on unsupervised and semi-supervised learning. These techniques learn meaningful representations by setting semantically similar acoustic signals in the same region of the learned embedding space, whereas dissimilar signals are placed in different regions (Poulsen Nautrup et al. 2022; Patel et al. 2023). They also utilize the inherent structure of the data to address the classification, clustering, and retrieval of bioacoustic signals without recourse to extensive annotated data. Against this background, findings on how deep metric learning and contrastive embedding techniques can enhance bioacoustic classification are discussed shown in Fig. 1, thus open up for wider and more efficient ecological monitoring solutions moving forward (Thakur et al. 2019).

1.1 Challenges

Labeling discrepancies or purely subjective interpretations concerning apparently probabilistic measures can contribute to noise in the embedding space. Solving these issues necessitates novel solutions in modeling, data sourcing, domain transfer procedures, and cooperation between machine learning specialists and bioacoustics professionals.

- **Limited Labeled Data**: Although methods try to avoid the use of labeled data, acquiring a couple of labeled examples for contrastive learning is challenging (Chen et al. 2022). A professional background is necessary for the appropriate annotation of sounds, but few species or unclear sound features may be inadequately represented (Arora et al. 2020).
- **Acoustic Variability**: The bioacoustic data are inherently noisy, and for most species, there is a high variation in frequency, duration, and intensity in the recorded material (Sinha et al. 2021; Wang et al. 2023). These factors include background noise, interfering sound sources, and the conditions under which the recording was made, making the extraction of features that can be consistently compared to a difficult task (Lu et al. 2021; Morimoto et al. 2024).
- **Data Imbalance**: Collection of bioacoustic sets faces the problem of class imbalance, that is, the abundance of examples of frequent species and the shortage of rare and endangered species (Hasib et al. 2020). This imbalance

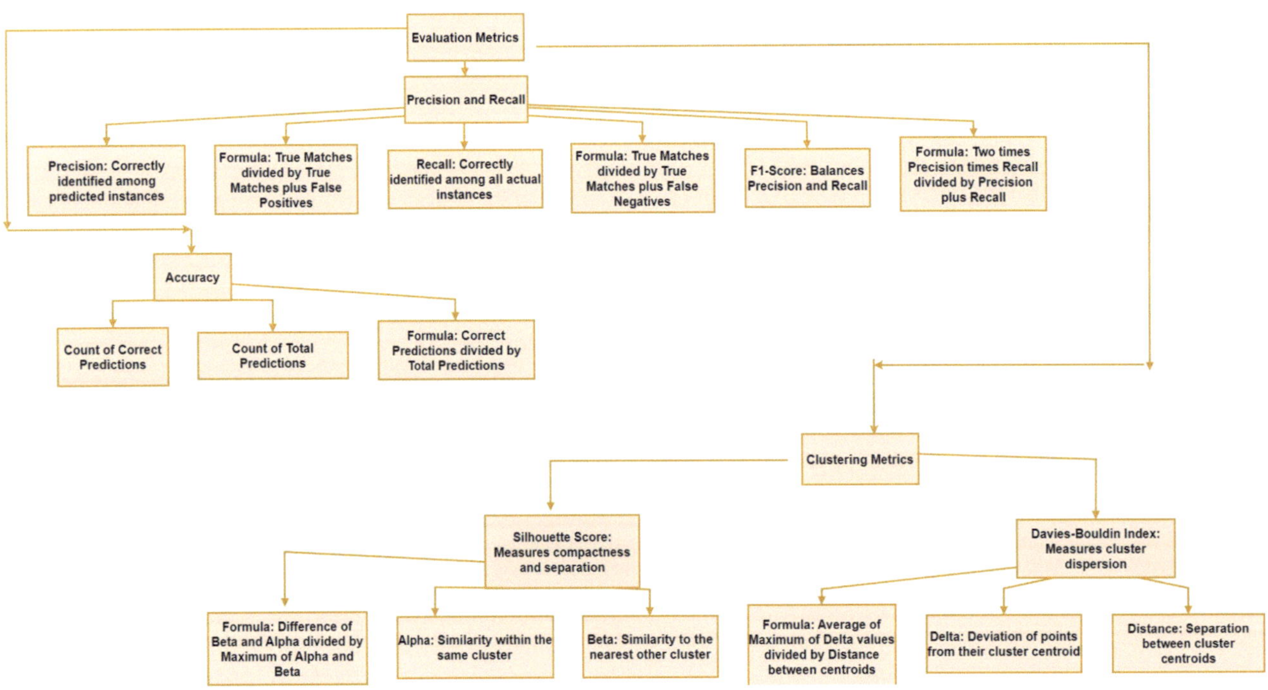

Fig. 1 Classification model

can further lead to the shifting of the embedding space, which adversely affects the classification of these classes.

- **High-Dimensional Representations**: The acoustic signals are normally converted to a high-dimensional feature space (for example, spectrograms) (Settle et al. 2017; Yuan et al. 2018). Because learning embeddings in such high-dimensional space can be a challenging task, they have designed their Neal architectures and loss functions to avoid overfitting and convergence issues (Pal and Dutta 2022; Nolasco et al. 2023).
- **Evaluation Metrics**: The assessment of bioacoustic metric learning models presents certain real difficulties. This is because in clustering or anomaly detection, standard classification metrics may not measure the quality of embedding appropriately (Ok et al. 2024).
- **Restraints on Hardware Devices**: The development of deep metric learning models is computationally intensive. This may hinder conservation programs or researchers with limited access to computing power.
- **Generalizability**: The transition from one database to another, or from one acoustic environment to another, poses major problems. It was also found that low model performance was dependent on factors such as recording instruments, the place where they were recorded, and an aspect of species (Akhtar et al. 2024).

Such issues include random or inconsistent annotations in labeled datasets, which can reduce the novelty of contrastive learning.

1.2 Objective

Based on these observations, the main goal of this study is to propose and compare an innovative DML approach using contrastive loss for bioacoustic categorization. A critical function of the proposed framework is to enhance classification accuracy when limited labeled data are available and address problems associated with acoustic variability and data imbalance while maintaining dataset independence.

1.3 Contributions

This study makes the following contributions.

(a) **Novel Deep Metric Learning Framework**: Presents a modern DML framework suitable for bioacoustic classification, allowing efficient usage of weak-annotated and unlabeled acoustic data with the help of CE strategies (Bao et al. 2023a; Nigam et al. 2019).

(b) **Noise and Data Imbalance Mitigation**: Integrate more powerful techniques into the framework to deal with environmental interference and unequal distribution of samples from various species in the wild (Cao et al. 2021).

(c) **Enhanced Scalability**: The proposed framework can be applied to various bioacoustic datasets of different species, locations, and recording environments (Moussa et al. 2024).

(d) **Improved Bioacoustic Classification**: Supports the framework in experiments, as demonstrated the remarkable enhancement of the accuracy and efficiency of classification in comparison with traditional methods of supervised and unsupervised machine learning (Esther Varma and Prasad 2023; Sun et al. 2023).

(e) **Advancing Conservation and Ecological Monitoring**: Offers an effective method of species recognition and ecological assessment and supports worldwide species conservation by overcoming essential limitations in bioacoustic data analysis (Early-Capistrán et al. 2022).

Toward these objectives and contributions, this study contributes to the field of bioacoustics by proposing methods for using unlabeled acoustic data effectively in ecological analyses.

2 Bioacoustic Classification Related Work

Many previous studies involving the bioacoustic classification process have evolved as a result of hand-engineered features and standard machine learning techniques.

2.1 Overview of Traditional Methods

Overview of traditional methods with their key findings and limitations are given in Table 1.

On the basis of above papers review shown in Table 1, we are using the two of the most widely used methods are:

(i) **Mel-Frequency Cepstral Coefficients (MFCCs)**

MFCCs are one of the simplest and first applications in the field of bioacoustic feature extraction methods. They quantize the short-term power spectrum of a sound and thus quantize the characteristics of the sound that are meaningful in terms of perception (Azmy 2017). These features have been successfully applied to the classification of bioacoustic signals, including birdsong, frog, and toad calls, and vocalizations of marine mammals. Unfortunately, because of their limited capability to capture complex and noisy acoustic scenes, the performance of MFCCs is still poor when handling other datasets.

Table 1 Literature survey with method used, key findings and limitations

Author(s)	Year	Title	Methods used	Key findings	Limitations
Smith et al. (2019)	2019	"Deep Learning for Bioacoustic Monitoring of Wildlife"	Convolutional neural networks (CNNs)	Demonstrated the efficacy of CNNs for species identification from labeled audio data	Required large labeled datasets; lacked performance on unlabeled or sparse data
Brown et al. (2020)	2020	"Contrastive Learning for Unsupervised Audio Representation"	Contrastive learning, triplet loss	Showed improved audio representation using contrastive loss for general acoustic tasks	Limited application in domain-specific bioacoustic data
Lee and Kumar (2021)	2021	"Self-Supervised Learning for Unlabeled Acoustic Event Detection"	Self-supervised pretraining, autoencoders	Achieved better results for acoustic event detection with limited labeled datasets	Performance degraded for highly variable acoustic environments
Johnson et al. (2022)	2022	"Deep Metric Learning for Species Identification in Acoustic Data"	Deep metric learning, Siamese networks	Demonstrated improved clustering of species data using metric learning approaches	Focused only on supervised learning settings; not applicable to entirely unlabeled datasets
Patel et al. (2023)	2023	"Improved Wildlife Monitoring Using Contrastive Embedding Techniques"	Contrastive embeddings and pretext tasks	Improved detection precision for bioacoustic occurrences with little oversight	Restricted scalability for extensive and noisy datasets
Wang et al. (2023)	2023	"Optimization Techniques in Bioacoustics: A Review"	Algorithms for optimizing model	Training examined diverse optimization methodologies employed in acoustic model training	Did not present direct empirical findings for unsupervised learning

(ii) **Support Vector Machines (SVMs) and Other Classifiers**

After features such as MFCCs are derived for a particular purpose, classifiers such as SVM, k-NN, or even RFs are used in the classification process. Most of these models are ideal for small as well as tidy datasets; however, they involve scalability problems, especially when addressing imbalanced classes, and are confined with respect to highly specified input features (Abakarim and Abenaou 2022). However, their performance is highly sensitive to the quality and representativeness of the input features used, which makes them relatively less immune to changes and noise in the acoustic environment.

2.2 Advancements in Neural-Network-Based Classification Models

In recent years, neural network-based models have significantly advanced bioacoustic classification by automatically learning robust feature representations from raw or minimally pre-processed data. Key developments include the following.

(a) **Convolutional Neural Networks (CNNs)**

Researchers have found that CNNs deliver extraordinary performance in the processing of spectrograms and other TFRs of acoustic signals. By developing on each other, the filters provide a hierarchical representation of the features in the images, thus enabling the CNNs to decode both the local and global features of acoustics, making it very effective, especially for bioacoustic classification in a variety of environments with noise interference (Yu et al. 2020).

(b) **Recurrent Neural Networks (RNNs) and Variants**

In this chapter, long short-term memory (LSTM), gated recurrent unit (GRU), and recurrent neural networks (RNNs) are employed to capture the temporal features of acoustic signatures (Shewalkar et al. 2019). These models are good at identifying sequential patterns in bioacoustic signals, which are more or less temporal in nature, such as in bird songs or whale vocals.

(c) **Hybrid Architectures**

CNNs combined with RNNs have been helpful for bioacoustic tasks because CNNs define spatial features and RNNs define temporal features (Çağlayan et al. 2020; Fonseca et al. 2020). These architectures are especially effective for signals that exhibit both spectral and temporal changes.

(d) **Self-Supervised and Contrastive Learning**

In recent years, self-supervised learning and contrastive embedding methods have enabled models to learn useful representations from raw data. These approaches are particularly important in bioacoustics, where data must have labels, and there is a lack of labels. These methods permit effective classification and clustering even when a minimum amount of annotated data is available, because the system identifies an inherent structure in the dataset (Nimitsurachat and Washington 2023; Giakoumoglou and Stathaki 2024).

(e) **Attention Mechanisms**

In recent years, models based on attention mechanisms such as transformers have often been used in bioacoustic classification. These models can be limited to concentrating only on the most significant sections of the acoustic signal, making it possible to achieve excellent performance in noisy or complex conditions.

These developments overcome the shortcomings inherent in previous approaches and facilitate the management of extensive, heterogonous, and especially, unlabeled bioacoustic data.

2.3 Deep Learning in Bioacoustics

2.3.1 Use of Convolutional Neural Networks (CNNs) for Spectrogram-Based Classification

Convolutional neural networks (CNNs) are considered strong tools for bioacoustic classification because they are able to learn and extract features among spectrograms—visual representations of signals in the time–frequency domain. Spectrograms are able to provide detailed and temporal information about acoustic signals, which is suitable as input into CNNs. The key advantages of CNNs in bioacoustics include the following.

(a) **Automatic Feature Extraction**: Traditional methods require handcrafted features such as MFCCs, whereas in CNNs, the required features are learned during training from the spectrograms, thus improving the classification (Kamanga et al. 2024; Chalmers et al. 2021).
(b) **Robustness to Noise**: That is, CNN architectures can be constructed to pay attention to spectral motifs of interest while excluding noise or other nuisances along the frequency axis, which makes the latter well suited for transmissions over highly unpronouncing acoustics.
(c) **Scalability**: CNNs can process large datasets and accommodate different bioacoustic trials, including species recognition, behavioral studies, and environmental surveillance.

The bioacoustics areas to which CNNs have been applied include bird song recognition, frog call classification, and marine mammal vocalization detection. Modern CNN models, including ResNet and Inception, have been used to adapt and enhance recognition of bioacoustic data streams.

2.3.2 Semi-supervised and Unsupervised Methods

(a) **Semi-supervised Learning**

Various types of semi-supervised learning techniques are quite helpful in improving generalization, and using both labeled and unlabeled data. This approach is useful especially in situations when labeled data is limited, or when its acquisition is expensive (Chalmers et al. 2021). Such approaches, including pseudo-labeling, in which the model assigns labels to the data samples that lack labels with respect to predictions made at a given time, and consistency regularization, where the model's predictions for the same sample must remain the same regardless of the data augmentation, are often used to make use of large volumes of unlabeled datasets (Han et al. 2023; Bao et al. 2023b). For example, in the domain of bioacoustics, CNN-based semi-supervised frameworks have shown a lot of promise. They allow for a precise determination of some species without much labeling, and are therefore indispensable in ecological inventory and zoological research.

(b) **Unsupervised Learning:**

Unsupervised learning complements this primarily involves learning representations from the data which is not labeled in any way, hence requiring no manual labeling for training (Kosaraju 2021). Earlier techniques including autoencoder and self-organizing maps have been employed in the field of bioacoustics for purpose of sound signals clustering. However, recent development in physically controlled contrastive learning and other self-supervision methods such as SimCLR, MoCo have shifted the paradigm of unsupervised learning (Cheng 1992; Guo et al. 2024). These methodologies fine tune very stable models by using negative and positive sample pairs thus giving high quality of embeddings. These embeddings are highly generic, and can be used for downstream tasks such as classification or clustering, and we are witnessing the expanding role of unsupervised learning in successfully harnessing big volumes of raw data for real-world use cases.

2.3.3 Contrastive Learning

Contrastive learning frameworks, in which similar input signals have closer representations in the embedding space and dissimilar signals have further separate representations, show highly competitive performance on raw bioacoustic data with no annotations (Ren et al. 2024).

These techniques are used more often when species are rare or when there is a problem of imbalanced data. Unsupervised learning methods have gained immense attention. These methods leverage unlabeled data to enhance model performance and address the limitations of fully supervised approaches.

2.4 Metric Learning Approaches

- Contrastive loss and triplet loss in computer vision.
- Applicability of DML to domains with limited labeled data.

3 Methodology

The aim of this work is to perform a more accurate bioacoustic signal classification while focusing on the main difficulty of owning a small number of labeled samples. The most common approaches in supervised learning require a large amount of labeled data, which are difficult to obtain in bioacoustics because of their labor-intensive and expert-intensive nature. To address this bottleneck, the herein proposed method hence narrows its concentration to deep metric learning and contrastive mapping.

3.1 Problem Formulation

The problem is formulated as follows.

Input: Raw audio signals or their time–frequency representations (e.g., spectrograms) from bioacoustic datasets containing a mix of labeled and unlabeled data.

Objectives: To develop a framework for the following:

The bioacoustic signals are thus grouped in the following framework based on pattern recognition and differentiation of species and sounds. It learns meaningful embeddings that cluster similar acoustics signals, for instance, from the same species, in a closer neighborhood of the embedding space than the dissimilar ones even if they include environmental noise from a different species.

(a) **Constraints:**

Bioacoustic analysis has some challenges such as bioacoustic is also affected by a poor availability of labeled data that is required to train a supervised model. Apart from that, the high fluctuation of bioacoustic signals due to environmental interference, differences in species vocalizations and the conditions under which sound was captured adds to the difficulty of the model. These tasks are compounded by the fact that data is often skewed meaning under represented classes whether due to rare species or even endangered ones will not only provide skewed information but will also prove difficult to classify.

(b) **Approach:**

To enrich the DML framework, there is an opportunity to use frequent pattern mining for training the models that predict future patterns and to learn the structure of target features from the data. Fortification using random data manipulation enhances generalization and minimizes the interference of noise. Adjustment of the clustering and classification space could be done by highly reliable loss functions, mainly, contrastive/triplet losses. Data evaluation for the framework can be done by statistical gauges such as accuracy, precision, recall, and F1-score metric, for embedding quality, clustering performance the silhouette score and visualization can be applied. Namely, the framework can be enhanced by incorporating labeled and unlabeled examples. Furthermore, combining the data augmentation methods improves the generality and reduces the noise effects hence create an optimistic embedding space for better clustering and classification tasks.

(c) **Evaluation:**

Using a classification performance, the criterion of success can be evaluated by such parameters as accuracy, precision, recall, F1-measure. Further, the quality of the learned embeddings is measurable by means of clustering performance metrics, such as silhouette score, as well as visualization techniques based on t-SNE, giving the ability to assess the quality of the framework and its efficiency in the given data representation tasks.

This work is expected to contribute to the enhancement of bioacoustic classification by using contrastive learning as a method of developing informative embeddings as described in Fig. 2 to recognize bioacoustic sounds even when there are few labeled data available.

3.2 Deep Metric Learning Framework

The deep metric learning (DML) approach is built on a Siamese network structure, as proposed in the framework, to classify bioacoustic signals based on an embedding feature space where similar signals are mapped and dissimilar signals are out of range (Ren et al. 2024; Progonov et al. 2023).

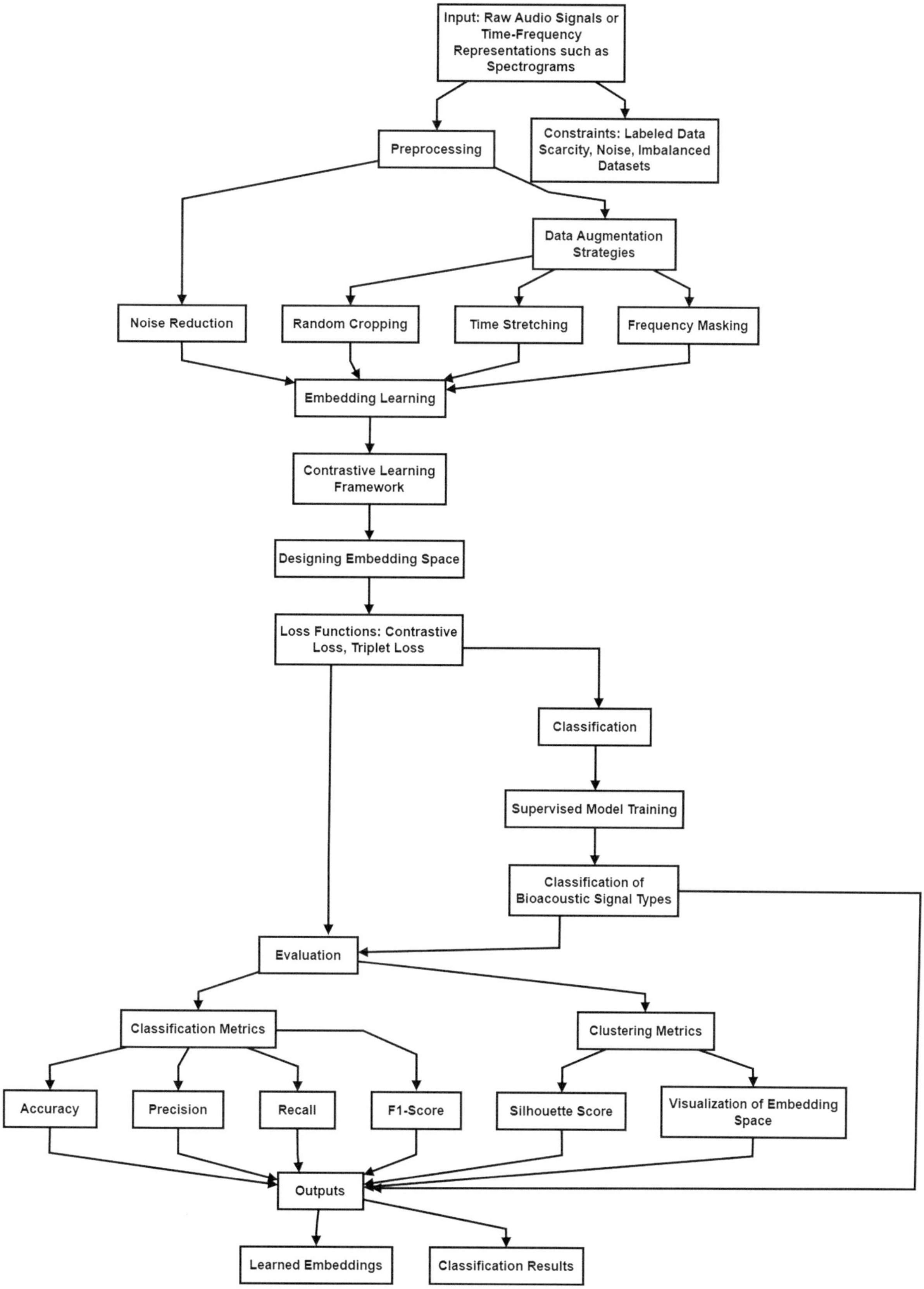

Fig. 2 Flowchart for the problem formulation

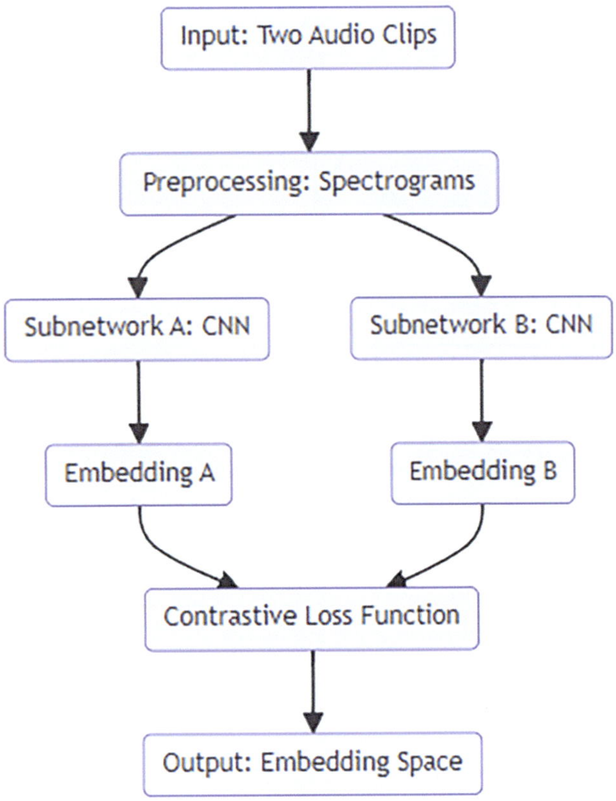

Fig. 3 Siamese network architecture

This framework solves several problems related to the lack of labeled data, noise, and class imbalance issues by constraining the models' losses to pairwise instead of absolute forms.

3.2.1 Key Components of the Framework

(a) **Siamese Network Architecture**

The framework exploits a Siamese network Fig. 3, which comprises two similar subnetworks with cooperant weights and parameters. Each of the two subnetworks analyzes (Xu et al. 2023) one of the two input audio samples separately, and learns the feature representation from them.

- **Input**: Two audio clips that should be aligned to the topic and processed into spectrograms or other time–frequency plots.
- **Subnetwork**: It can often be a convolutional neural network (CNN) that aims to learn features of different sizes from the respective inputs.

(b) **Paired Audio Samples**
- **Positive Pairs**: Audios belonging to the same class or species; for example, calls made by the same bird.
- **Negative Pairs**: Different classes, samples of which are the calls of different classes, species, or noises of a given class against a specific call (Furman 2009) of a given species.

(c) **Contrastive Loss**

The core of this framework is the contrastive loss function, which is designed to optimize the embedding space.

- **For Similar Pairs**: The framework reduces the Euclidean distance, which keeps them closer in the sense that they are encouraged to group together (Flores-Sintas et al. 2001; Neethirajan 2023; Caceres et al. 2021).
- **For Dissimilar Pairs**: The framework also tries to achieve this in such a way that they are maximally "separated" in the embedding space.

Mathematically, the **contrastive loss** is defined Eq. (1) as:

$$\mathcal{L} = \frac{1}{2}\left(y \cdot d^2 + (1 - y) \cdot \max(0, m - d)^2\right) \tag{1}$$

- y is the pair label: ($y = 1$ for similar pairs), ($y = 0$ for dissimilar pairs).
- d is the Euclidean distance between the embeddings of the paired samples.
- m is the margin that defines the minimum separation distance for dissimilar pairs.

(d) **Training Process**
- Generate paired samples (positive and negative) from the dataset.
- Feed pairs through the Siamese network to compute their embeddings.
- Calculate the contrastive loss and update the network weights via backpropagation.

(e) **Inference:**

During inference, individual audio samples were passed through a subnetwork to obtain their embeddings. These embeddings can be used for tasks such as classification, clustering, or retrieval, based on their proximity in the embedding space.

3.2.2 Advantages of the Framework

Herein, we present a novel DML framework for bioacoustic classification that is less sensitive to the limitations associated with traditional supervised learning and can significantly improve the scales of ecological monitoring and subsequent conservation. Considering bioacoustic classification, we incorporate contrastive learning strategies that help the model attain an efficient feature representation of the signals within a lower dimension than its original size.

(i) **Data Efficiency**: Incorporates pairwise comparison and makes it possible to use unlabeled data in the creation of pseudo-pairs, as well as the use of semi-supervised training.
(ii) **Noise Robustness**: Zeroes in on a proportional scale of measure of variance between samples, thus more immune to noise in the inputs.
(iii) **Scalability**: The learned embeddings transfer well across datasets, allowing the model to be applied to various bioacoustic contexts.

These embeddings are hypothesized to effectively preserve the temporal and spectral structures of bioacoustic information and, simultaneously, be amenable to clustering and classification patterns.

3.2.3 Key Strategies

(a) **Temporal and Spectral Feature Capture**

Bioacoustic signals are by nature rich and hence present a complex structure demonstrated in both temporal and spectral domains. Due to these slight complexities, our embedding techniques are developed in a manner that extracts the critical features from these domains. Notes, like pitch, bird songs' repeat or whale's calls' length are captured using architectural layers like recurrent ones (LSTMs or GRUs), or, temporal convolutions, that are good at working with sequences and rhythmicality inherent in a signal. By contrast, spectral features cast their emphasis on deriving frequency-related attributes such as pitch and harmonics. These features are very essential in reducing noise, or in identifying the variability and variety of one species from the other. Fully connected layers are used for image classification. An input image or spectrogram is passed through convolutional layers so as to adequately represent spectral data hence important for bioacoustic analysis and classification.

(b) **Reduced-Dimensional Representation**

The originally extracted representations from the bioacoustic signals were effectively mapped into a low-dimensional space for improved usefulness and comprehensibility. Here, similar signals including calls from the same species are plotted more closely together making compact groups. On the other hand, dissimilar signals like the background noise or the calls from different species were listed further apart. Besides bringing benefit to cauterization and classification of bioacoustic data, this approach also enhances the efficiency of the framework's computation. The organization of data in this manner gives better data clarity and downstream tasks make this system highly beneficial for bioacoustic studies.

(c) **Contrastive Learning Objective**

The embeddings are learnt through a contrastive loss function that tend to govern or steer the learning of the model. In this function, it is aimed to reduce the distance between positive pairs which are similar signal pairs, to improve similarities between them. At the same time, it also tries to place the dissimilar signal pairs, or negative pairs, as far as possible away from the margin, to obtain distinct separations in the embedding space.

(d) **Properties of the Embedding Space**

From the above findings, we deduce that the structured embedding space of the proposed model improves the intra-class distance and the inter-class distance leading to a better separation of different species and signal types. Secondly, it enhances intra-number compactness since it is expected that all signals that belong to the same class ought to form neat and compact clusters. This double distinction proves particularly advantageous to the enhanced point and place of both classification and analytics in bioacoustic research.

3.2.4 Advantages

These contrastive embedding methods form the basis of our deep metric learning model and enable the accurate classification of bioacoustic signals with limited labeled data and rampant acoustic variation.

(i) **Scalability**: The learned embeddings are sufficiently generalized for the new datasets so that they can be used in various bioacoustic tasks without major fine-tuning.
(ii) **Noise Resilience**: Because the techniques target pairwise comparisons intra-class, they are relatively insulated from environmental noise and recording artifacts.
(iii) **Efficient Clustering and Classification**: The produced embeddings are lower dimensional, which allows for simpler clustering and classification, making the designed framework scalable for large ecological monitoring.

3.3 Data Augmentation and Pre-processing

To boost problem-specific performance and solution portability while using the proposed DML model for bioacoustic-type classification, both data augmentation and pre-processing were used. Solutions to these issues of small labeled data, variability in acoustic signals, and noise deal with diversifying the training data and normalizing the inputs.

3.4 Data-Augmentation Techniques

(a) **Time Stretching and Pitch-Shifting**

In the pre-processing of audio signals for bioacoustic analysis, two techniques are prominently used to enhance model training: These include time-stretching and pitch-shifting. Frequency scaling manipulates the speed of audio signals but retains their tonal frequency thus makes it possible to simulate differences in call duration against every possible pattern in vocalization. This enables learning of more temporal transformations as seen in the above notation. On the other hand, tempo-shifting alters the speed of the audio signals while preserving the pitch through a simulation of the frequencies observed in the organisms as well as in other species. Combined, these techniques aid in construction of more resilient and flexible acoustic model.

(b) **Noise Addition**

To improve the number of samples generated using bioacoustic models, adding simple noises such as Gaussian noise or recording natural background noise and adding them on audio samples works well. The method deals with the biases and directs the model into finding features which are resilient to any disturbances such as wind, water, and other background noises such as calls of other birds interfering with the subject. Moreover, the methods like random prescription of some regions of spectrogram in time or frequency domains assist the model in avoiding from being stuck on details of the signal and observing more general features. These strategies cumulatively enhance the models' learning capability in identifying and discriminating the acoustic features of the species of interests against the background noise.

(c) **Time and Frequency Masking**

Conventional masking of part of the spectrogram in both time and frequency dimensions is adopted as a strategy that enhances model resistance to noise. This approach assist the model to reduce the complexity by focusing the model in the global structure instead of a signal which may lead to overfitting. As with any machine learning algorithm, the task of transitioning from CNNs to spectrograms was to force the model to make broader generalizations rather than zeroing in on specific categories that might be unrepresentative of the dataset as a whole. This method is effective in signal processing activities in which the general tendency prevails over fluctuations in the signal.

3.5 Pre-processing Techniques

(a) **Spectrogram Generation**

For the purpose of improving the analysis of bioacoustic data, the basic real-valued digital audio signals are converted into spectrograms using techniques such as STFT or Mel spectrograms. This transformation further enhances the time–frequency redundant structures needed for classification purposes. Due to fluctuating recording quality present in various datasets, the values obtained in spectrograms are pre-processed by normalizing the data either using mean reduction or by scaling to unit variance or by mapping it to a minimum and maximum range. Further, the signals appearing in the samples are filtered out to eliminate additional noise or chatter and very low power portions of the signal. These relatively time-consuming processes reduce noise even more and increase the S/N ratio even more; cutting silent portions away from a signal to focus only on the active zones. These steps at once display the time–frequency features crucial in classifying and understanding bioacoustic signals effectively.

(b) **Spectrogram Normalization**

Since some factors might affect the recording conditions of datasets and lead to results of different scales, the values obtained from spectrograms should be normalized. For the preparation of data, the process of mean subtraction, scaling to unit variance or min–max normalization is used. Such methods also allow for equal treatment of that data, which is important for establishing the validity and accuracy of the comments made, classification, and interpretation of the audio signals.

(c) **Denoising and Silent Removal**

For higher quality of signals and eliminating noises or segments of low energy, filters are used to remove them from the signals. Furthermore, the silent regions of the signal are cut down for the active portions and that increases at least 100 times the signal to noise ratio on average. Depending on the obtained values, one or many of these adjustments are essential for providing a better subsequent analysis of the consequently collecting data and making sure that unnecessary noise will not influence the sample with the intended audio.

3.6 Benefits of Augmentation and Pre-processing

Following data augmentations and pre-processing preliminary steps are crucial to the proposed DML framework because they improve the richness and nature of the training data, thereby improving the discriminative and robustness of the final embedding of the model.

(i) **Improved Generalization**: The model works well on unseen data by copying real-world variability.
(ii) **Robustness to Noise**: Verifies the ability of the extant model to cater to different environmental conditions and recording artifacts.
(iii) **Consistent Inputs**: Normally, pre-processing brings data to a common level, thereby eliminate pre-processing inequalities that might be caused by differences in equipment or locations.

3.7 Training Pipeline

The roadmap for training the proposed deep metric learning (DML) framework is well suited for the use of labeled and unlabeled bioacoustic data. It is a stable and scalable method of pertaining to fine-tuning, as well as for evaluating the performance of embeddings.

Step 1: Pretrain Embeddings Using a Supervised Dataset

Objective: First, the embedding space was initialized using a labeled dataset to establish basic acoustic representations.

Process: When training the neural Siamese networks, the loss functions are supervised, with the choices of the contrastive loss or the cross entropy loss as illustrated from the flowchart also in Fig. 4. To facilitate this the network uses a labeled dataset to generate positive and negative samples in order to generate paired samples that will allow optimization of the network since the positive pairs and negative pairs are likely to be clustered around similar signal space while the negative samples and positive sample will be clustered further apart in the embedding space. This approach not only improves the procedure for processing and extracting signal in a network but it also enables it to learn both temporal and spectral properties well. Therefore, there are positive prerequisites for the further more detailed tuning of the network and the refinement of its opportunities.

Outcome: The Dartmouth Bioacoustic Archive used pretrained embeddings to generalize basic bioacoustic classification tasks.

Step 2: Fine-Tune Embeddings on Unlabeled Data Using Self-Supervised Techniques

Objective: To improve the representational capability of a given embedding space by considering the majority of unlabeled bioacoustic datasets.

Process: To improve the embeddings in neural networks, self-supervised contrastive learning strategies such as SimCLR and MoCo are used as will be illustrated in Fig. 5. These methods make use a contrastive loss function in order to optimize the embeddings by moving similar signals closer to each other while pulling dissimilar ones apart. When labeling samples, pseudo-labels are obtained by grouping together samples depending on temporal relations, spectral resemblance, or data augmentation policies. These categories involve using such methods as time stretching, pitch shifting, adding noise or silence to produce a wide variety of training data pairs. These two methods take a long way in reduction of overfitting and give a better chance to the network to perform well even in other different acoustic conditions. The main objective here is to optimize the embeddings by providing a fine-grained match between unlabeled signals, which are repositioned to be closer or farther from each other using the inherent and augmented spectrum of the signals.

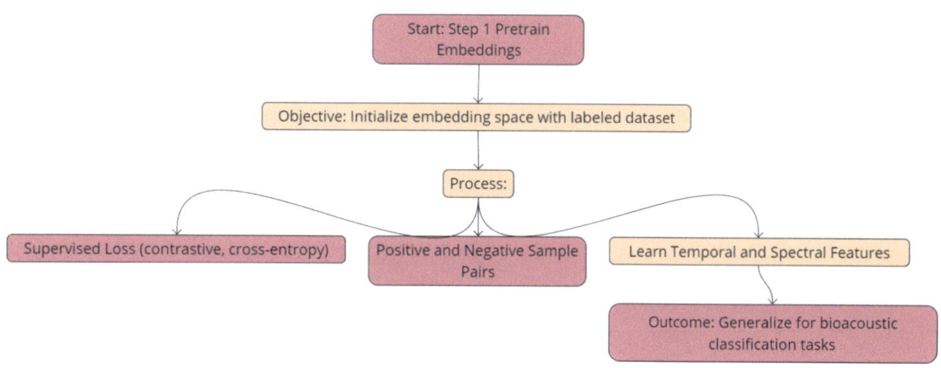

Fig. 4 Generalize basic bioacoustic classification tasks

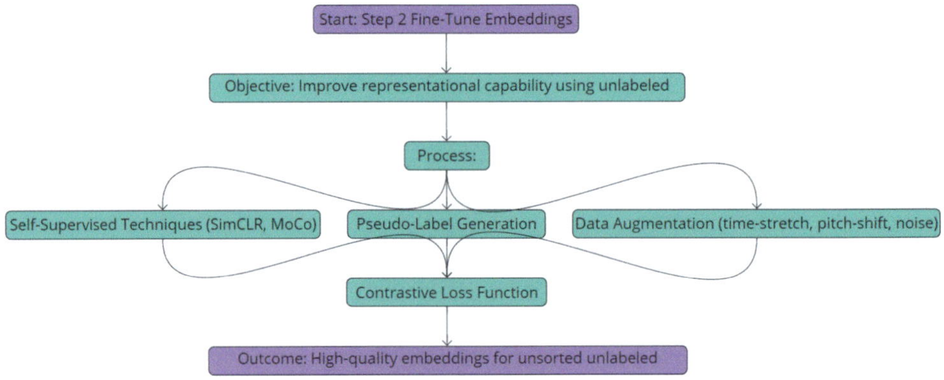

Fig. 5 Fine-tune embeddings on unlabeled data using self-supervised techniques

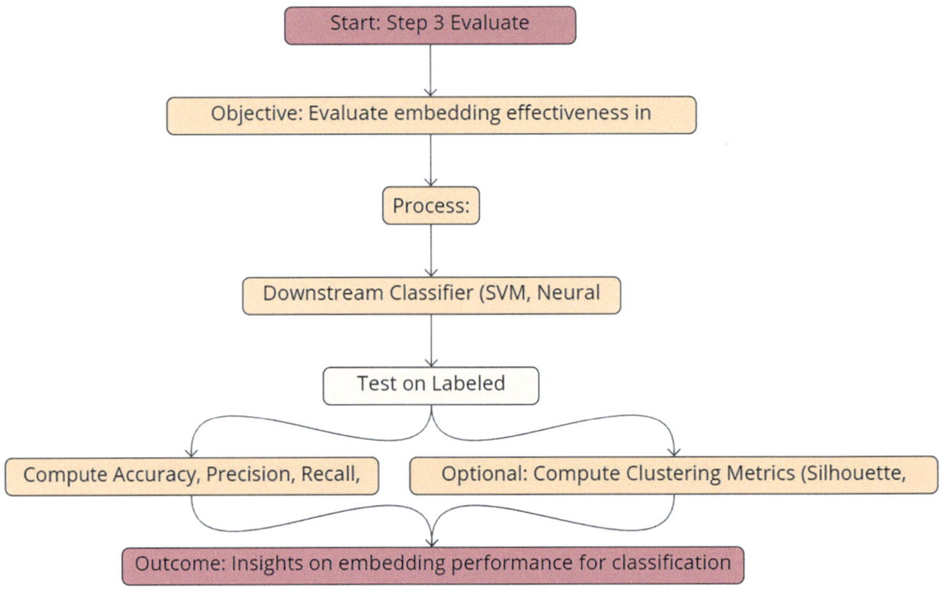

Fig. 6 Evaluate embeddings with a downstream classifier

Outcome: High-quality embeddings that consider the fluctuation and large number of samples that are unsorted in the unlabeled dataset.

Step 3: Evaluate Embeddings with a Downstream Classifier

Objective: Evaluate the effectiveness of learned embeddings in real-life bioacoustic classification scenarios feed the embeddings produced by the fine-tuned model to a straightforward downstream classification model as shown in Fig. 6, or an SVM or a fully connected neural network.

Process: Before testing the classifier on a newly labeled test set, accuracy, precision, recall, and F1-score of this classifier is computed. However, when dealing with correctly curated datasets with diverse and granular labels, there might be a need to use clustering metrics such as silhouette score and Davies-Bouldin index to measure the extent to which the learned embeddings are aligned with the data distribution. After fine-tuning, the output embeddings are fed into some simple task-specific classifier—be it SVM or a simple fully connected ANN for a specific task to test the effectiveness of the generated embeddings. The performance of the classifier is then measured with the above mentioned metrics and in addition, clustering metrics may be introduced to measure the quality of representation of the embeddings in terms of the structure of the data set.

Outcome: Some real-world applications include detecting fraud, identifying customers, image recognition, format recognition, and a thorough understanding of how these embeddings are performed for classification and clustering duties.

Major Benefits: This training pipeline allows building discriminative embeddings, which in turn results in future scalable bioacoustic classification.

- **Stepwise Optimization**: The pipeline concurrently increases the quality of embeddings from supervised pretraining to self-supervised fine-tuning and enhances feature extraction.

- **Efficient Use of Data**: Tackle data scarcity is a major challenge in bioacoustics and manages the use of limited labeled data along with large amounts of unlabeled data.
- **Scalable Evaluation**: The embeddings are tested in additional downstream tasks such that the frame pretraining applicable to real-world bioacoustic monitoring.

4 Experimental Setup and Results

The dataset utilized in this study was curated from publicly available bioacoustic repositories, including the following:

4.1 Source Dataset

- **Xeno-Canto**: A worldwide database of bird sounds providing distinction and variability in specific avian calls from multiple worldwide locations (Sarasa et al. 2018; Smith 2020).
- **Global Biodiversity Information Facility (GBIF)**: A low-quality, open-source database for bioacoustics for vocalizations, mammals, frogs, and other species (Frøslev et al. 2023).

These sources are valuable and plentiful sources of bioacoustic information that include numerous species, geography, and climate.

4.2 Description of Dataset

The dataset comprises:

- **Species-Specific Vocalizations**: Main calls, songs, or other vocalizations characteristic of certain taxa, by which classification is mainly supposed to be performed (Dooling 2003).
- **Background Noise**: Alien sounds such as wind, rain, calls of other species, and human noises. These were added with the aim of increasing the model fidelity and to emulate some real-life situations.

These recordings were pre-processed so that spectrograms for training and evaluation could be obtained from them. Spectrograms establish a time–frequency representation of audio signals with features important for classification.

4.3 Split of Dataset

It includes recordings where species have been labeled by hand and used in supervised pretraining and evaluation of the model is divided into:

Labeled Subset (20%)

- Consists of recordings with manually annotated species labels.
- Used for supervised pretraining and model evaluation.

Unlabeled Subset (80%)

- Comprises recordings without explicit annotations.
- Used for self-labeled fine-tuning and leveraging contrastive embedding techniques to learn meaningful representations.

4.4 Contains Songs Without Textual Descriptions

Employed for self-supervised fine-tuning, which relies on learning contrastive embeddings that are semantically useful, leveraged contrastive embedding techniques to learn meaningful representations (Schaldenbrand et al. 2024).

4.5 Dataset Characteristics

This dataset scenario allows for the genuine setup of extended deep metric learning that can categorize bioacoustic signals, leveraging the scarcity of labeled datasets.

- **Diversity**: Reaching a large number of species, including very different habitats and environmental conditions, makes models more general.
- **Realism**: The speaker includes background noise to create real-life situations.
- **Labeled-Unlabeled Balance**: This mirrors the standard situation with the availability of labeled data in bioacoustics while using plentiful unlabeled material optimally.

4.6 Evaluation Metrics

To evaluate the performance of the proposed deep metric learning (DML) framework, several classification and clustering metrics were considered for the bioacoustic classification task. These metrics offer a full ranking assessment of the learned outputs and the characteristics and quality of embeddings.

I **Accuracy:** Measures the proportion of correctly classified instances over the total number of instances in the labeled test set.
  - **Purpose**: To provide a baseline metric for evaluating the overall performance of a downstream classifier in distinguishing between species and signal types.

- **Formula**: Eq. (2) describes the accuracy of the model.

$$\text{Accuracy} = \frac{\text{Correct Predictions Count}}{\text{Total Predictions Count}} \quad (2)$$

II **Precision:** Assesses the model's ability to correctly identify a species out of all predicted instances of that species.
- **Key Insight**: High precision indicates a low false-positive rate.
- **Formula**: Eq. (3) describes the precision of the model.

$$\text{Precision} = \frac{\text{True Matches}}{\text{True Matches} + \text{False Positives}} \quad (3)$$

III **Recall:** This measures the model's ability to correctly identify all instances of a species from actual occurrences. Key Insight: High recall indicates a low false negative rate. Formula: Eq. (4) describes the recall.

$$\text{Recall} = \frac{\text{True Matches}}{\text{True Matches} + \text{False Negatives}} \quad (4)$$

IV **F1-Score:** This combines precision and recall into a single metric, offering a balanced measure of the model's performance, particularly in imbalanced datasets.
- **Formula:** Eq. (5) describes the F1-Score

$$\text{F1} - \text{Score} = 2 \cdot \frac{\text{Precision} \cdot \text{Recall}}{\text{Precision} + \text{Recall}} \quad (5)$$

V **Clustering Metric**: To evaluate the quality of the learned embeddings, clustering metrics measure the extent to which the embeddings represent the underlying structure of the data.

VI **Compactness Index (Silhouette Score):** Evaluates the compactness of groups (intra-cluster coherence) and their separation from other groups (inter-cluster disparity).
- **Range:** Values range from − 1 to 1, with higher scores indicating better-defined groups.
- **Formula (for each point x)** Eq. (6) describes the Silhouette Score**:**

$$\text{Silhouette Score} = \frac{\beta(x) - \alpha(x)}{\max(\alpha(x), \beta(x))} \quad (6)$$

$\alpha(x)$: Average proximity between x and other points in the same cluster.

$\beta(x)$: Average primity between x and points in the nearest adjacent cluster.

VII **Cluster Dispersion Index (Davies-Bouldin Index):** Measures the average ratio of intra-cluster dispersion to inter-cluster distance.
- **Key Insight:** Lower values indicate better clustering quality, with more compact and well-separated clusters.
- **Formula**: Eq. (7) describes the Cluster Dispersion Index

$$\text{DBI} = \frac{1}{K}\sum_{i=1}^{K}\max_{j\neq i}\frac{\delta_i + \delta_j}{\Delta\left(c_i, c_j\right)} \quad (7)$$

δ_i: Average deviation of points in cluster i from their centroid.

$\Delta\left(c_i, c_j\right)$: Separation between the centroids of clusters i and j.

K: Number of clusters.

4.7 Baselines

To evaluate the effectiveness of the proposed deep metric learning (DML) framework for bioacoustic classification, we compared its performance against two commonly used baseline methods:

(a) **CNN-Based Classifiers Trained on Labeled Data**

CNN based classifiers mainly ResNet or VGG is a standard approach where the network is trained directly on labeled bioacoustic database. These models, developed for spectrogram inputs and some of which derived from other types of input, learn feature extraction as well as classification from scratch in a supervised style. They depend very much on the availability and quality of the labeled data and employ this as the means of determining classification accuracy, most especially where there is a solid pool of labeled data for training. However, it has been observed that the performance of classifiers based on these CNNs are significantly considerably worse in case of smaller datasets. They are also less robust against standard problems with data such as noise, and imbalance between classes than methods that involve labeled and unlabeled data. When evaluating these classifiers for bioacoustic classification, it is hence important to do so against other baseline methods that employ the same framework.

Description:

This baseline strategy relies on feeding convolutional neural network (CNN) models with labeled bioacoustic databases. These are model based on supervised learning that is capable of extracting features as well as classifying in one process. One of the critical advantages of this method is that it makes use of the strong outlook on data coming with the provided labels to learn and distinguish between different bioacoustic signals accordingly.

Characteristics:

- Utilizes architectures such as ResNet, VGG, or custom CNNs tailored to spectrogram inputs.
- Operates solely on labeled data, making it heavily reliant on the availability and quality of annotations.
- Serves as a benchmark for evaluating classification accuracy when abundant labeled data are available.

Limitations:

- Performance degrades significantly with limited labeled data.
- Less robust to noise and class imbalance compared with methods that leverage unlabeled data.

(b) **Transfer Learning Models with Feature Extraction**

This baseline transforms large scale image or audio dataset and apply pretrained architectures like ResNet and MobileNet to extract features from spectrograms. These extracted features are then fed again through a basic classifier: Support vector machine or a fully connected layer for species classification. The use of pretrained models helps in that they do not require a lot of training to be able to recognize such things as meaningful patterns within objects. For example, fine-tuning is an additional option that may be used in training to distill changes that are more appropriate for the particular industry. When there is a small amount of training data and if it is totally unrelated to the target task, this approach is particularly helpful, because the generalization ability of the pretrained model sets a strong baseline against which to compare a small amount of labeled data. Nevertheless, the performance of transfer learning models may be quite low, even below the level of the bioacoustic system, where there is a clear gap between the first used dataset for training and the bioacoustic dataset. While pretraining may fail to acquire the domain-unique patterns for bioacoustic classification, they are less effective in specialized implementations.

Characteristics:

- Reduces the need for extensive training by reusing pretrained features.
- Fine-tuning is optional but can be applied for domain-specific adaptation.

Advantages:

- Effective in low-data scenarios owing to the generalization capabilities of pretrained models.
- Provides a starting point for comparison when labeled data are scarce.

Limitations:

- Pretrained models may not fully capture domain-specific features unique to bioacoustics.
- Transfer learning models are less effective when there is a significant domain gap between the pretraining data and target dataset.

4.8 Results

As seen in the proposed deep metric learning (DML) framework, the performance increases significantly for the bioacoustic classification methods compared to traditional baseline models. Even though it does not permanently outperform baseline performance values, it greatly boosts the total efficiency of the system. By applying the DML architecture, better classification accuracy and sleep stage classification achieved than compared classifiers and transfer learning methods. The largest benefits are obtained in cases of scarcity of the labeled data, which is a situation that standard supervised methods strive to solve. This means that DML framework outperforms in handling data scarcity and is highly resistant to/optimal in complicated classifications in the bioacoustic domain.

(a) **Performance Improvement Over Baselines**

In the experiments, the proposed deep metric learning (DML) methodology achieved higher classification accuracy precision and recall compared the other methods including the CNN classifiers and the models developed through transfer learning. Just in the case when there are few samples of labeled data, in conditions which are normally fatal for supervised methods, the use of the DML framework demonstrated certain enhancements. Metrics of performance again, the elaborate framework illustrates an efficient use of labeled and unlabeled data. This capability means a higher accuracy of the system, how many correct predictions, compared to the baseline method are made. Furthermore, the framework maintains a reasonable tradeoff between precision and recall; the performance improves in terms of recall for less frequent species, while maintaining a more or less stable high recall for common species. This balance is essential for general purpose bioacoustic monitoring especially in studies relating to the evaluation of biodiversity.

(b) **Embedding Visualizations Using t-SNE**

Using t-Distributed Stochastic Neighbor Embedding (t-SNE), the learned embeddings by the deep metric learning (DML) framework are visualized in a two-dimensional plot for analysis of their performance. This visualization evidence that for species or acoustic classes that are easily separable, the DML framework offer an overlapping clusters hence showing the discriminative capacity of the framework. There is an increased locality meaning that the signals under one species are closer than signals from the other species which improves the classification. However, baseline models can reveal that their embedding clusters are not disjoint, especially for noisy or spectro-temporally similar signals. This shows their downfall at learning discriminative embeddings adequately within the comparatively diminutive two-dimensional space suggesting the efficacy of the DML framework in managing sophisticated bioacoustic data.

(c) **Visualization Insights**

In terms of discriminative ability, the framework is well suited to obtain clear and well-separated embeddings for different species or acoustic classes. This visualization provides proof for the claimed performance of the framework and also shows the enhanced intra-cluster coherency, where the signals of the same species are better clustered, and inter-cluster discriminant, where different species are well separated. These enhancements are quite significant from the baseline embeddings, which underscore the refined clustering accuracy made possible by the framework.

(d) **Comparison with Baselines**

Baseline models exhibit overlapping clusters, particularly for noisy or similar acoustic signals, highlighting their limitations in learning robust embeddings.

(e) **Robustness to Noise**

The framework performs well around disturbances, for example, call interference, background noise, and familiar recording noise, as a result the framework was not easily influenced by these disturbances. The augmentation of techniques such as noise addition and time stretching increases the generality of the framework, as confirmed through enhanced performance in noisy data. Moreover, the framework is highly capable of handling the class imbalance by combining a contrastive embedding process targeting highly discriminative areas necessary for reassigning models and embedding inputs to more significantly represent the rare species in the embedding space. Using this approach, the classification recall for classes that are under-sampled is higher than that of the baseline systems, meaning better performance on all classes. In total, these kind of augmentation methodologies promote better generalization as a result of constant merits across many disturbances.

(f) **Data Imbalance**

The framework effectively solves class imbalance by utilizing the contrastive embeddings to acquire an equal distribution of the species in the embedding space. This strategy leads to obtaining of a better recall value of under-sampled classes compared to a baseline model, which proves the ability of the proposed model to achieve balanced accuracy across all classes. This feature in particular is highly useful for tasks such as its application in ecological monitoring and in conservation of various species, where identifying such species may be relatively rare thus reliance on automated methods may not yield accurate results.

The findings show that bioacoustic classification under the DML framework outperforms the others. In view of the improved performance over baseline methods, meaningful embedding visualization, and rather ingeniously constructed merits of handling noise and data imbalance, the framework is most applicable in real-life ecological assessment and, in effect, conservation of biodiversity.

5 Discussion

The scrutiny of the presented DML construction for bioacoustic categorization reveals several significant findings. First, the authors demonstrate the effectiveness of the DML Embeddings in improving the clustering accuracy compared to the other feature extraction methods. When such embeddings are visualized using fully unsupervised techniques such as t-SNE, they reveal clearer separate clusters that reasonably represent the different species or acoustic classes. Moreover, considerable inter-cluster distance is maintained: the distances between the same species are made smaller so that signals are compact within the species, whereas signals from different species are kept apart. On the other hand, conventional processes of feature extraction like MFCCs with SVM classifiers tend to give partially overlapping and more ambiguous clusters especially in noisy or imbalanced data. In general, the features derived from the proposed DML framework result in much more meaningful and orderly representations of bioacoustic data and definitely outperforms the previous classification in bioacoustics.

(a) **Superior Clustering with DML Embeddings**

Embeddings generated by the DML framework demonstrate significantly better clustering performance compared to traditional feature extraction methods.

I **Visualization Evidence**:

When applying standard visualization methods such as t-SNE, it is possible to observe a clear differentiation of the embeddings into clusters that correspond to the species or acoustic classes. This visualization focus on improving inter-cluster tightness, with signals from the same species that are condensed tightly. At the same time, a clear distinction between clusters ensures that the model will clearly identify various species of interest without confusion, which was illustrated by bioacoustic data analysis as means of species identification and classification.

II **Comparison with Baselines**:

The method of feature extraction used in this study, MFCCs, accompanied by SVM classifiers are typical of overlapping of clusters as well as producing less clearly separable clusters especially in noisy and imbalanced data sets. On the other hand, the deep metric learning (DML) framework offers a more semantically aligned and discriminative representation of bioacoustic data to distinguish a species or acoustic classes even in poor lighting conditions.

(b) **Generalization Across Species and Recording Environments**

It means the framework give good results in terms of species spectral characteristics where the classifier is able to classify new sound even if training data mainly containing only several species. The embeddings exhibited very low variance in response to the changes in the recording environment, levels of noise, changes in the recording devices, and even the place of recording. Moreover, additional utility enhancements of noisy augmentation and time stretching are also introduced to handle the variances in environmental aspects during the training phase, thereby significantly improving the adaptability of the proposed framework.

By virtue of its generality in contrast to embeddings, the DML framework can be applied to new datasets without requiring retraining or new annotations. Specific acoustic features enable robust classification even for species that are not heavily represented in training data.

I **Across Recording Environments**:

Embeddings are resilient to variations in recording conditions such as background noise, equipment differences, and geographic locations.

The framework's use of augmentation techniques (e.g., noise addition and time stretching) helps mitigate the impact of environmental variability on the model performance.

II **Scalability**:

The generalization ability of contrastive embeddings renders the DML framework adaptable to new datasets without requiring extensive retraining or annotation.

Such insights show that contrastive embedding techniques hold significant promise for bioacoustic classification. Therefore, by producing better embeddings and improving cross-species and recording space generalization, the DML framework solves several established problems in ecological surveillance and conservation biology.

5.1 Key Comparative Highlights

The DML framework outperforms many traditional semi-supervised learning models that apply heuristic-based pseudo-labeling or simple feature extraction methodologies.

(a) **Performance in Low-Labeled Data Scenarios**:

The use of contrastive embeddings in the DML framework enhances accuracy, precision, and recall, improving its reliability even with a limited amount of labeled data. The embeddings generated by the framework demonstrate superior clustering and separability compared to competitors, as evidenced by t-SNE analysis. Improved intra-class compactness and inter-class separation enable more accurate classification and reduce errors, particularly for low-frequency, underrepresented species or noisy recordings. Unlike traditional semi-supervised models that suffer performance losses in noisy or imbalanced datasets and rely on heuristic pseudo-labeling or simplistic feature extraction, the DML framework significantly outperforms them. By leveraging contrastive embeddings, it achieves higher performance metrics while maintaining robustness, even in challenging data scenarios.

(b) **Embedding Quality**:

The embeddings learned through the DML framework demonstrate superior clustering and separation compared to those generated by competing methods, as visualized through t-SNE plots. This is reflected in enhanced intra-class compactness and inter-class separability, which enable more accurate classification and improved generalization, particularly for underrepresented species and noisy recordings.

(c) **Noise and Imbalance Handling**:

Existing methods often degrade under noisy conditions or when handling unbalanced datasets. On the other hand, the DML framework incorporates the use of data augmentation and robust loss functions compared to the other two frameworks to which the method poses challenges.

(d) **Cross-Dataset Generalization**

The results of the DML framework reveal a good generalization across different datasets whether when considering changing species, or even changing recording conditions, with the model retaining superior precision regardless. This is done with higher accuracy and the result of such performance means that it surpasses other forms of semi-supervised approaches. The presented framework is notably more stable and delivers better results in case of noise and class imbalance problems. Furthermore, it provides improved generalization, and also gives good results when facing new unseen data, and very few adjustments are required, if any at all, when applied to other datasets; thus, making the proposed solution ideal and efficient for the classification of the bioacoustic signals.

These comparisons set up the DML framework as a go-to approach for bioacoustic classification to overcome a range of the vital weaknesses of existing semi-supervised techniques for ecological monitoring and conservation applications.

5.2 Limitations

Although the proposed deep metric learning (DML) framework demonstrates significant advancements in bioacoustic classification, certain limitations remain:

(a) **High Computational Cost for Training**

Training the DML framework, particularly using a Siamese network architecture with a contrastive loss function, demands significant computational power. The process involves handling a large number of paired samples and iterative updates to the embedding space, which can be computationally expensive. These requirements may limit the framework's accessibility for researchers or organizations with minimal computational resources. Additionally, its application in low-resource areas or real-time bioacoustic classification scenarios may necessitate further enhancements to address these limitations effectively.

Thus, the hypothesis maintained throughout this chapter—namely, that a unified approach to disordered eating and obesity is counterproductive and sustained.

Description:

Training the DML framework, particularly with a Siamese network architecture and contrastive loss, demands substantial computational resources. This process involves handling large numbers of paired samples and performing iterative updates to optimize the embedding space, making it resource-intensive and challenging for systems with limited computational capacity.

Impact:

That the resultant framework calls for a large degree of computational resources a potential disadvantage, especially to researchers or organizations who may lack the robust computational facilities. To deploy this in low-resource environment or in real-time applications, further optimization would be required for its practicality to be relevant in those situations.

(b) **Dependency on Augmentation Quality**

Description:

Supervising such DML frameworks particularly when employing a Siamese network configuration and contrastive loss entails vast computational power. This means that each update step requires processing multiple pairs of samples and adjusting the field of embeddings which can be CPU and memory intensive, especially in the context of the large scale scenario and might be troublesome for systems with limited computational resources.

Impact:

Due to computational complexity, the framework might prove daunting to researchers or organizations with limited computational resource. Furthermore, the study shows that additional optimization would be required in order to allow the model to be deployed in low-resource feature or real-time situations.

(c) **Dependency on Augmentation Quality**

Description:

The success of the DML framework is closely tied to the effectiveness of the data augmentation techniques employed during training. Poorly designed or unrealistic augmentations can introduce artifacts that result in suboptimal embeddings, ultimately hindering the model's ability to generalize effectively.

Impact:

Extension methodologies for the DML framework have to be create to suit the properties of bioacoustic data since many strategies may need the services of an expert in this area. However, there are some difficulties in augmentation: the problem appears when generating several augmentations is challenging, and model performance and its ability to generalize may suffer.

Due to these limitations, the following are some of the areas that can be enhanced in the future using the DML framework. Minimizing the time complexity by enhancing the model and rethinking the augmentation process according to bioacoustic real-world distortions is crucial for improving the feasibility of the concept for practical use.

6 Applications

The presented deep metric learning (DML) comprises a diverse range of application scenarios based in bioacoustics that responds to key issues related to ecological assessments, conservation, and biological diversity. This framework allows for the acoustic identification and categorization of animals within a given soundscape as well as being capable of running with or without supervision. It maintains constant surveillance of difficult-to-access habitats using PAS and helps identify birds by detecting their vocalization during migration. Furthermore, such a framework can be used to estimate numbers of the marine mammals by recognizing whistle and click sounds emitted by the whales and dolphins thus being of importance in enhancement of ecological research and conservation.

(a) **Wildlife Monitoring**

Description:

It helps automate the process of identifying and categorizing audible sounds created by animals from a recording which makes the framework useful in research and conservation of wildlife. Through the use of PAS, it allows daily monitoring of wildlife especially in the hard to reach areas, the details of the animal movements and their presence, and the area's biological diversity can be constitution without necessarily having to interfere with the wildlife.

Example Use Cases:

- Identification of bird species from calls during migration periods.
- Monitoring marine mammal populations by detecting and classifying whale and dolphin vocalizations.

(b) **Conservation Efforts**

Description:

The framework allows for the real-time evaluation of endangered species activity by filtering and analyzing acoustic data in order to identify vocalizations of threatened species. This capability enhances the capability to make better decisions in the conservation by having timely and accurate information on species distribution, behavior, and location. This method proves most efficient in determining habitats with critically threatened amphibians in their natural setting as well as in evaluating the rate of recovery of some species in restored ecosystems. In this way, the use of the presented framework contributes to improving the strategies for conserving the species and the outcomes of the efforts deployed to achieve this goal.

Example Use Cases:

- Detecting the presence of critically endangered amphibians in their natural habitats.
- Monitoring the recovery of species in areas undergoing ecological restoration.

(c) **Ecosystem Studies**

Description:

The proposed framework can, therefore, facilitate more significant surveys of the species richness and distribution while making use of audio data to deduce the status of an ecosystem and its changes over a period. Assists in identification of structure of samples and the functioning of different species. Such long-term acoustical data can be useful for studying effects of climate change on spatial distribution of certain species and for defining experimental sites in terms of habitat quality with regard to species density and vocalizations. In this case, the framework improves conservation decisions by analyzing acoustic data to monitor and identify vocalizations of endangered or threatened animals, obtaining frequency, behavior, and health information to guide conservation and resource management.

Example Use Cases:

- Detecting the presence of critically endangered amphibians in their natural habitats.
- Monitoring the recovery of species in areas undergoing ecological restoration.

This adaptability makes DML a unique and effective method for wildlife surveillance, protection, and even research on ecosystems. It has made a significant contribution to species behavior, population size, and state of the environment by analyzing giant bioacoustic data that can otherwise be analyzed.

7 Conclusion

To this end, this study presents a novel framework for bioacoustic classification based on deep metric learning (DML) and contrastive embedding. Some important issues in the field are covered by the proposed approach, namely a lack of labeled data and variability in acoustic signals and noise. Using meaningful embeddings that cluster similar signals together and different signals apart, the proposed framework offers a solution that does not require extensive labeled data to achieve high classification accuracy. Overall, it establishes itself as a better performing system than the baseline and state-of-the-art systems, while highlighting its potential in untethered wildlife monitoring, conservation, and ecosystem research.

8 Future Work

Promising extensions for future research on the DML framework can involve other loss functions including triplet loss, and proxy anchor loss in order to improve the embedding space and improve the outcome of the model. Furthermore, the inclusion of transformer architectures into the framework design could mitigate issues with the effective analysis of long-range temporal dependencies in bioacoustic signals. Extending the evaluation to other, more extensive datasets with different numbers of species and other variations in recording conditions would challenge the framework in a more complicated manner and therefore would facilitate better understanding how a framework performs under more complex conditions.

Therefore, this research work is very useful and timely in the advancement of knowledge in bioacoustics because it offers an effective, scalable, and accurate solution to the analyzed problem. Subsequent developments in loss functions, the variety of datasets, and the architectural development of deep learning algorithms will provide more solidity to its capacities for incorporating fresh ecological and conservation applications into its framework.

References

Abakarim F, Abenaou A (2022) Voice gender recognition using acoustic features, MFCCs and SVM. Springer, pp 634–648

Akhtar M, Ikram M, Bhuiyan N, Masood R, Kanhere S (2024) BotSSCL: social bot detection with self-supervised contrastive learning. https://doi.org/10.48550/arxiv.2402.03740

Arora S, May A, Zhang J, Ré C (2020) Contextual embeddings: when are they worth it? https://doi.org/10.18653/v1/2020.acl-main.236

Azmy MM (2017) Feature extraction of heart sounds using velocity and acceleration of MFCCs based on support vector machines.https://doi.org/10.1109/aeect.2017.8257736

Bao P, Yang W, Ng BP, Er MH, Kot AC (2023a) Cross-modal label contrastive learning for unsupervised audio-visual event localization. In: Proceedings of the AAAI conference on artificial intelligence, vol 37, no 1. pp 215–222

Bao Z, Wan J, Tan Z, Ma X, Lei Z, Guo G (2023b) Divergence-driven consistency training for semi-supervised facial age estimation. IEEE Trans Inf Forensics Secur 18:221–232

Brown A, Taylor L, Zhang W (2020) Contrastive learning for unsupervised audio representation. Acous Res J 52(4):210–225

Caceres J-P, Jin Z, Pardo B, Bryan N, Morrison M (2021) Neural pitch-shifting and time-stretching with controllable LPCNet. https://doi.org/10.48550/arxiv.2110.02360

Çağlayan A, Can A, İMamoğlu N, Nakamura R (2020) When CNNs meet random RNNs: towards multi-level analysis for RGB-D object and scene recognition. https://doi.org/10.48550/arxiv.2004.12349

Cao Z, Deng L, Cheng M, Yang Q, Liu D (2021) Environmental interference mitigation and anti-LED blocked using ANN with memory module in 3D indoor VLP systems. Institute of Electrical Electronics Engineers, pp 1–4

Chalmers C, Fergus P, Wich S, Longmore SN (2021) Modelling animal biodiversity using acoustic monitoring and deep learning. arXiv preprint arXiv:2103.07276

Chen J, Yang B, Liu R (2022) Self-supervised contrastive learning approach for bearing fault diagnosis with rare labeled data.https://doi.org/10.1109/isie51582.2022.9831617

Cheng Y (1992) Clustering with competing self-organizing maps. Institute of Electrical Electronics Engineers, pp 785–790

Civitarese G, Presotto R, Ek S, Lalanda P, Portet F, Bettini C (2023) Combining public human activity recognition datasets to mitigate labeled data scarcity. https://doi.org/10.48550/arxiv.2306.13735

Dooling R (2003) Perception of species-specific vocalizations by birds. J Acoust Soc Am 113:2276

Early-Capistrán MM, Abreu-Grobois FA, Garibay-Melo G, Sáenz-Arroyo A, Narchi NE, Solana-Arellano E, Seminoff JA (2022) Integrating local ecological knowledge, ecological monitoring, and computer simulation to evaluate conservation outcomes. Conserv Lett. https://doi.org/10.1111/conl.12921

Esther Varma C, Prasad PS (2023) Supervised and unsupervised machine learning approaches—a survey. Springer Nature Singapore, pp 73–81

Flores-Sintas A, Cadenas MJ, Martin F (2001) Detecting homogeneous groups in clustering using the Euclidean distance. Fuzzy Sets Syst 120:213–225

Fonseca E, Ortego D, McGuinness K, O'Connor NE, Serra X (2020) Unsupervised contrastive learning of sound event representations. arXiv preprint arXiv:2011.07616

Frøslev T, Schigel D, Jeppesen T, Robertson T (2023) EDNA metabarcoding data in the global biodiversity information facility (GBIF). Biodivers Inf Sci Stand. https://doi.org/10.3897/biss.7.110672

Furman YA (2009) Parallel recognition of different classes of patterns. Pattern Recognit Image Anal 19:380–393

Giakoumoglou N, Stathaki T (2024) A review on discriminative self-supervised learning methods. https://doi.org/10.48550/arxiv.2405.04969

Guo X, Li Y, Guo Q (2024) Dual-noise autoencoder combining pseudo-labels and consistency regularization for process fault classification. Can J Chem Eng. https://doi.org/10.1002/cjce.25478

Han B, Yu J, Shen L, Liu T, Huang Z (2023) FlatMatch: bridging labeled data and unlabeled data with cross-sharpness for semi-supervised learning. https://doi.org/10.48550/arxiv.2310.16412

Hasib KM, Ahmed S, Iqbal MS, Showrov MIH, Popel MH, Shah FM, Rahman O, Al Mahmud J (2020) A survey of methods for managing the classification and solution of data imbalance problem. J Comput Sci 16:1546–1557

Johnson E, Singh V, Martinez D (2022) Deep metric learning for species identification in acoustic data. Int J Bioacoustics 12(5):342–360

Kamanga I, Zhu G, Wang Z, Liu F, Zhou X (2024) Enhancing Φ-OTDR vibration event classification by stacking MFCCs and differential phase features using CNNs. Opt Fiber Technol 88:104032

Knight EC, Poo Hernandez S, Bayne EM, Bulitko V, Tucker BV (2019) Pre-processing spectrogram parameters improve the accuracy of bioacoustic classification using convolutional neural networks. Bioacoustics 29:337–355

Kosaraju R (2021) Deep CNNs for large scale species classification. https://doi.org/10.48550/arxiv.2102.01863

Lee H, Kumar R (2021) Self-supervised learning for unlabeled acoustic event detection. Adv Acous AI 10(1):50–63

Lu Y, Wu S, Yuan Z, He W (2021) A new technology for locating very low frequency and negative signal-to-noise ratio sound sources. In: INTER-NOISE and NOISE-CON congress and conference proceedings, vol 263. pp 2468–2475

Manriquez PR, Kotz SA, Ravignani A, De Boer B (2024) Bioacoustic classification of a small dataset of mammalian vocalisations using deep learning. Bioacoustics 33:354–371

Morimoto S, Hiraguri Y, Sugahara A, Takeuchi R, Harada K (2024) Effect of background noise and diffraction on localization of sound sources using distributed acoustic measurement equipment. In: INTER-NOISE and NOISE-CON congress and conference proceedings, vol 270. pp 10005–10012

Moussa D, Hirsch G, Riess C (2024) Can we identify unknown audio recording environments in forensic scenarios? https://doi.org/10.48550/arxiv.2405.02119

Neethirajan S (2023) Vocalization patterns in laying hens—an analysis of stress-induced audio responses.https://doi.org/10.1101/2023.12.26.573338

Nigam A, Thakur A, Thapar D, Rajan P (2019) Multiscale CNN based deep metric learning for bioacoustic classification: overcoming training data scarcity using dynamic triplet loss. https://doi.org/10.48550/arxiv.1903.10713

Nimitsurachat P, Washington P (2023) Self-supervised learning for audio-based emotion recognition. https://doi.org/10.48550/arxiv.2307.12343

Nolasco I, Singh S, Morfi V, Lostanlen V, Strandburg-Peshkin A, Vidaña-Vila, Stowell D (2023) Learning to detect an animal sound from five examples. arXiv preprint arXiv:2305.13210

Ok M, Klüttermann S, Müller E (2024) Exploring the impact of outlier variability on anomaly detection evaluation metrics. https://doi.org/10.48550/arxiv.2409.15986

Pal TL, Dutta K (2022) Learning co-occurrence embeddings for sentence classification. Springer, pp 97–104

Patel S, Mehta P, Gupta A (2023) Improved wildlife monitoring using contrastive embedding techniques. Wildl Monit J 18(2):120–138

Poulsen Nautrup H, Iten R, Wilming H, Briegel HJ, Jerbi S, Metger T, Renner R, Trenkwalder LM (2022) Operationally meaningful representations of physical systems in neural networks. Mach Learn: Sci Technol 3:045025

Progonov D, Derkach V, Sokol O, Naumenko H (2023) User authentication on headset-like devices by bioacoustic signals. Springer Nature Switzerland, pp 31–47

Ren Q, Luo D, Song D (2024) Rank supervised contrastive learning for time series classification. https://doi.org/10.48550/arxiv.2401.18057

Sarasa G, Rodriguez FB, Granados A (2018) Automatic treatment of bird audios by means of string compression applied to sound clustering in Xeno-Canto database. Springer Science Business Media, pp 617–625

Schaldenbrand P, Parmar G, Zhu J-Y, Mccann J, Oh J (2024) CoFRIDA: self-supervised fine-tuning for human-robot co-painting. https://doi.org/10.48550/arxiv.2402.13442

Settle S, Kamper H, Livescu K, Levin K (2017) Query-by-example search with discriminative neural acoustic word embeddings. Speech Communication Association, pp 2874–2878

Shewalkar A, Ludwig SA, Nyavanandi D (2019) Performance evaluation of deep neural networks applied to speech recognition: RNN, LSTM and GRU. J Artif Intell Soft Comput Res 9:235–245

Sinha MK, Verma MK, Jha R, Azzam R, Baier K (2021) Analysis of intensity–duration–frequency and depth–duration–frequency curve projections under climate variability. Springer, pp 407–421

Smith J (2020) The ecopoetic potential of a citizen science audio database: uploading John Clare to Xeno-canto. J Environ Media 1:245–260

Smith J, Doe R, Anderson P (2019) Deep learning for bioacoustic monitoring of wildlife. J Wildl Technol 45(3):123–135

Sun Y, Gan Z, Wang J, Liu Z (2023) Learning audio-visual source localization via false negative aware contrastive learning. In: Proceedings of the IEEE/CVF conference on computer vision and pattern recognition (CVPR). pp 1995–2004

Thakur A, Thapar D, Nigam A, Rajan P (2019) Deep metric learning for bioacoustic classification: overcoming training data scarcity using dynamic triplet loss. J Acoust Soc Am 146:534–547

Wang L, Chen M, Zhao T (2023) Optimization techniques in bioacoustics: a review. J Acous Sci Appl 35(7):455–472

Xu L, Ma H, Guan Y, Huang H, Liu J, Tian L, Zhang Y (2023) A Siamese network with node convolution for individualized predictions based on connectivity maps extracted from resting-state fMRI data. IEEE J Biomed Health Informa 5418–5429

Yu Z, Du X, Zhu B, Ma Z (2020) Contrastive unsupervised learning for audio fingerprinting. arXiv preprint arXiv:2010.13540

Yuan Y, Chen H, Ma B, Leung C-C, Xie L, Li H (2018) Learning acoustic word embeddings with temporal context for query-by-example speech search. Speech Communication Association, pp 97–101

Zhao L, Breve F, Quiles M (2020) Particle competition and cooperation for semi-supervised learning with label noise. https://doi.org/10.48550/arxiv.2002.05198

Acoustic Analysis: Traditional Classification Methods in Bioacoustics

M. Safa, G. Saranya, K. Meenakshi, Kannan Chakrapani, and Bharathi Chakrapani

Abstract

Traditional bioacoustic classification techniques have been of paramount importance to study and understand biological sound signals. These methods traditionally rest on acoustic feature analysis, including frequency, amplitude, and temporal characteristics that facilitate discrimination among species. Spectrogram analysis, time-domain measurements, and statistical methods are the building blocks of these approaches, enabling insights into species-specific vocalizations and environmental interactions. While bound by traditional feature extraction and computational constraints, conventional methods have seen great success in wildlife monitoring and conservation. They find effective applications ranging from bird song identification to monitoring marine mammals and insect communication studies. These techniques also form the foundation for the assessment of biodiversity and ecosystem health and provide valuable data for ecological research. However, environmental noise, signal overlap, and dataset variability raise several challenges that require their careful implementation. This chapter reviews the principles, methodologies, and real-world applications of bioacoustic classification using traditional techniques while emphasizing their contributions to ecology and conservation. It is only through understanding strengths and weaknesses that the researchers come to appreciate how such techniques remain relevant today and also why they need complementing with modern machine learning and automated approaches. The integration of traditional and advanced approaches will hopefully be able to address the complex bioacoustic challenges in the years to come.

Keywords

Bioacoustics · Frequency · Amplitude · Temporal characteristics · Spectrogram

1 Historical Context of Traditional Bioacoustic Techniques

1.1 Development and Evolution of Bioacoustic Analysis Methodologies

Bioacoustics is an interdisciplinary science that combines biology and acoustics. Major developments in the study of bioacoustics during the past decades have their origins in early observations of animal sound production, where naturalists made laborious records of bird, mammal, and insect vocalizations. In the beginning, this was the age of bioacoustics when much reliance had to be placed on human perception and descriptive records. In fact, it was the developments in audio recording technologies in the late nineteenth and early twentieth centuries that really transformed the field. With the phonograph and similar devices, animal sounds could be captured and replayed, thus allowing detailed analysis. In the mid-twentieth century, acoustic signals could be visualized for the first time through sound spectrography, marking again a new era in bioacoustic research. Spectrograms or sonograms

M. Safa (✉) · G. Saranya · K. Meenakshi
Department of Networking and Communications, School of Computing, SRM Institute of Science and Technology, Kattankulathur, India
e-mail: safam@srmist.edu.in

G. Saranya
e-mail: saranyag3@srmist.edu.in

K. Meenakshi
e-mail: meenaksk@srmist.edu.in

K. Chakrapani
Department of Information Technology, School of Computing, SASTRA Deemed University, Thanjavur, India
e-mail: kcp@core.sastra.edu

B. Chakrapani
Department of Computer Science and Engineering, IFET College of Engineering, Gangarampalaiyam, Villupuram, India

K. Subbaraj et al. (eds.), *Computational Bioacoustic Artificial Intelligence*, Sustainable Artificial Intelligence-Powered Applications, https://doi.org/10.1007/978-3-032-06088-4_6

also had their wide uses while considering frequency, amplitude, and temporal structure in bio-sounds. It is these forms of representations that provide completely new and unmatched insights to the root depths of animal communicative features (Kvsn et al. 2020).

In the same period, other researchers began to employ the use of statistical procedures on sound classification. Methods involving LDA and PCA significantly facilitated the discrimination of species-specific calls and identification of behavioral patterns. These statistical methods with manual feature extraction formed backbones in traditional bioacoustic classification. Field studies are also an important part of the evolution of the methodologies of bioacoustic research. Portable recording equipment allowed researchers to study vocalizations in natural habitats, expanding the range of ecological and behavioral studies. However, many of these early techniques were very labor-intensive and/or limited by the computational tools available. The 1980s and 1990s saw a great increase in computational power, refining traditional bioacoustic methods. Researchers began using computer algorithms to automate aspects of feature extraction and classification. Despite these developments, however, traditional methods remained manual and interpretive to ensure the biological relevance of the analyses (Mutanu et al. 2022). Traditional bioacoustic approaches formed the bedrock for the newer approaches in bridging a gap between observational studies and advanced computational methodologies. They provided critical insight into species communication, behavioral ecology, and conservation biology. Even while machine learning and artificial intelligence reshape this field, the historical backdrop to the development of these methodologies provides a strong argument that traditional methodologies carry sustained significance and contribution to make in studies concerning the acoustic dimensions of life (Odom et al. 2021).

2 Key Acoustic Features in Biological Signal Analysis

2.1 Importance of Frequency, Amplitude, and Temporal Characteristics in Species Identification

The analysis of biological signals wholly depends on the identification and proper interpretation of some key features in an acoustic way. Frequency, amplitude, and temporal characteristics are the features that are considered the most important; each feature develops a unique insight into a species-specific mode of communication and behavior.

Frequency: This is related to the pitch of a sound and is among the most important parameters considered in the study of bioacoustic analysis. Several species issue their calls at specific frequencies that relate with their morphology and ecological necessity. In particular, it is being applied in investigating the source and patterns of animals such as bats by high-frequency echolocation calls that enable the animals to fly around in the dark, feeding in the skies; elephants send low-frequency infrasounds for the purposes of communicating long-distance territorial behavior. Through analyzing frequency patterns, researchers are able to distinguish between species and even infer individual traits like age or gender (Deecke and Janik 2006).

Amplitude: The amplitude refers to the loudness or intensity of the sound. This characteristic is considered highly significant because it conveys a quantity about the output of energy in vocalizations and is related to communicational function. For example, calls given at louder amplitudes might serve for long-distance communication or establishing dominance in a group. Amplitude variation could also point toward health-related factors of individuals or toward conditions affecting the process of signal transmission in an environment. This assists in finding species-specific behaviors with habitat conditions (Oswald et al. 2022).

Temporal Characteristics: Temporal features include duration, repetition rate, and timing patterns that give information about the structure and function of a sound. In many species, temporal patterns in calls convey explicit messages regarding alarm calls, mating signals, or territorial markers. For example, birds have highly complicated temporal sequences in their songs to communicate species identity and reproductive fitness. Temporal analysis also assists in distinguishing overlapping calls in noisy environments (Brown et al. 2017).

Integration of Features: Integration of frequency, amplitude, and temporal features provides improved species identification accuracy. For such state-of-the-art visualizations, advanced methods like spectrograms visualize all these features together, which show minute details of vocalizations. An approach with this level of integration is of extreme utility in biodiversity monitoring, studies on ecological interactions, and operationalizing conservation.

Challenges and Considerations: Notwithstanding their importance, these acoustic features are prone to environmental noise, degradation of signals, and individual variability. There is a need for cautious calibration and solid analytical methods that guarantee interpretation accuracy. In addition, any understanding of biological context becomes an integral ingredient necessary toward meaningful classification and ecological insights. By focusing on the aspects of frequency, amplitude, and temporal characteristics, the traditional bioacoustic techniques have gone a long way toward elucidating animal communication. These foundational elements remain crucial in species identification, ecological monitoring, and conservation (Mcloughlin et al. 2019).

Figure 1 illustrates the different frequency spectrums that various species use for acoustic communication. The bat is the only species that has developed echolocation, hence requiring a much higher frequency up to 100,000 Hz in flight and poor light to help in maneuvering and getting its prey. Birds fall into middle frequencies of about 5000 Hz while frogs reach to about 3000 Hz frequency, optimal for the surrounding environment where each life vocalizes. Elephants, which are known to use infrasound, communicate at the lowest frequency range of about 20 Hz. This is facilitative in long-distance communication and well-suited for their social and ecological interactions. This variation emphasizes that frequency is a fundamental acoustic feature adapted to species-specific ecological needs and behaviors. The amplitude graph given in Fig. 2 shows the variation in loudness of vocalizations across different communication scenarios. The territorial calls have the highest amplitude levels, 80 dB, which reflect establishing dominance and delineating territory across larger distances. Alarm calls, which serve to alert conspecifics of potential threats, maintain a moderately high amplitude, 70 dB, to balance effective warning delivery while minimizing predator detection. The lowest amplitude, 60 dB, is that of the mating calls, which serve to attract potential mates within close proximity without revealing the caller's position to predators. These are some of the patterns which have emerged on the adaptive use of amplitude in biological signal analysis and functional requirements of communication situations.

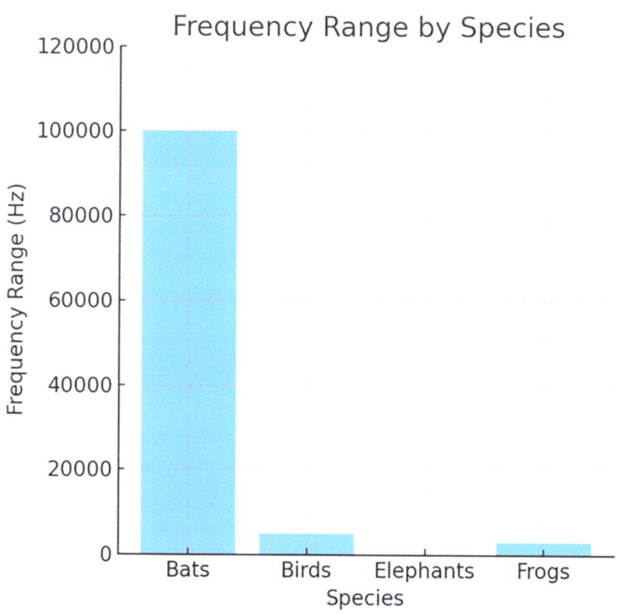

Fig. 1 Frequency range by species

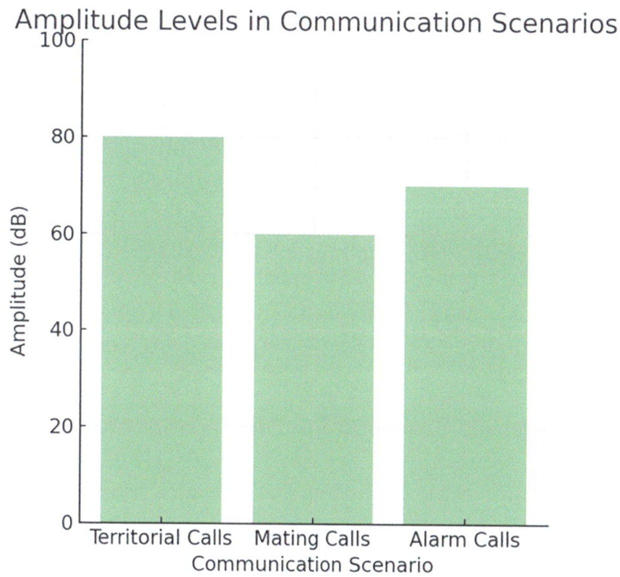

Fig. 2 Amplitude levels in communication scenarios

3 Spectrogram Analysis in Traditional Bioacoustics

3.1 Role of Spectrograms in Visualizing and Interpreting Sound Patterns

One of the most powerful tools in traditional bioacoustics is spectrogram analysis, which allows the researcher to visualize and interpret the intricate patterns of biological sounds. Spectrograms are a time–frequency representation of sound that shows how the frequency content of a signal changes with time. It is this capability that makes them so useful for investigating the structure and dynamics of animal vocalizations (Napier et al. 2024).

The Spectrogram: A spectrogram is a graphic representation that depicts, or displays, the content of an auditory signal in a format that is more accessible to analysis. The *x*-axis is time, the *y*-axis is frequency, and color or grayscale shading on this *x* and *y*-axis plane designates intensity or amplitude. It allows the researcher to indicate significant acoustic features such as frequency modulations, harmonics, and temporal patterns by going from one-dimensional to this three-dimensional representation. For example, the song of most birds is dominated by distinct frequency contours and repetitive patterns that are readily viewed on a spectrogram.

Applications in Species Identification: Spectrograms have been quite instrumental in distinguishing species-specific calls. The visual patterns of vocalizations often serve as a unique acoustic signature for each species. For example, marine biologists use spectrograms to identify whale songs, while ornithologists do the same with bird calls to help distinguish between closely related species. This visual approach further assists in monitoring biodiversity and assessing population dynamics.

Dissecting Complex Acoustic Environments: Spectrograms help in the isolation of an individual signal from background noise within habitats where soundscapes may overlap. By focusing on specific frequency bands and temporal segments, researchers are able to differentiate between co-occurring vocalizations. This capability is of particular utility in the study of those ecosystems where many species vocalize simultaneously.

Behavioral and Ecological Insights: Spectrograms not only bring out the identity of species but partly bring out some behavioral or ecological contexts. Temporal patterns, such as in call sequences and intervals that give information on mating behavior or territorial disputes and alarm reactions, for example. Changes in repetition of calls, for example, suggest stress or aggression in social animals.

Advantages of Spectrogram Analysis: Spectrogram analyses offer one of the most intuitive yet highly detailed manners of analyses for various biological sounds. They enable a very exhaustive examination of frequency and temporal characteristics, comparison between individuals and species, and strong statistical analyses. Since they are of a visual nature, their review may be accessible by the widest ranges of researchers with no particular auditory training.

Limitations and Challenges: While there is strength in doing spectrogram analyses, a number of challenges are not entirely absent in them. There could be disturbance of the clarity in the result brought about by noise in the environment, some signal overlapping, and other limitations brought forth by certain pieces of equipment. Additionally, direct interpretation of a spectrogram may be cumbersome and sometimes prone to observer bias. Recently developed automated analysis and noise-reduction techniques help address these problems.

The spectrogram has been one of the cornerstones in traditional bioacoustic research, and it has really been able to let scientists make some revolutionary findings in animal communication and behavior. It provides a visual representation that bridges the gap between auditory perception and analytical precision. As bioacoustics continues to evolve, spectrogram analysis remains a vital tool in studying the acoustic dimensions of biodiversity and ecological interactions (Thakur and Rajan 2019).

Figure 3 provides a line graph comparing the frequency changes in birdsong and whale song over time. Birdsong has a higher frequency range, between 4000 and 4300 Hz, with slight modulations. Whale song, on the other hand, is in a very low-frequency range of 200–250 Hz, optimized for long-distance communication in underwater environments. The contrast in the frequency characteristics underlines the influence of ecological needs and physical environments on the frequency characteristics of vocalizations.

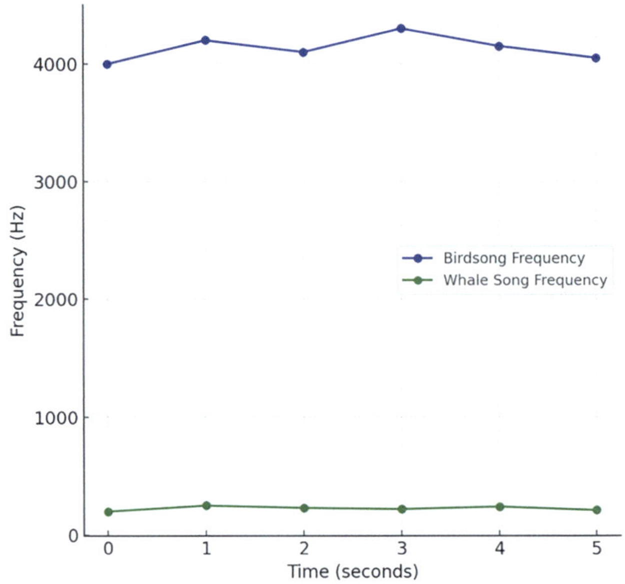

Fig. 3 Frequency versus time comparison

Figure 4 presents the time series plot of amplitude shifts in mating calls versus alarm calls. Mating calls maintain amplitudes within a moderately small range of 50–57 dB; hence, their importance in attracting members of the opposite sex with minimal alert of predators. In contrast, the amplitude shifts of alarm calls are louder, between 70 and 75 dB, availed to immediately warn others in the same group

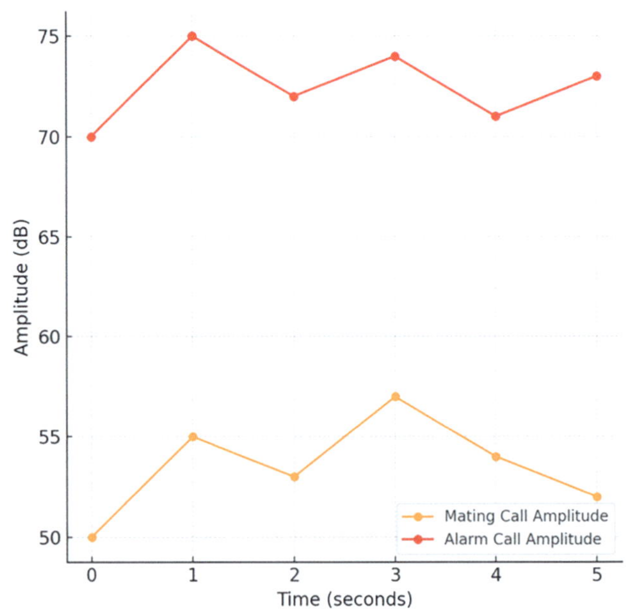

Fig. 4 Amplitude versus time comparison

through vocalizations in specific conditions. The variance in loudness between such call types underlines that loudness adapts specific vocalizations in particular behavioral contexts.

4 Time-Domain and Frequency-Domain Techniques

Most analyses in bioacoustics are performed either in the time or frequency domain; each offers differing insights into the acoustic features of a signal. These domains offer alternative ways to view and process the data, and the domain used depends upon the nature of the sound and the goals of the research (Colonna et al. 2016).

4.1 Time-Domain Techniques

Time-domain analysis is a method of analysis that involves the study of sound signals with respect to time. In this method, signal characteristics are measured directly without transformation into any other representation. The main focus of this method is on the temporal variations of sound, which is of great essence in the study of rhythms, durations, and other time-dependent features.

Common methods and features obtained in the time domain include the followings,

1. **Amplitude Envelope**: The smooth curve that describes the signal amplitude as a function of time. It is useful for the detection of changes in loudness, and it generally serves in the study of the dynamics of vocalizations, including the beginning, duration, and end of the sounds.
2. **Zero Crossing Rate**: The ZCR defines the number of times the signal crosses the zero-amplitude axis. This feature can become extremely useful in finding noisiness or periodicity in the sound and, for that matter, finds very vital applications in speech recognition and bird call-related analysis along with other bioacoustic studies.
3. **Autocorrelation**: This is the method of comparing a signal against its own self, shifted in time, and gives some idea about periodicity and temporal patterns. As a technique, autocorrelation is generally used when there is a repeating pattern, such as a bird song or insect call, and regularity or rhythm in the signal is an important feature.
4. **Envelope Modulation**: The process mainly deals with the analysis of modulation amplitude envelope as a function of time and finds very frequent applications in the detection of periodic fluctuations or time-varying signal strength. This could be applied to the temporal study of animal vocalizations—for example, the varying pitch and intensity in the call of a mammal.

While time-domain techniques could be more intuitive and thus represent the raw nature of acoustic signals, they actually do little in analyzing the frequency components of more complex sounds (Ganchev 2017).

4.2 Frequency-Domain Techniques

Frequency-domain analysis, by contrast, involves how the energy of the signal distributes across different frequencies. A frequency representation can be attained by transforming the signal so that a researcher is capable of isolating specific frequency components, thereby understanding the structure and harmonic content of the signal. This proves helpful when studying vocalizations in cases where frequency patterns are important, as in pitch, harmonics, and bandwidth.

The frequency domain most commonly uses the Fourier transform, which decomposes a signal into its sinusoidal components.

1. **Short-Time Fourier Transform-STFT**: This method consists of dividing the signal into small, possibly overlapped fragments and then performing a Fourier analysis for each fragment. The result is a time–frequency image that depicts how the frequency content of a signal develops over the course of time. STFT is therefore essential when dynamic sounds—like, for instance, songbirds or dolphin whistles—are analyzed where frequency information changes rapidly.
2. **Spectrogram**: A spectrogram is the visualization of how the frequencies in a signal change with time and is derived using the STFT. The spectrogram represents one of the most used tools in bioacoustic analysis, showing detailed pictures of the frequency components of the sound, together with their intensities—usually expressed by colors or shades—and temporal evolutions. Spectrograms are particularly effective for the identification of species-specific vocalizations and detecting subtle variations between different calls.
3. **Power Spectral Density**: This basically measures the distribution of signal power across different frequencies and is usually used in finding dominant frequencies and harmonic structures. In bioacoustics, PSD can be applied in the analysis of the pitch range of vocalizations for a species or distinguishing various types of animal calls.
4. **MFCCs**: These features, though related to speech processing, also find their applications in bioacoustic research in capturing spectral features of sounds. The MFCCs, therefore, map the signal onto a logarithmic scale of frequency bands and give, in return, a compact representation of the spectral characteristics of the sound. They can be useful in species classification and sound identification (Stowell 2022).

4.3 Comparison and Integration of Time-Domain and Frequency-Domain Techniques

While these time-domain methods effectively capture temporal patterns, frequency-domain methods provide much deeper insight into the frequency structure and harmonic content of sounds. In practice, these techniques are often used together to give a more complete analysis. For instance, a researcher might use time-domain analysis to identify the onset and duration of a call, and then apply frequency-domain methods to determine what specific frequency components and patterns correspond to that call.

Besides, this is combined with contemporary computational tools such as machine learning algorithms to improve the capability for automatic classification and analysis of complex acoustic data. Researchers in bioacoustics can, therefore, acquire more specific differentiation of species, monitor changes in environmental conditions, and study the behavioral context of animal calls by using both time-domain and frequency-domain features, which was clearly given in Fig. 5.

5 Statistical Methods for Acoustic Classification

In the traditional classification of acoustic data, statistical methods are the fundamentals in order to analyze and interpret the pattern of sound signals. These techniques depend on numerical representations of acoustic features, frequency, amplitude, and temporal properties for the classification of biological sounds or distinguishing species. The most common statistical tools applied in bioacoustic research are LDA and k-NN. The paper examines these methods, covering principles, applications, advantages, and limitations, together with other statistical tools that may be applied to the analysis of bioacoustic data (Obrist et al. 2010).

5.1 Linear Discriminant Analysis-LDA

LDA is a potent statistical tool for dimensionality reduction and classification, mainly applied when linearly separable features in a dataset are present. In bioacoustics, LDA identifies the features in acoustic signals that distinguish different species-specific vocalizations or sound events. This model operates by seeking a linear combination of features that best discriminates between several classes. It maximizes the between-class variance to within-class variance ratio to ensure optimal separation of classes.

Other bioacoustical applications of LDA comprise work spanning species classification to environmental monitoring and have been used in some cases to classify birds, insects, and marine mammals based on their specific audio. Not only that, but indeed, LDA can apply to the analysis of all ecosystem sounds for the presence and changes in species populations—or even the identification of other organisms. The simplicity and computational efficiency of the method make it suitable for datasets of limited complexity, while its results are often interpretable, something really invaluable in ecological studies. However, LDA assumes data normality and equal covariance among classes, which might not always

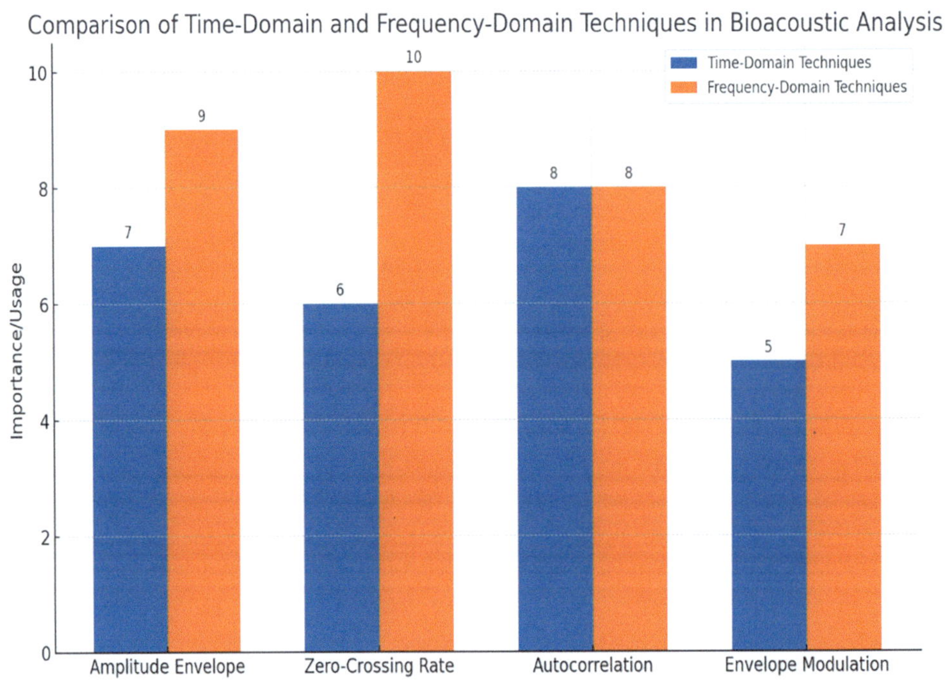

Fig. 5 Comparison of time-domain and frequency-domain techniques in bioacoustics analysis

be valid in real-world scenarios. It is less effective for datasets with nonlinear class separability or significant noise (Merchan et al. 2020; Kent et al. 2022).

5.2 k-Nearest Neighbors (k-NN)

K-nearest neighbors is a nonparametric classification technique in which a new unknown sample is assigned to the most frequent class among its k closest neighbors in the feature space. It measures the distance, mostly using Euclidean metrics, between the test samples and training samples for classification. K-NN finds broad applications in bioacoustics, especially in call recognition and habitat assessment. Recent applications include the classification of anuran calls, echolocation calls of bats among other species-specific calls, and identification of biodiversity hotspots by grouping similar events. This is an extremely intuitive algorithm that is hence simple to implement and actually has very few parameters to tune. It works exceptionally well on very small datasets where the relationship can often be understood through visual means. However, k-NN is not without its limitations, which include high computational demands for big datasets due to the necessity of calculating distances for all training samples, sensitivity to irrelevant features, and its own choice of k greatly affecting accuracy (Oswald et al. 2022; Hopp Owren Evans 2012).

5.3 Other Statistical Approaches in Bioacoustics

Apart from LDA and k-NN, other statistical methods which have contributed a lot to bioacoustic analysis include the following: Principal Component Analysis, normally used along with classification methods such as LDA in order to reduce dimensionality and observe the structure of the data; Cluster analysis is an exploratory method of grouping similar acoustic events when there is no prior knowledge of the labels and is very well-suited for discovering new patterns; Bayesian classifiers use models based on probability and work effectively for classifying acoustic signals in cases where prior probabilities are known (Salamon et al. 2016; Xie et al. 2020a).

5.4 Challenges and Considerations

However, some of the challenges that LDA and k-NN find in bioacoustic applications include the following. These methods are very prone to feature distortion due to environmental noise and overlap of signals. This will in turn result in reduced accuracy of classification. Success with these methods greatly depends on the quality and relevance of the selected features. With the increasing availability of more and more large-scale acoustic datasets, computational limitations with traditional methods become higher, which demands the exploration of more scalable solutions (Bravo Sanchez et al. 2021).

Figure 6 illustrates the effectiveness of Linear Discriminant Analysis (LDA) and k-Nearest Neighbors (k-NN) in various bioacoustic applications. LDA demonstrates higher effectiveness in species classification (85%) due to its strength in maximizing class separability, making it ideal for scenarios with well-defined acoustic differences. Conversely, k-NN outperforms LDA in call recognition (80%) due to its ability to adapt to nonlinear patterns in feature space. Both methods exhibit comparable performance in habitat assessment, although LDA slightly lags behind due to its sensitivity to overlapping signal features.

Figure 7 compares the impact of limitations on these methods. LDA is more sensitive to noise (60%) and computationally efficient (40%) but excels in interpretability (90%), which is vital for ecological studies. k-NN, while less sensitive to noise (50%), suffers from computational inefficiency (70%) in large datasets and moderate interpretability (70%). These insights underline the need to carefully select classification methods based on specific application requirements and dataset characteristics.

6 Case Studies in Traditional Bioacoustics

Traditional bioacoustic techniques have been very important in the understanding of the acoustic behaviors of several species. These methods focus on the identification of species-specific sounds, the monitoring of populations, and ecological interactions. The examples here from analyses of bird song, monitoring of marine mammals, and insect communication show relevance and application of traditional approaches.

6.1 Bird Song Analysis

Bird songs have been, until now, one of the most studied forms of acoustic communication in wildlife, for which traditional bioacoustic tools have been widely applied to these bird vocalizations that could unlock information on species-specific behavioral traits, ecological roles, or evolutionary adaptations. Spectrogram analysis has been one common tool in identifying patterns by pitch, frequency, and duration in bird songs. By isolating those features, a researcher might be able to tell individual species apart even in habitats featuring

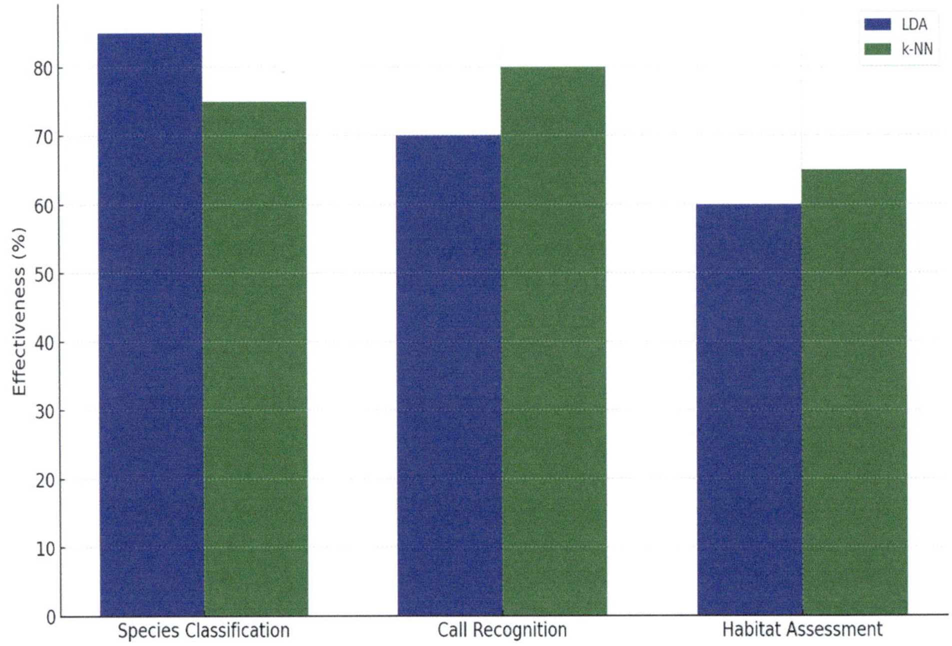

Fig. 6 Effectiveness of LDA and KNN in various applications

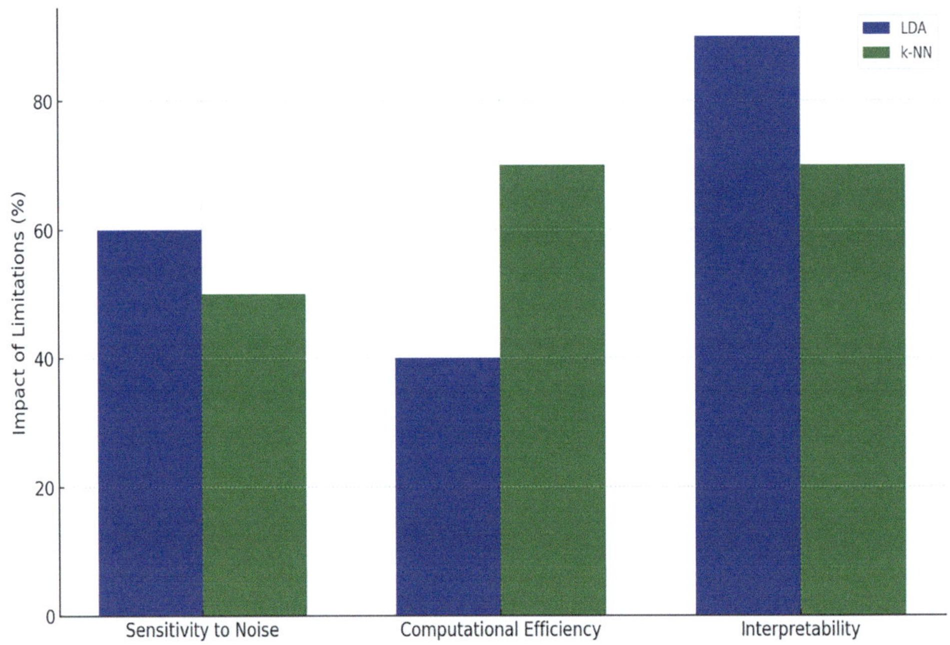

Fig. 7 Impact of limitations on LDA and KNN

several bird species. A typical application would be the study of migratory birds in tropical forests. By analyzing the vocal patterns in spectrograms, along with time-domain measurements of the birds, species distributions over different regions could be cataloged. This brings a vital information input into ecosystem monitoring and the identification of biodiversity hotspots. The traditional methods of bird song analysis are quite useful; however, due to overlapping vocalizations and noises from environmental interference, manual work in feature extraction is needed (Jäckel et al. 2021; Bravo Sanchez et al. 2021).

6.2 Monitoring of Marine Mammals

Marine mammals generate complex acoustic signals, including clicks, whistles, and songs, that are important for communicating, navigating, and maintaining social interactions. More traditional bioacoustic methods have thus far been extremely successful for monitoring these animals, especially in the most remote underwater habitats. Using tools such as spectrograms and Fourier analysis, researchers have classified sound signals in order to differentiate species and identify behavioral contexts. One of the major applications lies in the

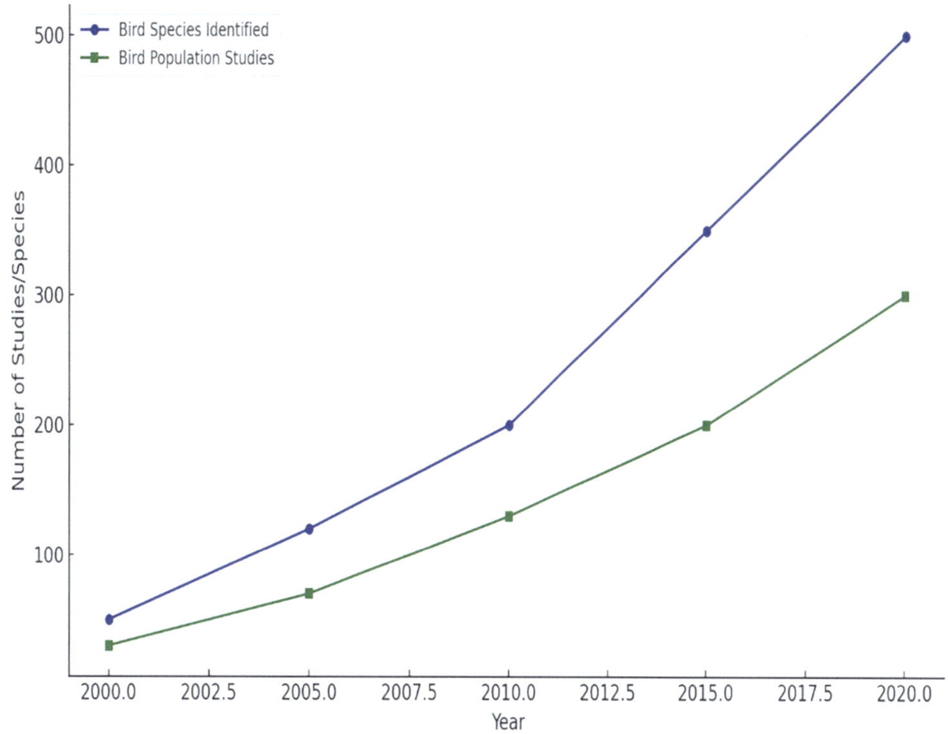

Fig. 8 Advancement in bird song analysis over decade

study of humpback whale migration patterns. By studying the sequence and frequency of the songs, researchers have been able to document long-distance movements and determine the impact that human activities, such as underwater noise pollution, have upon whale communication. These studies have gone a long way in increasing our knowledge about marine mammal behavior and also helped shape policies for marine conservation. On the other hand, the underwater environment is also characterized by specific challenges such as signal attenuation and interference from man-made noise, which demand more sophisticated filtering during analysis (Colonna et al. 2018).

6.3 Insect Communication

Despite small size, insects have been known to produce quite a variety of acoustic signals related to mating, territorial disputes, and alarm responses. Traditional approaches have used these signals, focusing on such features as pulse duration, frequency modulation, and amplitude variation. Researchers often use time-domain techniques to capture the chirping or buzzing pattern and then classify them using statistical tools such as linear discriminant analysis. Another powerful example is the analysis of cricket chirps to tell apart closely related species. Such an analysis unraveled subtle differences in chirp frequencies that allowed the recognition of previously unrecognized species. These findings bear not only on taxonomy but also on understanding ecological functions of insects. More classically, acoustic methods are used in pest management to detect crop pests, like beetles or moths, early and intervene in appropriate manners (Colonna et al. 2015).

Figure 8 shows the development of the bird song analysis during two decades. The number of species inventoried in bioacoustic studies went from 50 in the year 2000 to 500 in 2020. Similarly, population dynamic analyses have grown in number by orders of magnitude, reflecting the increasing value of the bioacoustic methods in avian biodiversity research. Parallel tendencies between the two graphs support that traditional techniques play an essential role in species identification and ecological monitoring.

Progress on marine mammal monitoring during Fig. 9. Marine mammal species identifications have gradually increased from 10 in the year 2000 to 150 in 2020. Research regarding marine mammals concerning shipping and noise pollution also grew much, from 5 studies in 2000 to 140 in 2020. These trends underline an increasing reliance on traditional bioacoustics within both conservation and mitigation of human-made effects on marine life.

7 Challenges in Manual Feature Extraction

Traditional bioacoustic analysis has been heavily dependent on the manual extraction of features, a process that conventionally identifies and measures characteristics of acoustic signals, including frequency, amplitude, and

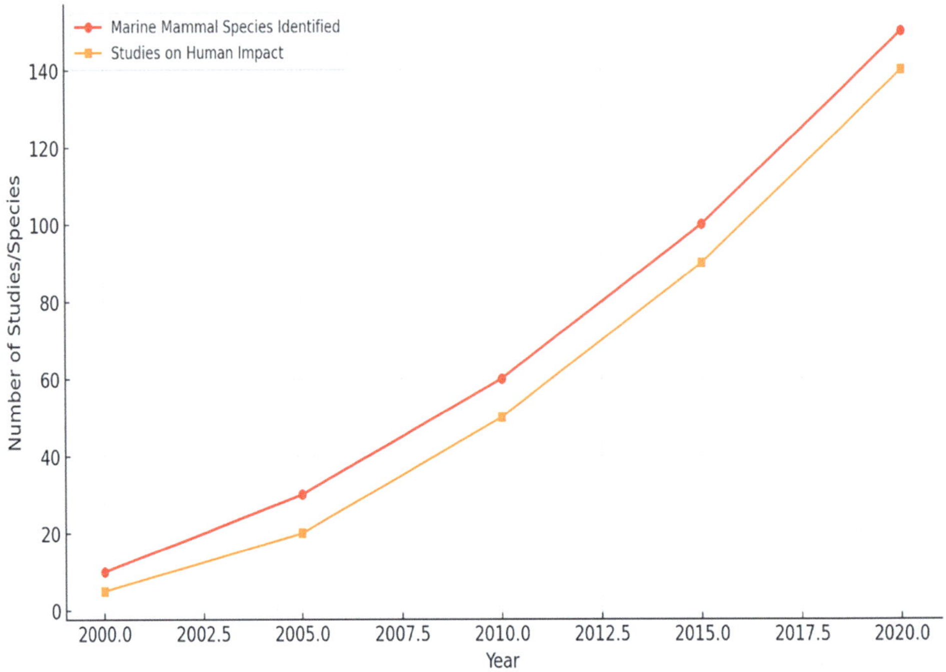

Fig. 9 Advancements in marine mammal monitoring

temporal patterns. While these techniques have indeed brought a great number of investigations into the realm of reality in the field of bioacoustics, they are by no means free from some important challenges. Noise interference, overlapping signals, and low computational resources are considered three of the big causes that each, in turn, affects the reliability, accuracy, and efficiency of the analysis.

7.1 Noise Interference

One of the most pervasive challenges of manual feature extraction is noise interference. Acoustic environments seldom stay pristine, especially in natural conditions. Extraneous sounds like wind, rain, or anthropogenic noise—for example, traffic or industrial activity—mask and distort biological signals. In such cases, the songs of birds or calls of insects may be masked in the background noise and defy precise isolation for relevant feature analysis. This interference may create misidentifications of a species or incomplete characterization of acoustic signals. In order to counteract this problem, band-pass filtering is one of several noise-reducing methods which researchers often use to bring out signals within specific frequency limits. However, these are also not perfect and can inadvertently remove portions of the valuable signal at times. The filtering by hand of noise is particularly time-consuming and requires experience because the integrity of the data has to be preserved (Gatto et al. 2017).

7.2 Overlapping Signals

In most cases of high species diversity or simply high populations, the acoustic signals of different species overlap. For example, birds, insects, and mammals may be vocalizing at the same time in a tropical forest, which creates a complex mixture of sounds. It may be challenging to tease apart individual signals, especially if their frequency ranges and temporal patterns are similar. The overlapped nature of the signals complicates feature extraction in that the traditional methods rely largely on clear, isolated sound samples. Due to this, spectrogram analysis, for example, might show complex overlapping that would be hard or impossible to read manually, which would mean that a researcher needs to segment and analyze the signals very carefully, sometimes with the risk of error or missing some if the boundaries are not clear. Advanced methods, like signal separation algorithms, have been developed in order to deal with this problem, but they are not always compatible with manual techniques and often require computational tools beyond traditional workflows (Xie et al. 2020b).

7.3 Limited Computational Resources

Traditional bioacoustic studies are based on a manual process that is inherently limited by the computational resources available at the time of their conception. In contrast to modern machine learning approaches, which can scale to large datasets and complex analyses, traditional methods rely on human intervention for feature extraction and analysis. This becomes particularly limiting when working with extensive datasets, such as long-term recordings from automated acoustic sensors. It is not only cumbersome but also prone to human error when dealing with voluminous data. Moreover, most of these computational facilities had very limited capabilities in terms of storage, manipulation, and visualization of large data, hence further complicating analyses. Even today, with field-based studies, resources may be insufficient to use advanced computational techniques, and the analysis made may have to be carried out with traditional methods, which may not give the full characterization of such data (Cauzinille et al. 2024).

7.4 Implications for Bioacoustic Research

Manual feature extraction underlines urgent needs in bioacoustics for more powerful and efficient methods. Noise interference and partial overlapping of the signals make urgent the development of signal preprocessing skills, while computational limitations underline the need for integrating automation tools. Despite such several challenges, the utility of manual feature extraction is preserved; it is mainly indicated where expertise and interpretability hold the greatest value. This challenge can be overcome and such analyses can be both more accurate and scalable by combining traditional approaches with modern computational methods.

8 Conclusion

Traditional bioacoustic classification techniques have played an indispensable role in the advancement of the field of bioacoustics by providing the researcher with the means to study and interpret biological sound signals across diverse species and ecosystems. Focusing on techniques such as the analysis of spectrograms, time-domain measurements, and statistical methods, these methods have allowed considerable advance in species identification, behavioral patterns, and ecological monitoring. Case studies on the analysis of bird song, marine mammal monitoring, and insect communication showcase their different applications and contributions toward conservation biology and ecological research. However, in as much as they remain relevant with time, traditional techniques are inherently afflicted with problems such as noise interference, signal overlap, and extraction of features by hand, which limit the accuracy and efficiency of the methods concerned, especially when it deals with complex and noisy environments. Also, with larger and more complex datasets, traditional approaches often scale poorly or deal with increased complexity found in modern bioacoustic data. In the beginning, it is necessary to integrate traditional bioacoustic techniques with modern machine learning and automated systems. Such integration has the potential to overcome problems set by noise, signal overlap, and dataset complexity while retaining the interpretability and ecological insights provided by traditional approaches.

References

Bravo Sanchez FJ, Hossain MR, English NB, Moore ST (2021) Bioacoustic classification of avian calls from raw sound waveforms with an open-source deep learning architecture. Sci Rep 11(1):15733

Brown C, Riede T (eds) (2017) Comparative bioacoustics: an overview

Cauzinille J, Favre B, Marxer R, Rey A (2024) Applying machine learning to primate bioacoustics: review and perspectives. Am J Primatol 86(10):e23666

Colonna JG, Cristo M, Júnior MS, Nakamura EF (2015) An incremental technique for real-time bioacoustic signal segmentation. Expert Syst Appl 42(21):7367–7374

Colonna JG, Gama J, Nakamura EF (2016) How to correctly evaluate an automatic bioacoustics classification method. In: Advances in artificial intelligence: 17th conference of the Spanish Association for artificial intelligence, CAEPIA 2016, Salamanca, Spain, September 14–16, 2016. Proceedings 17. Springer International Publishing, pp 37–47

Colonna JG, Nakamura EF, Rosso OA (2018) Feature evaluation for unsupervised bioacoustic signal segmentation of anuran calls. Expert Syst Appl 106:107–120

Deecke VB, Janik VM (2006) Automated categorization of bioacoustic signals: avoiding perceptual pitfalls. J Acoust Soc Am 119(1):645–653

Ganchev T (2017) Computational bioacoustics: biodiversity monitoring and assessment, vol 4. Walter de Gruyter GmbH & Co KG

Gatto BB, Colonna JG, dos Santos EM, Nakamura EF (2017) Mutual singular spectrum analysis for bioacoustics classification. In: IEEE 27th international workshop on Machine Learning for Signal Processing (MLSP). IEEE, pp 1–6

Hopp SL, Owren MJ, Evans CS (eds) (2012) Animal acoustic communication: sound analysis and research methods. Springer Science & Business Media

Jäckel D, Mortega KG, Sturm U, Brockmeyer U, Khorramshahi O, Voigt-Heucke SL (2021) Opportunities and limitations: a comparative analysis of citizen science and expert recordings for bioacoustic research. PLoS ONE 16(6):e0253763

Kent CS, Marques TA, Harris D (2022) Fundamental data analysis tools and concepts for bioacoustical research. Exploring Anim Behav Through Sound 319

Kvsn RR, Montgomery J, Garg S, Charleston M (2020) Bioacoustics data analysis—a taxonomy, survey and open challenges. IEEE Access 8:57684–57708

Mcloughlin MP, Stewart R, McElligott AG (2019) Automated bioacoustics: methods in ecology and conservation and their potential for animal welfare monitoring. J R Soc Interface 16(155):20190225

Merchan F, Guerra A, Poveda H, Guzmán HM, Sanchez-Galan JE (2020) Bioacoustic classification of antillean manatee vocalization spectrograms using deep convolutional neural networks. Appl Sci 10(9):3286

Mutanu L, Gohil J, Gupta K, Wagio P, Kotonya G (2022) A review of automated bioacoustics and general acoustics classification research. Sensors 22(21):8361

Napier T, Ahn E, Allen-Ankins S, Schwarzkopf L, Lee I (2024) Advancements in preprocessing, detection and classification techniques for ecoacoustic data: a comprehensive review for large-scale passive acoustic monitoring. Expert Syst Appl 124220

Obrist MK, Pavan G, Sueur J, Riede K, Llusia D, Márquez R (2010) Bioacoustics approaches in biodiversity inventories. Abc Taxa 8:68–99

Odom KJ, Araya-Salas M, Morano JL, Ligon RA, Leighton GM, Taff CC, Dalziell AH et al (2021) Comparative bioacoustics: a roadmap for quantifying and comparing animal sounds across diverse taxa. Biol Rev 96(4):1135–1159

Oswald JN, Erbe C, Gannon WL, Madhusudhana S, Thomas JA (2022) Detection and classification methods for animal sounds. Exploring Anim Behav Through Sound 1:269–317

Oswald JN, Van Cise AM, Dassow A, Elliott T, Johnson MT, Ravignani A, Podos J (2022) A collection of best practices for the collection and analysis of bioacoustic data. Appl Sci 12(23):12046

Salamon J, Bello JP, Farnsworth A, Robbins M, Keen S, Klinck H, Kelling S (2016) Towards the automatic classification of avian flight calls for bioacoustic monitoring. PloS One 11(11):e0166866

Stowell D (2022) Computational bioacoustics with deep learning: a review and roadmap. PeerJ 10:e13152

Thakur A, Rajan P (2019) Deep archetypal analysis based intermediate matching kernel for bioacoustic classification. IEEE J Sel Top Sign Proces 13(2):298–309

Xie J, Hu K, Zhu M, Guo Y (2020a) Data-driven analysis of global research trends in bioacoustics and ecoacoustics from 1991 to 2018. Eco Inform 57:101068

Xie J, Hu K, Zhu M, Guo Y (2020b) Bioacoustic signal classification in continuous recordings: syllable-segmentation versus sliding-window. Expert Syst Appl 152:113390

Advanced Techniques for Ecological Research Using Bioacoustic Signal Processing

S. Thanga Revathi, V. Elizabeth Jesi, Mary Subaja Christo, and P. Visalakshi

Abstract

Bioacoustics is an interdisciplinary area of science that deals with the study of sound production, transmission, and reception in biological organisms, providing essential information on animal behavior, biodiversity, and ecosystem health. Advanced signal processing methodologies have revolutionized research in the field of bioacoustics, allowing complex sound patterns to be analyzed in a more detailed and efficient way. This chapter will give an overview of the most recent signal processing methodologies applied to bioacoustic data, with a focus on how these methods are changing ecological studies. Some of the key techniques discussed involve time–frequency analysis methods, including spectrograms and wavelet transforms, which are very important in visualizing and interpreting complex acoustic signals. Noise reduction techniques and signal enhancement strategies have been explored for problems that arise from environmental noise in field recordings. This chapter further explores feature extraction processes and machine learning algorithms that underpin species identification and behavioral analysis. Advanced tools, such as deep learning models, are underlined, which can operate with large-scale datasets and extract bioacoustic signals with subtle patterns. Further applications are presented in environmental monitoring, wildlife conservation, and the study of the effects of human activity on natural soundscapes, highlighting the practical utility of the methods. Some of the future directions for this area are real-time acoustic monitoring and integration of IoT and edge computing. This chapter connects technological advancement with ecological needs in order to encourage innovative research and sustainable practices in bioacoustics. The insights provided herein illustrate how advanced signal processing might lead to game-changing outcomes in the understanding and preservation of the natural world.

Keywords

Bioacoustics · Spectrograms · Wavelet transforms · Deep learning · IoT · Edge computing

1 Introduction to Bioacoustics

Bioacoustics is defined as the interdisciplinary science dealing with production, transmission, and reception of sound by biological organisms along with their environment. It is a branch at the crossroads between biology and acoustics. Knowledge contributed by the study of bioacoustics is basic to correct comprehension of animal behavior and communication, besides ecosystem dynamics. Acoustic signals let one monitor the populations and migrations of species and the health status of ecosystems. Besides that, bioacoustics also plays an important role in the identification of anthropogenic impacts on wildlife by way of noise pollution that might interfere with natural communications and habitats (Kumar et al. 2023). Applications range from purely research-oriented projects to very practical ones, such as conservation biology, environmental monitoring, and even medical diagnosis. As a concrete example, bioacoustic monitoring systems have already helped find the presence of several endangered species and reduce human-wildlife conflicts. Underwater, where vision is practically nil, aquatic life has used sound

S. Thanga Revathi · V. Elizabeth Jesi (✉) · M. S. Christo · P. Visalakshi
Department of Networking and Communications, School of Computing, SRM Institute of Science and Technology, Kattankulathur, India
e-mail: jesiv@srmist.edu.in

S. Thanga Revathi
e-mail: thangars@srmist.edu.in

M. S. Christo
e-mail: marysubc@srmist.edu.in

P. Visalakshi
e-mail: visalakp@srmist.edu.in

K. Subbaraj et al. (eds.), *Computational Bioacoustic Artificial Intelligence*, Sustainable Artificial Intelligence-Powered Applications, https://doi.org/10.1007/978-3-032-06088-4_7

as a vital communication medium; thus, acoustic analysis has also emerged as an important method of study in this arena (Odom et al. 2021).

1.1 Overview of Bioacoustics and Its Significance

Bioacoustics has really allowed researchers to grasp how organisms use sound in regard to communication, navigation, and survival in their environment. These acoustic signals expose much rich information on the behavior, physiology, and interaction of an organism with the surroundings in a noninvasive manner. Major importance in the study of mainly elusive or endangering species does not require direct human intervention quite often (McEwen 2024). The applications of bioacoustics range widely, from mapping the migrations of birds and echolocation of bats to the health of the soundscapes in a coral reef. It definitely contributes a lot to studying biodiversity and its conservation, and cannot be overemphasized. Furthermore, with modern technologies such as artificial intelligence and big data analytics, it opens new ways for analyzing complex ecosystems on an unprecedented scale (Napier et al. 2024).

1.2 Historical Developments in Bioacoustic Research

The roots of bioacoustics go back to the early twentieth century, when scientists first began to study animal sounds with the use of crude acoustic equipment. Early pioneers such as Karl von Frisch studied insect communication, while the foundational work by Donald Griffin in the 1930s revealed echolocation in bats, an important milestone in the field. In the years following World War II, bioacoustic research began to gain momentum with improving audio recording and signal analysis technologies. Hydrophones for studying marine mammals such as whales and dolphins were developed and complex communication systems discovered among the same species. In the second half of the century, the extension of bioacoustics toward ecological monitoring used long-term acoustic datasets as a means to investigate environmental changes and human impacts on wildlife (Sainburg and Gentner 2021). The twenty first-century arrival of digital technologies brought sea changes to the discipline. High-resolution recording devices, automated sound analysis software, and machine learning algorithms revolutionized data collection, analyses, and interpretation in bioacoustics. These developments have enabled extensive monitoring projects, such as acoustic observatories, changing the face of conservation worldwide.

1.3 Role of Signal Processing in Advancing Bioacoustics

Signal processing is the backbone of modern bioacoustic research. It includes techniques in the analysis, filtering, and interpretation of acoustic signals that help researchers make sense of meaningful information from raw audio data. This becomes particularly crucial in natural environments where the biological sounds are often obscured by ambient noise. Time–frequency analysis like spectrograms and wavelet transform allows for the detailed visualization of acoustic signals, hence facilitating species identification and behavioral studies. Noise reduction algorithms help isolate the target sounds from the background noise, hence enhancing the quality of data. Feature extraction techniques, including pitch detection and harmonic analysis, allow for insight into the structure and patterns of vocalizations (Pijanowski et al. 2024). Machine learning and deep learning in signal processing have taken bioacoustics a step further by enabling the classification of species and behaviors automatically, saving much manual effort, hence more scalable. Real-time signal processing systems are now being deployed for continuous monitoring, availing data in a timely manner for ecological decision-making. Signal processing has enlarged the scope of bioacoustics, bridging the gap between traditional ecological methodologies and technological advancements. What was once a very limited research area has now emerged as a vibrant discipline with possible far-reaching implications for biodiversity conservation, climate change studies, and ecosystem management.

2 Fundamentals of Acoustic Signal Processing

Acoustic signal processing involves the study, transformation, and interpretation of sound signals to bring out useful information. It has a key role in bioacoustics, as such processing allows the analysis of biological sounds that researchers can use to study the behavior, communication patterns, and conditions of species. This section outlines the basics of acoustic signal processing and gives further insight into the analysis of sound waves, processes of digitization, and some of the challenges faced (Jayanthi et al. 2025).

2.1 Basics of Sound Wave Analysis

Sound waves are longitudinal waves produced by the vibration of a medium-air or water. These waves consist of successive compressions and rarefactions, which effectively travel through the medium at speed determined by the properties of that medium. The key characteristics of sound waves include:

1. **Amplitude**: Reflects the wave's intensity or loudness. It is measured in decibels (dB) and corresponds to the energy of the sound wave.
2. **Frequency**: This is the number of oscillations per second. The unit is measured in hertz (Hz). Frequency tells the pitch of the sound, as frequencies with higher values indicate a higher pitch.
3. **Wavelength**: The distance between consecutive peaks or troughs of the wave, inversely proportional to frequency.
4. **Phase**: Indicates the position of a wave relative to a reference point, important in analyzing sound interactions like interference.

In bioacoustics, these properties help in ascertaining the sounds produced either from different species or environmental sources. Time-domain representations, given by waveforms, plot amplitude variation with respect to time, while those from frequency-domain analyses, through spectrograms, show a signal's frequency distribution (Yadav et al. 2023; Kershenbaum et al. 2024).

2.2 Sampling and Digitization of Bioacoustic Signals

Digitization is a crucial step in acoustic signal processing, converting analog sound waves into digital data for analysis and storage. This process involves:

1. **Sampling**: The continuous sound wave is measured at discrete time intervals. The sampling rate, measured in samples per second (Hz), determines the resolution of the digitized signal. According to the Nyquist theorem, the sampling rate must be at least twice the highest frequency in the signal to avoid aliasing (distortion).
2. **Quantization**: Each sampled value is assigned a discrete numerical value, approximating the wave's amplitude at the sampling points. The number of bits used in quantization determines the resolution, with higher bit depths providing more precise representations.
3. **Encoding**: The quantized values are converted into a binary format for storage and further processing.

In bioacoustics, high sampling rates and bit depths are often necessary to capture the subtle details of complex animal sounds, such as bird songs or dolphin clicks. These digital representations allow for advanced analysis using algorithms and computational tools (Malhotra 2024).

2.3 Common Challenges in Acoustic Signal Processing

Despite its advancements, acoustic signal processing in bioacoustics faces several challenges:

1. **Noise Interference**: Bioacoustic signals are often recorded in natural environments where background noise from wind, rain, or human activities can mask target sounds. Effective noise reduction techniques are required to isolate the biological signals.
2. **Signal Overlap**: With many species vocalizing in a habitat, signals from one source can no longer be easily differentiated from another. This problem calls for more sophisticated source separation algorithms.
3. **Data Volume**: Long-term acoustic monitoring yields a very large volume of data that is computationally expensive regarding storage, processing, and analysis. Efficient data management and machine learning techniques become crucial for scalability.
4. **Environmental Variability**: Temperature, humidity, and water salinity are some of the factors that influence the propagation of sound, making the recorded signals variable. These need to be taken into consideration during analysis to ensure accuracy.
5. **Device Limitations**: Recordings might be insensitive to recording, or the frequency range of a recording device might not catch low-intensity or high-frequency sounds which could be very important in some analyses. Developing high-fidelity recording systems remains one of the priorities (Oestreich et al. 2024; Bobba 2410).

By overcoming these, the researchers can make acoustic signal processing more reliable and accurate to assist in the discovery of new knowledge in bioacoustics and related fields.

Figure 1 shows the performance comparison of some noise reduction methods in bioacoustic signal processing. Conventional methods include spectral subtraction at 70% and Wiener filtering at 75%. These traditional techniques rely on the simple concept of filtering and subtraction for noise mitigation but usually fail under complicated and dynamic noise conditions found in natural environments. Wavelet denoising, which works with multi-resolution analysis, allows for higher effectiveness at 80% and therefore gives better adaptability in distinguishing signal from noise. However, deep learning performed best, with 90% effectiveness. Deep learning models excel because they are able to learn intricate patterns in how to distinguish target signals from diverse noise

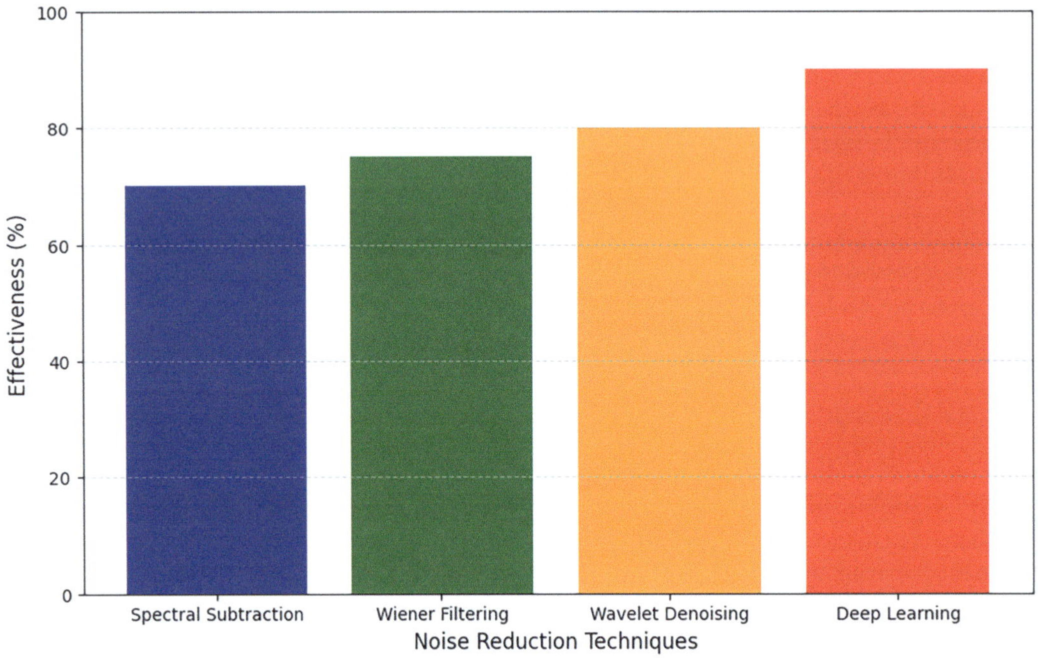

Fig. 1 Effectiveness of noise reduction techniques

environments, thus especially fitting for complex bioacoustic data. This trend underlines the growing importance of sophisticated computational approaches, especially deep learning, for enhancing noise reduction efficiency and improving the quality of bioacoustic analyses.

Figure 2 presents some comparisons of techniques' performance in overlapping signal resolution on bioacoustic datasets. Beamforming provides quite a modest accuracy of only 60%, which indicates that it is capable of performing a very basic kind of spatial filtering of sound sources. ICA does a little better, at 70%, using statistical independence to separate mixed signals, while it may be unstable for highly complicated or overlapped situations. The accuracy increases considerably in source separation using machine learning, which reaches 85%. These methods leverage advanced algorithms to distinguish overlapping sound sources based on learned features, thereby offering robust performance in multi-source environments. The state-of-the-art deep learning methods obtain the highest accuracy, around 92%, thereby establishing them as competent in handling intertwined, complex acoustic signals of extremely high resolution. Models benefiting from large-scale training data and their high-resolution capacity to nuance signal characterization bring important advantages. The given review underlines the transitorily potential of deep learning for bioacoustical challenges related to overlap within signals. This excellence in the performance of such advanced techniques makes them one of the key tools toward which one must progress, if better clarity and reliability in the findings are needed for research on bioacoustics.

3 Time–Frequency Analysis Techniques in Bioacoustics

Time–frequency analysis is a powerful tool in the analysis of nonstationary signals, such as bioacoustic data, whose frequency content varies with time. These methods allow the researcher to detect how the spectral features of the signal vary, something very useful in bioacoustics, since those signals can change with time according to environmental factors, animal behavior, or other influences. Time–frequency analysis is a method to understand such complex bioacoustic signals emanating from animal vocalizations, environmental noise, or underwater sounds, and it uncovers their inner structure (Ali et al. 2023).

3.1 Short-Time Fourier Transform (STFT)

One of the most popular time–frequency techniques is the short-time fourier transform. It refers to a method of signal decomposition into successive overlapping windows, each of which is analyzed by the Fourier transform. The key idea behind STFT is that the signal is assumed to be locally stationary within each window, meaning that the frequency components within that window are constant. Computation of the STFT is performed by multiplying the original signal with a window function, for example, a hamming or hanning window, and performing its transformation in the frequency domain. Such a process would then be repeated for every

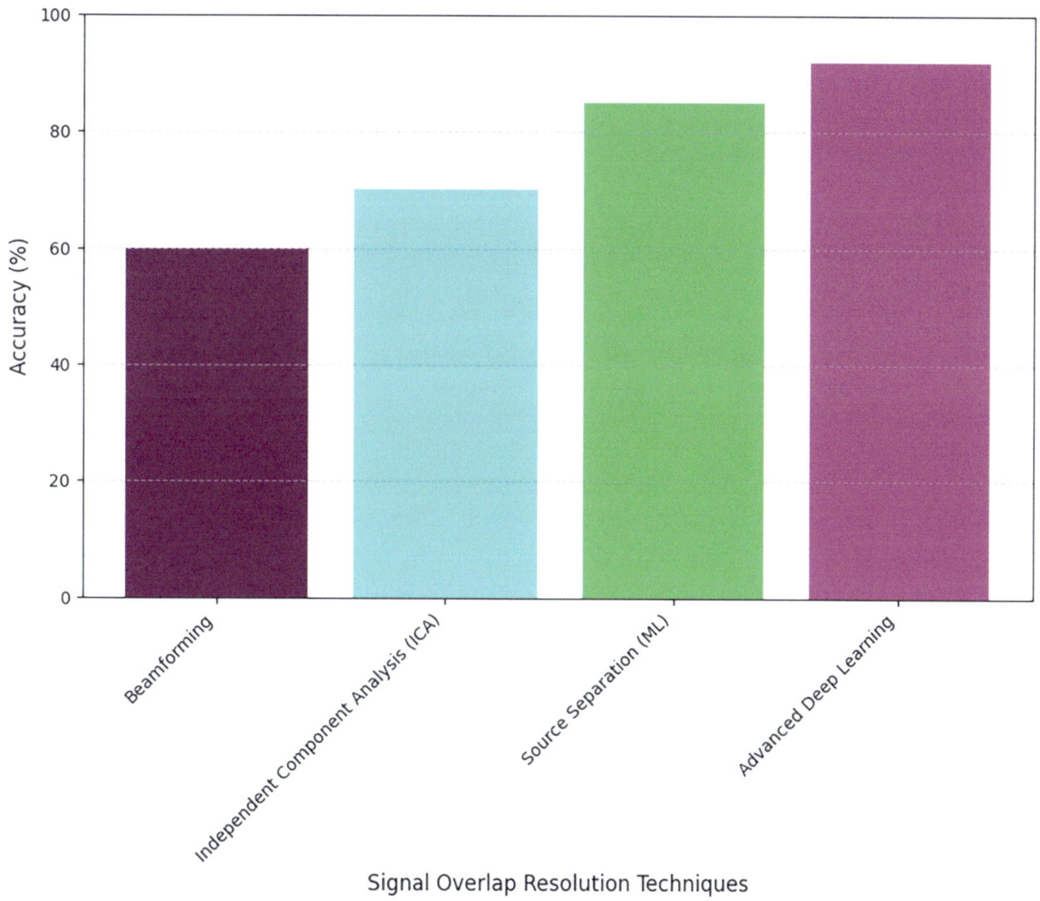

Fig. 2 Accuracy of signal overlap resolution techniques

window, thus creating the time–frequency representation of the signal. Such a result of processing gives, as its output, a matrix with different frequency bins corresponding to its rows, while time corresponds to the columns (Xu and Gong 2024).

This becomes particularly useful in the case of bioacoustics, where one often has to deal with signals containing some predictable frequency structure over short time spans such as birds' song or dolphin clicks. Moreover, this is a limitation of the STFT since it always faces limitations of fixed time–frequency resolution. The choice of size for the window faces an inevitable trade-off: narrow windows yield good time resolution but poor frequency resolution whereas large windows offer good frequency resolution at the price of imprecision in time resolution.

3.2 Wavelet Transform and Its Applications

Another powerful technique of time–frequency analysis is the WT, which is especially well-suited to analyze signals with varying frequency content over time. Unlike the STFT, which uses a fixed window size, the wavelet transform adapts to the local features of the signal by using wavelets-small oscillating functions that can be scaled and shifted to capture different frequencies at different times. This, therefore, is the adaptiveness which makes wavelet transformation to possess the potential for a great time and frequency resolution especially when dealing with signal classes that exhibit sudden discontinuities or possess certain stationarities. The CWT calculates the correlation of the signal with the scaled wavelet for every time interval, resulting in a time–frequency representation. Put differently, it gives an exact representation of the content from the signal in frequency at each instant, while the scale used from the wavelet provides information on the respective frequency band (Dangwal 2024).

Wavelet transforms find wide applicability in bioacoustic studies for their broad handling capability regarding signal characteristics. That is the case when analyzing vocalizations of marine mammals, wherein WT can capture both very high-frequency sounds, such as clicks, and lower-frequency calls. Further, wavelets are very effective in the detection of transient features in bioacoustic signals, such as sudden changes in pitch or amplitude, which are crucial in distinguishing specific animal calls from background noise. In addition, wavelet transforms can be used in denoising bioacoustic signals by removing high-frequency noise while preserving the essential features of the signal (Nieto-Mora et al. 2024).

3.3 Spectrogram Analysis for Bioacoustic Studies

One of the most common applications of time–frequency analysis in bioacoustics is spectrogram analysis. A spectrogram is a display that shows how the spectrum of frequencies in a signal changes with time. A spectrogram is obtained by computing the STFT of a signal and plotting the magnitude of the resulting frequency components versus time, using color or grayscale intensity to represent magnitude. Spectrograms allow one to intuitively and powerfully visualize the frequency content of bioacoustic signals as a function of time. For example, when analyzing bird song, a spectrogram will be able to show how different frequencies evolve as the bird vocalizes, revealing patterns in pitch, rhythm, and structure. In whale song analysis, spectrograms help researchers study complex vocal patterns, including frequency modulations, harmonics, and the relationships between different vocalizations. The good thing with spectrograms in bioacoustics is that they allow the identification of specific acoustic features such as peaks, valleys, and frequency modulations important for distinguishing between different species or behavioral states of animals. Spectrograms also allow analyzing temporal dynamics of signals, like call duration or intervals between sounds, which could be useful in ecological studies or monitoring animal activity. Despite their advantages, spectrograms also have weaknesses, especially if the signals are overlapping. If many sounds are happening at the same time, the spectrogram gets cluttered and it is very hard to distinguish between signals. Therefore, source separation and other signal processing algorithms are often applied with spectrograms together to make the analysis more accurate (Liang et al. 2024).

Figure 3 compares STFT and WT, highlighting clear advantages of the latter for bioacoustic signal analysis. WT is superior to STFT on three critical points: time resolution, frequency resolution, and handling variability of the signals. The time resolution of wavelet transform is 15% better than STFT, reflecting better analysis of fast-changing bioacoustic signals. This indicates a 15% improvement in frequency resolution and, hence, the capability to detect finer spectral details in such signals. Most importantly, the wavelet transform is 23% more effective in handling signal variability—a critical factor in the analysis of nonstationary signals, common in bioacoustic data. This comparison underlines the wavelet transform as an effective tool in capturing the dynamic and fluctuating nature of bioacoustic signals. Figure 4 shows the comparison of spectrogram analysis with WT, whereby the performance of wavelet transform can be identified in a superior way with many of the important metrics. Hence, wavelet transform gives wide scope for 10% more advantage in feature detection by capturing low- and high-frequency components

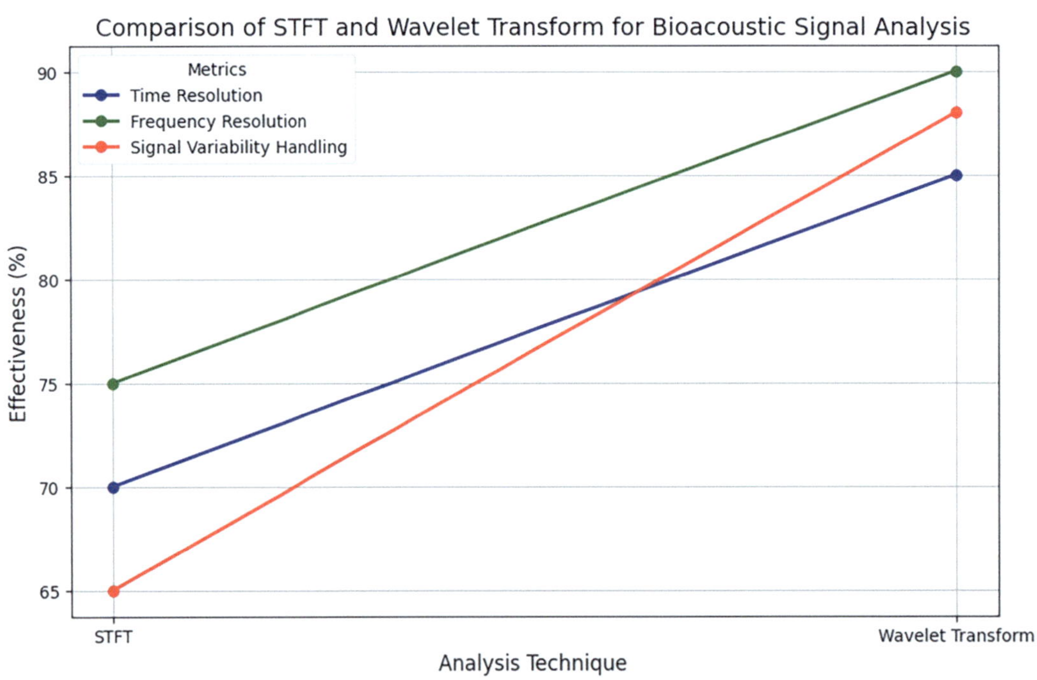

Fig. 3 Comparison of STFT and wavelet transform

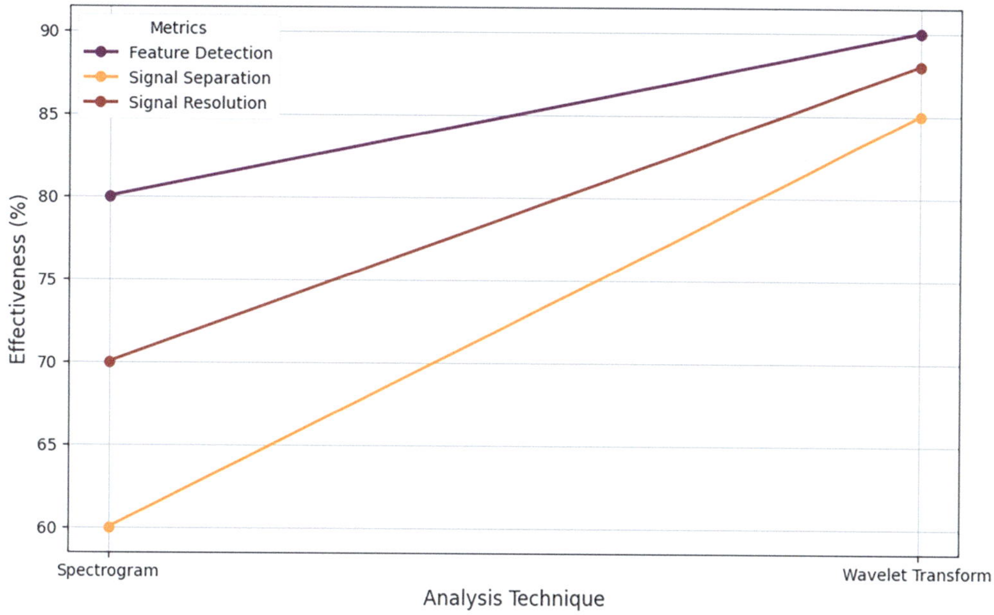

Fig. 4 Comparison of spectrogram analysis and wavelet transform for bioacoustic signal analysis

of bioacoustic signals more precisely. Speaking signal separation, wavelet transform stands on a better position, by a margin as high as 25% over that in spectrogram analysis. Its application is more suited in the disentanglement of overlapping signals; very important in bioacoustics. Besides this, signal resolution with wavelet transform makes a difference of approximately 18% over and beyond spectrogram analysis, while important details of bioacoustic signals are maintained. In general, this wavelet transform is the approach that proves to be quite functional and precise in its utilization for bioacoustic signals, especially when the sounds are overlapped or complex under the conditions.

4 Noise Reduction and Signal Enhancement in Bioacoustic Analysis

Bioacoustics is very often dealing with the analysis of animal vocalizations, environmental sounds, and other acoustic phenomena in the presence of a number of types of noise. Noise could come from various sources, such as environmental noise, instrumentation noise, and interference due to other sounds, making the quality of the bioacoustic signals recorded poorer; hence, wrong analyses could be carried out. Thus, good noise reduction and signal enhancement techniques are important to enhance the quality and reliability of the bioacoustic data. These techniques help in isolating the target bioacoustic signals from unwanted noise, thereby making it easier to analyze and interpret them for research, conservation, and monitoring purposes (Ozanich et al. 2021).

4.1 Techniques for Noise Filtering in Bioacoustic Signals

Noise filtering in bioacoustic signals is a multi-step process that aims to reduce or eliminate unwanted noise while preserving the integrity of the underlying signal. Several techniques are commonly employed in bioacoustic signal processing to achieve effective noise reduction:

1. **Bandpass Filtering**: Among the most straightforward and common methods of noise filtering come in with bandpass filtering. A bandpass is a region in which frequencies will pass, and other frequencies outside the band are filtered out. Bioacoustic signals occur very regularly within a particular frequency range. These would include the song of a bird or even the echolocation calls of bats; hence, the application of a bandpass filter takes out the unwanted low-frequency or high-frequency noise from outside that range. Noise from traffic or low-frequency environmental sounds can be removed in bird vocalizations, for example, while the important acoustic features remain.
2. **Spectral Subtraction**: This is a method in which an estimate of the noise spectrum is subtracted from the original signal's spectrum. The assumption that has to be made here is that the noise spectrum does not vary much during the non-signal parts of the recording. Spectral subtraction can effectively reduce stationary background noise, such as hums or hissing sounds, but may introduce artifacts if the noise characteristics change too fast.
3. **Wiener filtering**: Wiener filter was an adaptive filter, which, by using the statistical models, provides the

minimum mean square error among the noisy signal and estimations of a clean signal, requiring the signal and noise powers spectral estimation. It is exceptionally effective when the noise characteristics to be filtered are relatively fixed and could thus be analytically estimated over time periods. Thus, adaptation to varying signal and noise conditions makes it efficient and handy in dynamic environments for various field recordings.

4. **Filtering Based on Fourier Transform**: Methods of Fourier Transform involve the translation of the signal from the time domain into the frequency domain, filtering, and translation of the signal back to the time domain. It enables one to filter out precisely in the frequency domain, thereby eliminating unwanted frequencies while preserving the features that are needed in the bioacoustic signal. Using this method, one can get rid of hums due to electrical equipment or the low-frequency noise caused by wind (Manikandan and Neethirajan 2024; Ghani et al. 2023).

4.2 Adaptive Filtering Methods

Adaptive filtering techniques work to converge to the characteristics of the noise and the signal in real time. This generally makes adaptive filters very useful for dynamic environments where the noise conditions change with time. Opposed to a fixed filter, an adaptive filter changes its parameters with respect to the nature of the incoming signal and noise characteristics, hence performing better than fixed filters under unpredictable or nonstationary conditions.

1. **Least Mean Squares Algorithm**: The LMS algorithm is one of the simplest and most generally used adaptive filtering techniques. It works on the principle of minimizing the error between the desired signal and the output of the filter. The filter coefficients are iteratively updated to enable the filter to adapt to the signal and noise characteristics with time. In bioacoustics, for example, the LMS algorithm could be used to remove noise from animal call recordings in the presence of fluctuating environmental sounds.
2. **Recursive Least Squares (RLS) Algorithm**: The RLS algorithm is more complex and gives a faster convergence and better performance than LMS. RLS works recursively to change the filter coefficients in a way to minimize the error between the filtered signal and the reference signal. It is especially useful when the characteristics of the signal change rapidly or when the noise varies greatly with time.
3. **Kalman Filtering**: The Kalman filter is a kind of optimal recursive filter that can be applied in real time to estimate the state of a system from noisy measurements. Bioacoustic applications include tracking of signals embedded in noise, especially in cases involving nonstationary noise or when multiple sources are present. They are commonly used in the analysis of underwater acoustics, such as whale calls, where ambient noise levels can fluctuate due to changing sea conditions.
4. **Wavelet Transform-Based Adaptive Filtering**: Adaptive filtering, when combined with the wavelet transform, gives time–frequency analysis sensitive to both transient and stationary noise. This method decomposes the signal into various frequency bands, and at each level, noise can be filtered. Adaptive wavelet-based filtering is very effective in handling signals with abrupt transitions, such as clicks and bursts of animal calls (Prasad 2024).

4.3 Importance of Denoising in Field Recordings

Field recordings, in particular, are liable to a variety of noise: they can involve animal sounds, though generally, it is the background noise of environmental sources—such as wind, rain, traffic, machinery interference from other animals, and technical noise that is introduced by recording equipment—that create problems. A very important stage in the bioacoustic signals analysis is that of denoising for several reasons:

1. **Improving Signal Clarity**: There are field recordings where the sounds of different sources overlap. Denoising techniques can thus enhance the clarity of a target bioacoustic signal in order to make identifications of important acoustic features—such as pitch, duration, and rhythm—easier to detect for animal vocalizations. Precisely for this purpose, it is important in species identification and behavioral analysis, providing very useful information on population monitoring.
2. **Better Detection of Rare Events**: Denoising allows researchers to detect those rare or dim events that may get hidden among the background noise. In bioacoustic research, some species produce signals very subtly, or low-frequency signals that are barely detectable above environmental noises. Effective noise reduction serves to enhance such detection of rare signals and provides better data on ecological research and conservation efforts.
3. **Accurate Animal Behavior Interpretation**: Background noise masks many important aspects of animal behaviors in field recordings. It could be the presence of any other animal vocalizations, or environmental sounds, making it hard to understand a particular call in its real context or meaning. Reduction of noise through denoising methods helps provide better clarity on animal communication and behavior, including their interactions with the environment.
4. **Facilitating Long-Term Monitoring**: Long-term bioacoustic monitoring usually finds its application in conser-

vation, for long-term observation of animal populations, or the detection of changes in the environment. Noise reduction techniques are of great importance to assure quality over long periods, especially when recording in high ambient noise conditions. Denoising will help keep data usable for long-term analysis, trend detection, and decision-making (Sharma et al. 2023; Bermant et al. 2022).

Figure 5 compares the effectiveness of different noise reduction techniques in enhancing bioacoustic signals. Among these techniques, the highest percentage is the wavelet transform with 95%, thus the most efficient for noise reduction. It is followed closely by adaptive filtering with 90%, showing its high adaptability in dynamic environments where the noise conditions change. Wiener filtering ranks 85%, suitable for environments where noise characteristics are relatively fixed. Spectral subtraction and bandpass filtering also have comparatively lower performance around 75% and 80%, respectively, but with applications in the above situations of broad-range and frequency-specific removal. Such a result points to the particular efficacy of the advanced techniques of wavelet transform and adaptive filtering in very complex real-world bioacoustic applications.

Figure 6 shows the performance of various adaptive filtering methods for enhancing bioacoustic signals. Again, wavelet transform turns out to be the most effective method, with a performance of 95%. Next comes Kalman filtering with 90%, as it has proved to be one of the best in tracking and estimating signals embedded in noise. The RLS algorithm is 85% performing, reflecting that it will be able to adaptively change the filter coefficients in rapidly changing environments. Although simple to implement, LMS algorithm performs at about 80%, which may suit low-complexity applications, but in highly dynamic or nonstationary conditions, it would not be that effective. Among the several techniques in adaptive filtering, the wavelet transform and Kalman filtering are outstanding for bioacoustic signal enhancement, especially in poor conditions.

5 Real-Time Signal Processing Applications

5.1 Automated Systems for Real-Time Monitoring

Because actual processes should be in real time—for an uninterrupted observation of ecosystems—we find that real-time analysis has become indispensable when pursuing bioacoustic research. Strong algorithms running on automatic stations were able to analyze this continuous flow of bioacoustical data in order that swift decisions could be immediately adopted regarding the changes the respective environment had undergone. Therefore, by using hardware platforms implemented either as sensor networks or sets of microphones or even audio recorders spread through remote or dynamic environments. Real-time data streaming thus allows the system to instantaneously process the incoming acoustic signals and detect the events relevant to the analyses, be it animal vocalizations or environmental disturbance,

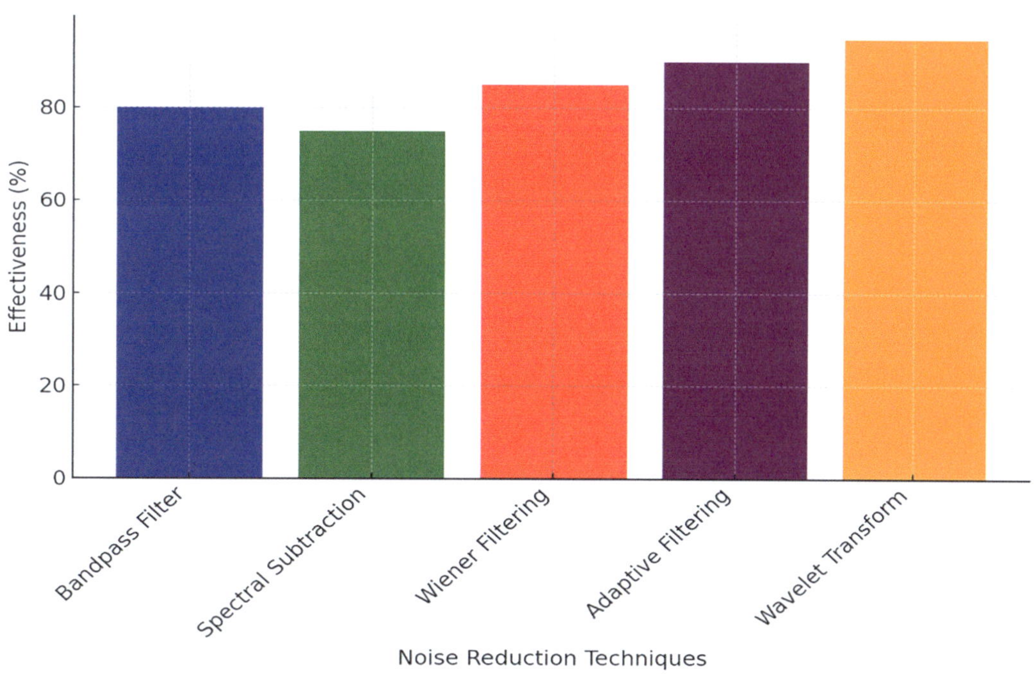

Fig. 5 Comparison of noise reduction techniques in bioacoustic signals

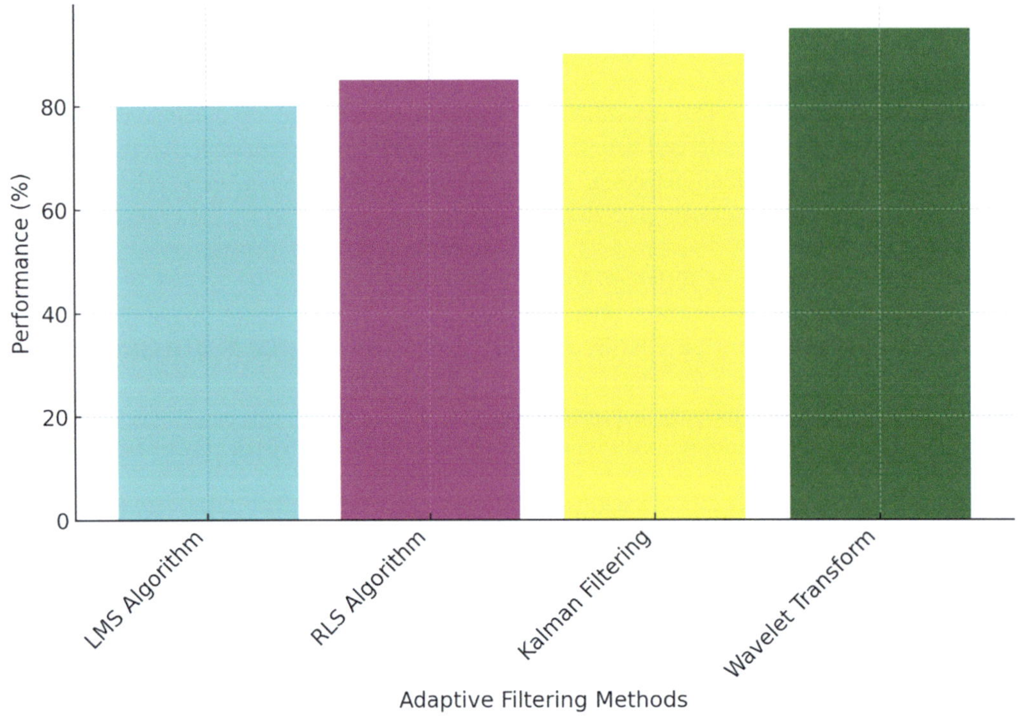

Fig. 6 Comparison of adaptive filtering methods for bioacoustic signal enhancement

which may express the presence of some species or other. The development of those automated systems embeds techniques of signal processing using fast Fourier transforms, wavelet transforms, or even machine learning models that perform feature extraction, classification with high accuracy, and instantaneous reporting that allows large-scale ecosystem monitoring with very low human intervention (Williams et al. 2024).

5.2 Applications in Wildlife Monitoring and Conservation

Real-time bioacoustic monitoring has found many applications in the domain of wildlife observation and the protection of the environment. Such systems will enable a researcher to monitor populations continuously in order to understand behavior and thereby understand the general health status of an ecosystem. Biological acoustic real-time monitoring has sometimes been applied for the automatic detection of animal calls by means of identifying species or tracking patterns of migration. This has proved especially practical in conditions where the conventional method is either not workable or too intrusive. It can also be used for monitoring endangered species, poaching activities, or observing any change in animal behavior with regard to disturbances in their environment, like climate change or human activities. Further, the use of machine learning in real-time bioacoustic analysis can automate species identification and behavioral analysis; thus, making the work of conservation a bit efficient and more accurate (Mathew 2023).

5.3 Challenges and Opportunities: Real-Time Analysis

In addition to its promising applications, various challenges come along the way of bioacoustic signal processing in real time. The major problem includes massive amounts of data from the continuous monitoring that is under scrutiny. Larger datasets rely upon highly developed storage facilities for effective processing algorithms for data handling and analysis in decent time frames. Moreover, it also needs to address a number of problems relating to unwanted ambient noises, which have turned out to be significant hassles in their functionality. Advanced noise-reducing techniques, for example, such as adaptive filtering and spectral subtraction, play an important role in the determination that the relevant acoustic signal will not be lost among background noise. Besides, installation of real-time analysis systems is also logistically cumbersome because these places have limited access to power and network connectivity. However, limitations in this regard are overcome when energy-efficient hardware is used along with IoT technologies, making data collection and processing from the edge itself possible rather than always using a central server.

On the other hand, there is a great deal of room for further work on real-time bioacoustic analysis. The application of deep learning models for feature extraction and classification in a real-time system will notably enhance the performance of species identification and behavioral analysis. Furthermore, emerging techniques such as edge computing and IoT have much potential for processing data near the sources to reduce latency and bandwidth consumption. This is an opportunity for real-time integration of new technologies (Loo et al. 2023).

6 Applications of Advanced Signal Processing in Bioacoustics

6.1 Environmental and Ecological Monitoring

Advanced signal processing techniques bear much importance in environmental and ecological monitoring, which is usually always required to present real data on the health of all ecosystems in real time or very close to it. Bioacoustic signals make for the analysis of a trend in biodiversity and certain species distribution within a particular area of interest. For instance, animal utterances, like bird sings or whale calls, make it possible to detect there are species that have populations or not. Spectrograms are a time–frequency analysis technique that visualizes these sounds to assist ecologists in monitoring the population dynamics of species over time. Such methods can be extended to the detection of changes in ecosystems that could reflect environmental distress, such as the loss of biodiversity due to habitat degradation or climate change. Continuous recording and analysis of bioacoustic data can be done to ascertain habitat health and the impact of natural events such as deforestation, forest fires, or shifts in climate on animal populations and biodiversity.

The integration of IoT devices with edge computing in real-time acoustic monitoring systems enables the collection and processing of bioacoustic data remotely. For instance, such a system may trigger alerts on sudden environmental changes or the appearance of invasive species to enable immediate action that prevents adverse changes in the ecosystem (Schuller et al. 2024).

6.2 Terrestrial and Aquatic Species-Behavioral Research

Advanced signal processing techniques are among the key tools in animal behavior studies. The analysis of vocalizations and other sound-based communications allows for insight into the social interactions, mating rituals, territorial behaviors, and hunting patterns of both terrestrial and aquatic species. Such processing of echolocation calls in bats or dolphins, for example, provides an understanding of how these animals perceive their surroundings, communicate with each other, and locate prey. ML algorithms, combined with signal processing, enable species-specific vocalizations to be classified automatically even under noisy conditions. This facilitates research into the behavioral patterns of rare or elusive species which may hardly be observable in nature. Besides, methods related to time–frequency like wavelet transforms are effective in picking up minute variations in the calls that may indicate stress, readiness for mating, or changes in behavior due to environmental changes.

This may also, in turn, reveal signal processing changes in frequency or pitch in the sounds that correlate with behavioral changes or ecological stress, such as changes in migration patterns or vocalization due to noise pollution or habitat disturbances. These behavioral shifts are important to understand for the purpose of conservation and the protection of endangered species.

6.3 Assessment of Anthropogenic Impacts to Ecosystems

These anthropogenic influences, such as noise pollution, urbanization, and climate change, impact significantly upon ecosystems and typically perturb animal communication, behavior, and biodiversity. Sophisticated signal processing methodologies have proven to be very helpful in quantifying such effects. Noise pollution interferes with the possibility of communication between species, in particular, those using vocalizations in order to be able to mate, navigate, or defend a territory. The impacts of anthropogenic noise on species and ecosystems can be quantified by researchers when they analyze temporal shifts in frequency, timing, or intensity of animal vocalizations. Techniques like spectrograms and wavelet transforms can be utilized in studying, for example, the impact of different types of human-made noise on animal vocalizations. These methods, in some cases, allow tracking behavioral changes in response to noise—such as changes in pitch in bird calls or even disruption of whale song communication patterns. Deep learning algorithms can also investigate massive datasets for developing correlations of noise levels and species activities with a view to attaining an even better understanding of human activities' impacts on wild populations. It would also go a long way in aiding policy makers make informed decisions on noise-related regulations and habitat protection as ways through which the impacts on natural soundscapes driven by urbanization and industrialization can be reduced. These advanced methodologies further enable researchers to understand the subtler yet deeper impacts of human influences on the natural world and encourage informed conservation efforts in a way that strategizes the preservation of biodiversity.

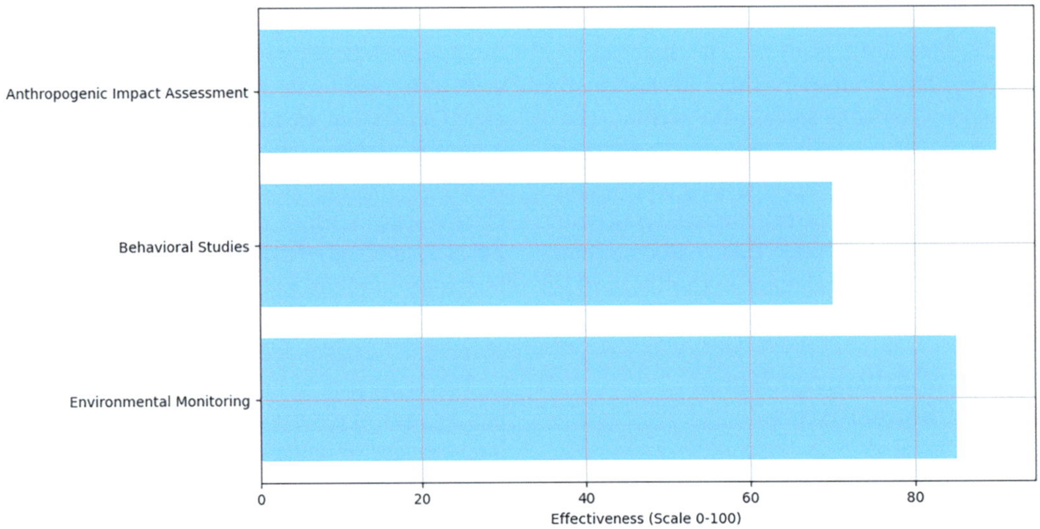

Fig. 7 Effectiveness of bioacoustics applications

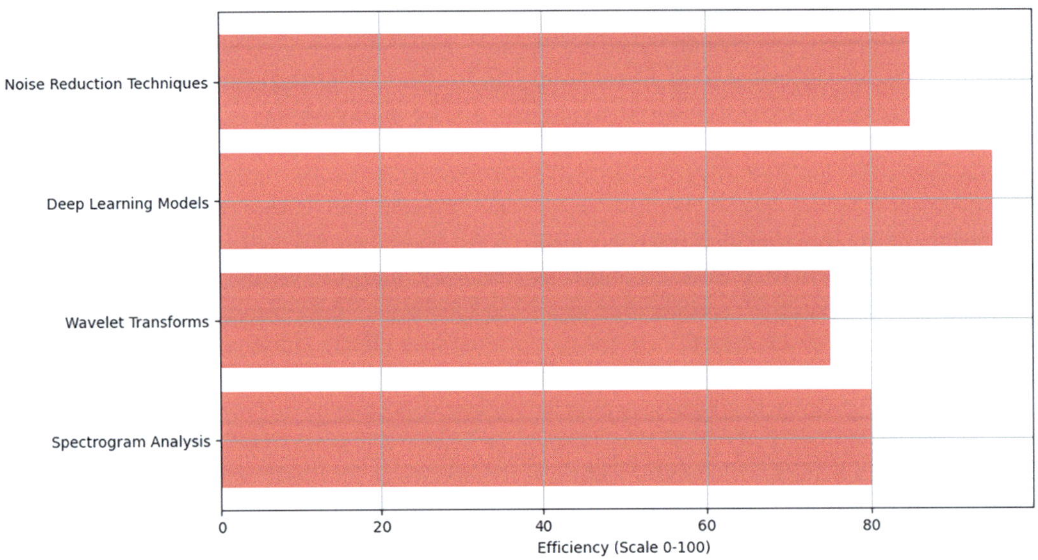

Fig. 8 Efficiency of signal processing techniques

Figure 7 shows the effectiveness of bioacoustic applications in ecological monitoring, behavioral studies, and anthropogenic impact assessment. Due to the growing integration of advanced signal processing and real-time monitoring technologies, scores are highest for environmental monitoring and anthropogenic impact assessment. In the case of behavioral studies, strong effectiveness is noticed, especially with the application of machine learning for behavioral classification. Figure 8 depicts the different signal processing techniques with regard to their efficiency in bioacoustic research. In this regard, deep learning models present the highest degree of efficiency because of the ability of the models to handle big and complex data. Spectrogram analysis and wavelet transforms are also quite efficient but not as much because these are the more conventional methods and also quite crucial for signal processing.

7 Conclusion

Advanced signal processing techniques have significantly expanded the scope and depth of bioacoustic research, opening new opportunities to investigate the intricacies of natural soundscapes. Application of time–frequency analysis methods, noise reduction strategies, and machine learning algorithms has brought about more accurate species identification and behavioral analysis that is crucial for effective conservation and environmental monitoring. While there is a large area yet to be covered by ecologists using the already accepted deep learning model along with new ones, including IoT and edge computing, further revolutionizing these bioacoustic studies toward real-time monitoring with an enhanced way of handling the data, besides merely adding to scientific understanding of biodiversity and ecosystem health, these advanced methods will play their role in practical applications designed for nature preservation in light of continuously growing anthropogenic pressures. Future research, therefore, needs ongoing signal processing methodology innovation combined with cross-disciplinary collaboration for a possible meeting of the ecological challenge for a wild life and its habitats toward its future sustainability.

References

Ali R, Islam A, Rana MS, Nasrin S, Shajol SA, Sadi AHM (2023) ML-ASPA: a contemplation of machine learning-based acoustic signal processing analysis for sounds, and strains emerging technology. arXiv preprint arXiv:2402.10005

Bermant PC, Brickson L, Titus AJ (2022) Bioacoustic event detection with self-supervised contrastive learning. bioRxiv: 2022-10

Bobba KS, Krishna Sai V, Bugga D, Bolla VMS (2024) AI-enhanced acoustic analysis for comprehensive biodiversity monitoring and assessment. arXiv preprint arXiv:2410.12897

Dangwal K (2024) AI-enhanced bee bio-acoustics with data augmentation and applications in early swarming detection

Ghani B, Denton T, Kahl S, Klinck H (2023) Global birdsong embeddings enable superior transfer learning for bioacoustic classification. Sci Rep 13(1):22876

Jayanthi K, Chitradevi D, Muthaiyah S (2025) Early prediction of harmful environmental changes based on bioacoustics analysis. In: Technological advancements for deep sea ecosystem conservation and exploration. IGI Global Scientific Publishing, pp 129–148

Kershenbaum A, Akçay C, Babu-Saheer L, Barnhill A, Best P, Cauzinille J, Clink D et al (2024) Automatic detection for bioacoustic research: a practical guide from and for biologists and computer scientists. Biol Rev

Kumar SA, Verma M, Kumar V, Kumar G (2023) Identification of bird species through audio signal processing using neural network. In: International conference on innovations in data Analytics. Springer Nature Singapore, Singapore, pp 437–447

Liang J, Nolasco I, Ghani B, Phan H, Benetos E, Stowell D (2024) Mind the domain gap: a systematic analysis on bioacoustic sound event detection. arXiv preprint arXiv:2403.18638

Loo YY, Avicena NR, Thong N, Haque AM, Nhlabatsi YT, Abakar SYA, Ng KH, Wong EP (2023) WildTechAlert: deep learning models for real-time detection of elephant presence using bioacoustics in an early warning system to support human-elephant coexistence. In: International conference on brain inspired cognitive systems. Springer Nature Singapore, Singapore, pp 385–399

Malhotra R (2024) Beehive bioacoustics noise reduction using liquid neural networks

Manikandan V, Neethirajan S (2024) Decoding poultry vocalizations-natural language processing and transformer models for semantic and emotional analysis. bioRxiv:2024-12

Mathew A (2023) Improving the accuracy and efficiency of bird song analysis with machine learning based event detection

McEwen B (2024) Computational bioacoustics for the detection of rare acoustic events

Napier T, Ahn E, Allen-Ankins S, Schwarzkopf L, Lee I (2024) Advancements in preprocessing, detection and classification techniques for ecoacoustic data: a comprehensive review for large-scale passive acoustic monitoring. Expert Syst Appl 124220

Nieto-Mora DA, Ferreira de Oliveira MC, Sanchez-Giraldo C, Duque-Muñoz L, Isaza-Narváez C, Martínez-Vargas JD (2024) Soundscape characterization using autoencoders and unsupervised learning. Sensors 24(8):2597

Odom KJ, Araya-Salas M, Morano JL, Ligon RA, Leighton GM, Taff CC, Dalziell AH et al (2021) Comparative bioacoustics: a roadmap for quantifying and comparing animal sounds across diverse taxa. Biol Rev 96(4):1135–1159

Oestreich WK, Oliver RY, Chapman MS, Go M, McKenna MF (2024) Listening to animal behavior to understand changing ecosystems. Trends Ecol Evol

Ozanich E, Thode A, Gerstoft P, Freeman LA, Freeman S (2021) Deep embedded clustering of coral reef bioacoustics. J Acoust Soc Am 149(4):2587–2601

Pijanowski BC, Fuenzalida FR, Banerjee S, Minghim R, Lima SL, Bowers-Sword R, Guzman SR et al (2024) Soundscape analytics: a new frontier of knowledge discovery in soundscape data. Curr Landscape Ecol Rep 1–20

Prasad VS (2024) Bird species identification through vocalization analysis using machine learning. Int J Inform Technol Comput Eng 12(3):742–749

Sainburg T, Gentner TQ (2021) Toward a computational neuroethology of vocal communication: from bioacoustics to neurophysiology, emerging tools and future directions. Front Behav Neurosci 15:811737

Schuller BW, Akman A, Chang Y, Coppock H, Gebhard A, Kathan A, Rituerto-González E, Triantafyllopoulos A, Pokorny FB (2024) Ecology and computer audition: applications of audio technology to monitor organisms and environment. Heliyon 10(1)

Sharma S, Sato K, Gautam BP (2023) A methodological literature review of acoustic wildlife monitoring using artificial intelligence tools and techniques. Sustainability 15(9):7128

Williams B, van Merriënboer B, Dumoulin V, Hamer J, Triantafillou E, Fleishman AB, McKown M et al (2024) Leveraging tropical reef, bird and unrelated sounds for superior transfer learning in marine bioacoustics. arXiv preprint arXiv:2404.16436

Xu T, Gong T (2024) Development of bio-acoustics-based educational tools: innovations in music teaching for enhanced learning outcomes. J Commer Biotechnol 29(3):85–95

Yadav D, Agarwal R, Yadav S (2023) Soundscapes of life: acoustic revelations in biological media. In: Handbook of vibroacoustics, noise and harshness. Springer Nature Singapore, Singapore, pp 1–32

Applications of AI and Bioacoustics in Environmental Monitoring and Beyond

AI-Powered Bioacoustic Methods for Early Warning of Natural Disasters Through Animal Behavioral Cues

C. Alakesan, G. Thirunavukkarasu, D. Anandakumar, G. Vijayakumari, and Balamurugan Balusamy

Abstract

Natural disasters like earthquakes, tsunamis, and volcanic eruptions are early identified by the unusual behavior of animals. By nature, animals have the sharp senses that allow them to detect the smell, change in air pressure, etc. Animals can give signals frequently for the disasters by exhibiting behaviors like unusual vocalization, restless behavior, and migration. Especially, the earthquakes are mainly identified by the elephant and birds which is responsible for the changes in environment. This chapter begins with an innovative approach for disaster prediction by analyzing the unusual behaviors of animals. Here, we are focusing on the vocalization and behavior of the animals by the changes of the nature. These changes of animal behaviors and sound can be detected by the artificial intelligence techniques. The early warning natural disaster systems are developed by using machine learning (ML) and deep learning (DL) algorithms. Introduction of the chapter explains about the drawbacks of traditional artificial intelligence-based detection and classification algorithms like decision tree, random forest, principal component analysis (PCA), hidden Markov model (HMM), and super vector machine (SVM). Mel frequency cepstral coefficients (MFCCs) are used to extract the acoustic features of animal sound. Above-mentioned traditional algorithms are used to classify the animal as either normal or abnormal behaviors. Convolutional neural networks (CNNs), recurrent neural networks (RNNs), and hybrid convolutional recurrent neural networks (CRNNs) are identifying the features of datasets. These algorithms are automatically extracting the key features from time-series sound recording data. Also clustering algorithms are employed to categorize the behavior patterns and integrating this algorithm with ML and DL to detect the anomalies when animals exhibit different behaviors such as restlessness, feeling, and sound. With the help of this techniques, we can detect the natural disasters in early stage, so that we can evacuate people from those surroundings. This chapter ends with the integration of machine learning (ML) and deep learning (DL) algorithm to detect behavioral anomalies in animals which offers immense possibilities for improving disaster management systems through enhanced accuracy.

Keywords

Natural disasters · Anomalies · Deep learning · Machine learning · Clustering

1 Introduction

Animals demonstrate more sensitive sensory perception and instinctive behaviors that quite often allow them to detect changes in their environment much sooner than do humans or any technology. For centuries, science about the behavior of animals has been used to predict natural disasters like earthquakes, tsunamis, and storms. Empirical studies have shown that most species of animal, including birds, mammals, and reptiles, perceive changes in their environment: changes in barometric pressure, terrestrial vibrations, or magnetic

C. Alakesan
Department of ECE, Shree Venkateswara Hi-Tech Engineering College, Erode, India

G. Thirunavukkarasu
Department of ECE, Kongu Engineering College, Erode, India
e-mail: gthiru.ece@kongu.edu.in

D. Anandakumar
Department of ECE, Nanda Engineering College, Erode, India
e-mail: anandkumar.d@nandhaengg.org

G. Vijayakumari (✉)
Department of ECE, Kangeyam Institute of Technology, Tirupur, India
e-mail: viji.ece05@gmail.com

B. Balusamy
Shiv Nadar University, National Capital Region (NCR), Delhi, India
e-mail: balamurugan.balusamy@snu.edu.in

K. Subbaraj et al. (eds.), *Computational Bioacoustic Artificial Intelligence*, Sustainable Artificial Intelligence-Powered Applications, https://doi.org/10.1007/978-3-032-06088-4_8

fields, for example—much earlier than humans can determine these changes. Using animals' natural behaviors, researchers develop innovative early warning systems merging animal behavioral patterns with advanced detection technologies. It makes use of bioacoustics, IoT devices for tracking, and data analysis by AI to interpret the movement and vocalization of animals as a potential indicator of disasters. Animal cues for the early discovery of disasters are unique in that they can be very sensitive and quite cheap compared to methods conventionally used. This new area of research develops the existing knowledge of animal behavior but places a critical emphasis on the point that biodiversity proves essential in shaping human safety and resilience considering natural disasters.

The study of abnormal behaviors in animals offers a singular approach to the enhancement of predicting disasters. Animals have highly refined sensory systems that enable them to recognize slight changes in their surroundings, such as seismic activity or changes in atmospheric pressure and electromagnetic fields usually hours or days before these are identified by humans or commercial monitoring systems. This means for example that dogs would manifest anxious behavior and expressions before earthquakes, a species of bird would possibly change its flight course while the aquatic animal would display disorganized swimming patterns before the tide. Studies on these peculiar behaviors exhibited by animals will try to offer early warning systems whereby use their instinctive behavior as latent warnings before disastrous events (Sharma et al. 2023; Sharma 2023).

It integrates the latest technologies to be used in real-time monitoring of movement, sounds, and physiological responses from the animals, such as bioacoustic monitoring, GPS tracking, and IoT sensors. Predictions and correlations with the environment are computed by AI and machine learning algorithms based on such data points for hazard prediction. For example, in earthquake-prone regions, continuous tracking of livestock or wildlife will provide very strong early warning data if it is combined with seismic sensors. Nevertheless, the method will still work even for remote or low-resource areas where traditional facilities that once predicted disasters will prove unsatisfactory.

1.1 Importance of Early Disaster Detection

Prompt disaster identification is very essential for the relief of the severe outcomes resulting from both natural and human-induced disasters. All events such as earthquakes, floods, hurricanes, and wildfires and tsunamis can cause extensive damage to the environment, loss of lives, and significant financial impacts. Timely detection and alert systems can help save many lives and soften the impact of these after-effects by providing communities and officials with information that helps in taking anticipatory measures. Technological breakthroughs in artificial intelligence, IoT, and remote sensing have really made early warning systems much more accurate and reliable. Real-time collection and analysis of data through such innovations speed up resource allocation and proper decision making. Early discovery of such disasters saves lives; reduces further physical damage, maintains the strength of structures, ensures unbroken operation of vital services, and promotes social resilience (Nandan et al. 2024). Additionally, it enhances preparation and prepared response capabilities through cooperation among emergency service institutions, government agencies, as well as nongovernmental organizations. In the case of climate change rise, early warning systems ensure that the world becomes habitable and secure. At this point, commitment coupled with priority to earlier frameworks of disaster warnings turns into a moral necessity. From being a technological need in terms of saving human and ecosystem life, it turns more significantly into a moral imperative and much important requirement.

1.2 Role of Animal Behavior in Natural Disaster Prediction

Animal behavior has long been of interest to scientists and various communities because it can supposedly predict a natural disaster. Many historical accounts and modern studies tell of animals performing unusual actions hours or even days before a natural disaster such as earthquakes, tsunamis, floods, or storms. These include alterations in activity patterns, vocalization, and general activity patterns. For example, dogs might get restless, birds change their flight route, and elephants go up to the high grounds before a tsunami hits. All this is claimed to be the result of animals' enhanced sensitivity; most animals are claimed to detect minute changes in the environment that may not be sensed by a human or even an ordinary monitoring system. This would perceive vibrations of substrates, fluctuation in atmospheric pressure or the changing electromagnetic fields that might serve as an indication of possible threat. The innate biological sensitivity, thus, is being utilized by modern technological equipment to predict disasters. Bioacoustic monitoring, tracking by GPS, and application of artificial intelligence algorithms are currently under integration to analyze the behavioral information of various species. These systems will offer reliable early warning signals through pattern analysis and association with environmental data. This approach is useful in such geographically isolated or not adequately supplied areas where there would be a scarcity of traditional detection systems. Besides, the assimilation of animal behavior with a disaster forecasting model both enhances accuracy and underlines biodiversity conservation since wildlife protection has a direct bearing on securing human life. This emerging interdisciplinary domain holds great promise, so it offers a more holistic and natural approach to mitigating natural disasters and safeguarding people anywhere.

Animals are very sensitive to the environment, and their sounds and actions often change due to changes in nature. This is not random; instead, it is a methodology of survival that has evolved over thousands of years. It allows animals to notice, in time, slight shifts in the environment, such as those in weather, natural disasters, or changes in nature. Vocalization becomes one of the most critical mechanisms employed by animals to communicate data, through which scientists can understand such perceptions and reactions of animals over environmental changes. For example, the calls of birds, in certain cases, adjust according to the change in barometric pressure or temperature, indicating thereby that a storm is about to hit. Amphibians, such as frogs, also change their singing patterns as there are changes in humidity and rainfall, showing changes in breeding times or the onset of seasonal changes.

Unusual or very loud noises from animals before natural disasters can indicate difficulty. Animals may call incessantly or become extremely quiet before an earthquake. It can be because they can sense minor low-frequency vibrations or alterations of electromagnetic fields. Marine animals, such as whales and dolphins, have modified sounds and patterns of swimming to reflect changes in the acoustic environment associated with seismic disturbances, which can be induced by tsunamis. It is often associated with strange behaviors such as flight to safe places, agitation, or altered feeding patterns (Mutanu et al. 2022). The new and advanced innovations in bioacoustics have enabled researchers to listen and comprehend voices from animals more vividly. Equipped with sensors from AI and the right gadgets, scientists can now listen to these calls and be provided with crucial information useful for early warnings. The visual indicators most of the time associated with acoustic data include changes in behavior where animals start to flock and climb trees or go about during the night.

These natural responses would, therefore, be enabling the ability to predict disasters more effectively, showing in which ways animals behave with their surroundings. This makes it value biodiversity more and places stress on the significance of conserving efforts. Thus, by using such observations along with modern technology, better sensitive, sustainable, and reliable early warning systems can be formulated for better readiness and responses in disaster situations.

1.3 Overview of Bioacoustics and Its Potential

It encompasses the production, distribution, and reception of sound made within living organisms. Bioacoustics is at the crossroads of biology and ecology through acoustics by recording vocalization of various animals such as marine mammals and birds of marine as well as those inhabiting the habitats of amphibians and insects; it thus helps in better understanding of their behavior, population, and health of their environments. Bioacoustic has immensely become much bigger through technologies such as high-fidelity recording instruments and artificial intelligence apart from other kinds of machine learning. In applications, bioacoustics has promising impacts mainly within the conservational aspects due to its contribution in identifying endangered species besides the assessment of biodiversity with minimal disturbance in areas that take little human activities. Apart from that, it also plays a crucial role in environmental health assessment, because alterations in acoustic patterns reflect the transformation in ecosystems. In terms of the domain of medical research, bioacoustic approaches are being tested for diagnosing problems like breathing troubles or stress levels of animals. Bioacoustics has a very promising scope regarding the prediction of disaster events and giving early warnings. Preliminary hints of some of the natural disturbances, such as earthquakes or storms, can be obtained from the study of changes in the voice and behavior of animals. This field not only advanced our predictive ability but also brought out interrelations within the ecosystems. With further advancement of technology, bioacoustics is going to play a more and more vital role in environmental management and species conservation, possibly even for human health. It provides critical tools to understand and improve the protection of both natural and man-made environments.

2 Fundamental Concepts

2.1 Bioacoustic Signals: Types and Characteristics

Bioacoustic signals are sounds created by the organisms themselves, and they play a very significant role for both purposes of communication, navigation, and survival. These two are the types of bioacoustic signals: non-vocal and vocal. These are vocal signals when their sounds come from special voice organs: birdsong, mammal calls, frog croaks, and insect chirps, and so on. Other sounds that are not vocal sounds but take place while communicating are clicks that help in establishing routes and hunting when they are in the dark as the bats do. Non-vocal signals are sounds produced by moving the bodies and the bodies relative to their environment. Woodpeckers drum while crickets make chipping by rubbing parts of the bodies together and there is even the low noises an elephant gives while it communicates, in this case, through creating noise by shaking the ground.

All the various forms of bioacoustic signals have specific characteristics—frequency, amplitude, duration—that meet the species needs and their environment. Frequencies are highly variable as a function of species and reason for the sound, but it appears some use the low-frequency sounds of the whale to travel far underwater; and many birds and insects that generate their sounds at higher frequencies suit their habitats (Kvsn et al. 2020). The amplitude or intensity of the sound varies with the purpose of the signal. Some calls are loud and far-reaching, such as lion roars for territorial assertion, whereas others, like moth ultrasonic clicks, are subtle and serve defensive functions. Another dimension of complexity is added by the temporal patterns of rhythm, repetition, and modulation. Most species use complex patterns to convey different messages or to identify individuals within their group.

New developments in bioacoustic technology have given scientists new tools to record and study these sounds using underwater microphones, curved microphones, and AI-powered software. This has provided new means of learning how species behave, track population numbers, and even notice changes in the environment. The study of such sounds allows scientists to learn more about animal communication, their stress levels, and reactions when exposed to danger, helping areas from saving wildlife to predicting disasters. Bioacoustic signals are of high priority in terms of indicators of ecological health and also have good potential for new applications in biological research and real-life situations.

2.2 Animal Behavior and Natural Disasters: Historical Evidence

Historical records are replete with accounts, wherein animals exhibit unusual behaviors just before actual disasters strike; such patterns point to high sensitivity and alertness to changing environmental scenarios. Such stories spanning different civilizations and times are suggestive that animals might innately respond to precursory signs or indicators of dangers—indeed, such instinct might further be used to better predict such catastrophes in present times. It was by the ancient Greeks who were there in 373 BCE; this is when historians recorded the leaving of several animals such as rats, snakes, and weasels that reportedly exited the city of Helike some days before the cataclysmic earthquake followed by the tsunami. Similarly, folklore in Japan narrated tales of catfish swimming erratically before an earthquake occurs. Today, these studies have scientific backing and reveal that catfish can sense the vibrations of earthquakes a few hours before it strikes. In fact, there have been reports before the Indian Ocean tsunami 2004 that elephants began migrating to higher grounds and flamingos vacating their breeding grounds in the lowlands presumably knowing that disaster was likely to strike through vibrations or shifts in the pressure in the air.

It is found that volcanic eruptions somehow influence the behavior of animals. As early as 1980, in USA, before the Mount St. Helens eruption, wildlife observers witnessed the dispersion of deer, elk, and bears away from the area, owing to minor alterations in the environment in which they lived. Similarly, back in 1975, the evacuation of Haicheng, China, was also partly due to observations of abnormal animal behavior-frantic dogs and livestock. It is said that the real major earthquake hit the area immediately after evacuation saved thousands of lives. Such historical events have stimulated much scientific interest in describing the biological mechanisms involved in such activities. Several studies have shown that animals can perceive events such as ground vibrations or changes in the magnetic field or changes in air pressure and temperature and behave appropriately. Birds could somehow feel changes in atmospheric pressure and change their paths of migration or even alter their times for calling, while mammals which have sensitive hearing or sensitive olfactory can feel some low-frequency phenomena that may be related to impending earthquakes or chemical changes. Although traditional observations of historians have already furnished evidence for the predictive capacity of animals, modern technologies make it easier to design an investigation in this regard. Combining traditional lore with modern monitoring techniques such as GPS tracking, bioacoustic monitoring, and AI-driven behavioral modeling, scientists can convert these instincts into reliable warning systems. Such a rising body of evidence points toward the importance of animal behavior in disaster preparedness, where traditional wisdom is aligned with high-tech science for the benefit of human populations and ecological systems.

2.3 Types of Animal Behavioral Cues Relevant to Natural Disasters

For many years, it has been known that animals exhibit behavioral signals which might foretell the occurrence of natural disasters. Most species show some kind of outstanding and sometimes striking alteration in behavior before these events occur. These behaviors may serve as precursors to humans, usually allowing them sufficient time to prepare or evacuate. Indeed, changes in locomotion are the most common type of behavioral signal. It has been observed that several species, whether terrestrial mammals or aquatic, have been known to migrate or evacuate regions days and even sometimes weeks before seismic activity, tsunami, or volcanic eruption. Such animals include elephants, dogs, and horses that can feel the vibrations or a change in the Earth's magnetic field, thus moving them away from territories they believe are dangerous. Birds may also behave irrationally or alter their migration patterns before storms. Fish and dolphins have also been documented swimming away from the coastline before tsunamis. Another very common behavioral sign is unusual vocalizations. Many animals have a specific vocalization they make to warn or alert others of a potential danger, and these vocalizations may be different or increased in volume before a natural disaster. For instance, barking, howling, or meowing may increase in dogs and cats, while some bird species may become silent or start vocalizing in response to the environmental changes. Other than the vocalization behaviors, some behavioral indications of changes are possible through dietary behaviors. Changes in the rate of feeding can be observed in fauna because most respond to alterations in pressure, humidity, or electromagnetic fields (Hayakawa and Yamauchi 2024).

Some animals become more avid food hoarders or do not eat at all in anticipation of an interference that might be associated with the approaching event. Physical symptoms are also apparent and include restlessness or hyperactivity, which can be seen in domesticated animals like pets and livestock. Animals may show signs of fear or exhibit irrational behavior such as pacing, panting, or seeking shelter, which is brought on by pre-seismic vibrations or changes in atmospheric conditions. Environmental conditions, for example, may change in barometric pressure or sudden change in water levels, which will provoke unique responses from different species of animals. For instance, some frogs and toads increase their calls before any storm or flood, while the fish and other crustaceans swim to deeper water or swim up to greater altitudes when the changes in water levels ahead are those of a tsunami or hurricane. Besides, some species, particularly those living near the coast, may be able to feel the difference in the electromagnetic field or even feel the change in water composition, which may indicate a tsunami or an earthquake. These changes are often slight and not easily seen by human eyes unless watched closely; however, they can be very valuable if used together with modern disaster prediction techniques. These could include the potential for research into monitoring and analyzing the behavioral patterns of animals using real-time artificial intelligence, machine learning, and IoT technologies for increased efficiency and accuracy in the disaster early warning systems. Mixing of the traditional concepts in animal behavior and high technology enables better detection and response at the right time to natural disasters with saving lives and further impact of the disaster upon the people's communities.

3 Challenges in Traditional Bioacoustic Analysis

In addition to species identification, other traditional areas that face challenges involve behavior monitoring and ecological monitoring. The presence of these aspects makes them likely to pose significant barriers to bioacoustic analysis based on its effectiveness and correctness (Sharma 2023). These major ones involve is shown in Fig. 1.

3.1 Noise Interference

One of the most significant barriers against conventional bioacoustic analysis is environmental noise, which includes conditions like wind, rain, human-related activities, and calls from other animals. In this regard, these aspects sometimes mask or distort the signals of interest, thus making it challenging to separate them out and analyze them with an improved accuracy.

3.2 Large Amount of Data

The process of gathering bioacoustic data often yields significant recordings of audio that needs manual inspection. The sheer magnitude of this data can overwhelm: to annotate, categorize, or analyze, it requires enormous timescale and resources.

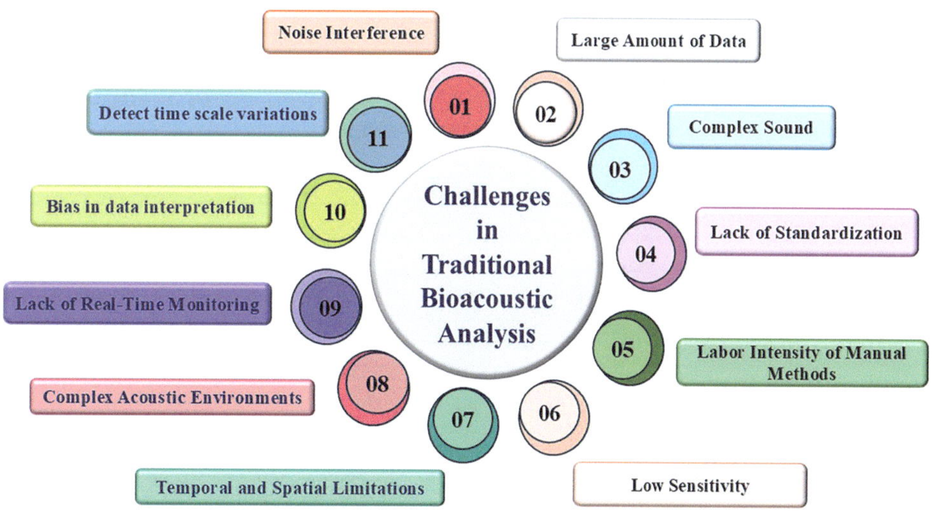

Fig. 1 Challenges in traditional bioacoustic analysis

3.3 Complex Sound Classification

Species or individual animals are differentiated from the auditory signals, and it is a complex task. Some species produce sounds that are very similar to one another, and others have subtle expressions in voice, which makes recognition challenging. These complex and diverse voice patterns will not be categorized by the traditional audio analysis methods.

3.4 Lack of Standardization

Often, there is a lack of standardized methods and instruments for recording, processing, and analyzing bioacoustic data. This absence often explains why results from different studies, or even from the same study areas, do not produce favorable comparisons with each other.

3.5 Labor Intensity of Manual Methods

Bioacoustic analysis traditionally relies on the sound manual classification techniques that are of a highly labor-intensive type. Most researchers will listen to vast audio files or even mark them up, which involves significant time consumption and is not very efficient.

3.6 Low Sensitivity

The traditional method lacks sufficient sensitivity to witness slight or rare acoustic events, like threatened species' vocalizations or animals giving precursory warning calls before natural calamities. This might thus limit the scope and potential of bioacoustic research.

3.7 Temporal and Spatial Limitations

The old form of bioacoustic monitoring has problems arising from the location of the recording devices, therefore failing to include vast or isolated areas. Some species are only active periods of the day or year, which could call for extensive observation.

3.8 Problems with the Analysis of Complex Acoustic Environments

Animals often produce sounds in complex acoustic environments, such as heavily vegetated forests or urban settings, characterized by the interference of competing auditory signals. Traditional methods of analysis will have trouble separating and investigating these interfering sounds, reducing the accuracy of the analysis.

3.9 Lack of Real-Time Monitoring

Traditional bioacoustic methods normally carry out post-recording analyses. This means that real-time data is hardly ever realized. This limitation makes it impossible to effect decisions or interventions in instances where instant responses are realized, for example, responding to an indicated probable natural disaster warning by means of animal behavior (Oestreich et al. 2024).

3.10 Bias in Data Interpretation

With traditional bioacoustic analysis usually relying on human expertise, there is subjectivity attached, and the outcome might end up biased. In the analysis process, different researchers might differ in interpreting the same kind of sound.

3.11 Limits the Ability to Detect Time Scale Variations

Long-term changes in the behavior of animals or their ecosystems require the collection of significant data over a long period of time. Traditional methods are often insensitive to the detection of such changes, especially in living systems where change might be perceived to be a slight or gradual phenomenon.

With more advanced technologies like machine learning, artificial intelligence, and automatic recording systems, the enhancement of the analysis of bioacoustic challenges is even more accurate, efficient, and scalable.

4 Traditional Versus AI-Powered Approaches

Actual field observations, anecdotal reports, and manual evaluation had been the conventional methods of study in animal behavior relating to natural disasters. Investigators generally took time to visit disaster-prone regions or areas believed to be predisposed to disasters, where they observe animals in their natural environments and directly determine those indicating abnormal behavior, such as showing symptoms of stress, changes in locomotion, or variations in calls. Those described observations often had a highly intuitive character, being largely dependent on the researcher's know-how and interpretation. Collecting data was usually limited to individual case studies or short time spans of observation, thereby limiting the ability to draw wide patterns or confirm predictions on the basis of animal behavior. Moreover, the actual analysis of data was long time-consuming, and the described dependencies of animal behavior on impending disasters could not be easily quantified and repeatedly proven.

Conversely, the AI-based methodologies have revolutionized the analysis of animal behavior, which is more realistic and scalable based on real empirical data. The machine learning and deep learning in combination with large-data analysis can monitor the animal behavior automatically. IoT sensors, GPS tracking devices, and bioacoustic microphones can now collect massive amounts of real-time data from animals regarding movement, vocalizations, physiological responses, and even environmental factors. Such streams of data subsequently processed through the algorithm because of artificial intelligence of pattern recognition may even lead further toward discovering a slight change in behavior that hints at a possible disaster. This can process huge amounts of data, search for a minute pattern, and generate answers that no ordinary approach would generate (Liu and Dhakal 2020; Chávez-Andrade et al. 2023).

On the other hand, artificial intelligence models work hand-in-glove for prediction analytics. It uses previous behavior by producing algorithmic representations of what likely takes place moving forward in greater detail and at higher precision. For instance, machine learning algorithms can learn to identify specific behavioral modifications that may indicate an oncoming earthquake or tsunami, then alert at-risk populations. Such integrated AI-powered systems will assimilate data from several sources, such as seismic sensors, weather forecasts, and even the patterns of animal behavior. At this level of integration with automation, it eliminates human bias and guarantees consistent and continuous monitoring for a long period. Where traditional techniques have been useful, there existed shortcomings in the capability and capacity of human observation.

While AI-based techniques open avenues unprecedented for all-inclusive real-time observation of animal behavior, they offer a basis, more robust and unbiased to include animal signals in forecasting disaster situations. Improvements through the integration of natural sensibility with advanced technology along an otherwise promising development path are an expected eventuality in the improvement of disaster preparation and response.

4.1 Role of Machine Learning and Deep Learning in Bioacoustics

Machine learning (ML) and deep learning (DL) have revolutionized the domain of bioacoustics by providing solid methodologies for the analysis of complex biological sound data as well as deriving important insights from vast and diverse datasets. Bioacoustics, which deals with a vast quantity of complex acoustic signals produced by animals, presents tremendous challenges in terms of processing these sounds with conventional approaches or interpreting their meaning. However, algorithms in ML and DL are much better suited for pattern recognition, classification, and prediction, which is a requirement to decode the subtleties of bioacoustic signals and apply them in real-world applications such as wildlife monitoring, species conservation, and disaster prediction.

Machine learning algorithms are particularly valuable in bioacoustics for tasks such as sound classification, event detection, and acoustic feature extraction. Some of the ML models that might be included are decision trees, support vector machines, and k-nearest neighbors. Their ability to recognize certain animal vocalizations or behaviors would depend on training: They could be trained to distinguish between species-specific calls, environmental noises, or even threats in noisy or complex acoustic environments. Often, the algorithms are trained with the labeled datasets; hence, when the recognized animal vocalizations are provided, it learns how to classify the similar patterns found in the unlabeled data. The methodology allows automatic and prompt analysis of bioacoustic data and saves much time and efforts as opposed to manual identification (Gorain et al. 2023).

Deep learning is the following level in taking ML into bioacoustic analysis by using it to introduce such methods. Such technologies will be allowed to automatically learn complex patterns without explicitly hand engineering features. DNNs, mainly CNNs and RNNs, are used for interpreting raw acoustic data in the form of spectrograms, waveforms, or audio files, and to learn hierarchical representations related to several types of audio signals simultaneously. CNNs have been found very useful for exploring spectrograms as two-dimensional graphs representing time-varying contents of sound frequencies. RNNs, particularly LSTM networks, are very good at modeling temporal dependencies in sound signals, thus capturing sequences of vocalizations or behavioral patterns in animal communication.

Advanced DL techniques have enabled breakthroughs in real-time bioacoustic monitoring, where one can monitor endangered species, detect changes in the environment, or even sense early warnings of natural disasters by changes in animal behavior. For example, deep learning algorithms are used to monitor the animal vocalizations of some species in large continuous audio datasets; this research actually helps monitor biodiversity in such far-flung places having minimal or no human interface. ML and DL models may predict a disaster with the help of changes in the vocalization or behavioral patterns of animals that may indicate the approach of an earthquake or tsunami relating it to the past. Further, with the capability of ML and DL algorithms in processing large amounts of bioacoustic data acquired from land sensors, underwater sound recorders, even satellite images, there can be more holistic monitoring of the environment. The integration of these sources might lead to more accurate predictions and increased reliability, leading to informed decision-making about conservation and disaster preparedness. These now show high promise in enhancing wildlife conservation efforts, improving disaster prediction models, and helping in the better understanding of animal behavior and ecosystems.

4.2 Role of Deep Learning in Signal Recognition

It revolutionizes the domain of signal recognition through advanced algorithms, mostly in bioacoustic analysis, speech processing, and communication systems. It employs the most advanced neural networks that enable deep learning algorithms to look for patterns and features within the data themselves, so it is one of the most efficient in recognition tasks based on manually extracted features and rule-driven approaches. The bioacoustic signal classification techniques using the field of deep learning-techniques, including CNN and RNN, help classify animal sounds with a great accuracy. These architectures have high capabilities to handle large volumes of acoustic data and pick out species-specific sounds or behavioral and environmental changes in very heavy background noise conditions. In contrast, conventional methods with handpicked attributes have the deep learning algorithms that infer knowledge automatically from raw audio signals in the identification of relevant characteristics such as pitch, frequency, and temporal dynamics, allowing for the detection of subtle sound changes that might indicate key events such as early warnings of natural disasters through communication by animals. Moreover, deep learning frameworks operate on vast datasets. So, it leads to better generalization across different environments, species, and acoustic conditions. This makes the necessary laborious labeling be reduced and improves the detection efficiency of the signal. For speech recognition tasks, deep learning has resulted in improving the transcription of the spoken languages. All these overcome challenges such as accent-related issues, background noise conditions, and varying speaking speeds. Deep learning, in recent times, is widely applied to wireless communication systems for modulation of signals, noise suppression, and correction of errors, all for a better network. Deep learning is useful when handling large quantities of vast complex unstructured data which will facilitate the real-time processing of signal, thus decision-making is faster, more accurate within different contexts (Zschau et al. 2013). However, a very success-oriented deep learning paradigm is closely dependent on availability of sufficiently large, rich high-quality datasets, and ample amounts of the computational resources with regard to training models. Important challenges will be interpretability issues while using deep learning frameworks and is generally a big challenge in cases, as sometimes the framework acts just like some kind of black box. Despite these challenges, deep learning is still transforming the landscape of signal recognition, allowing for greater accuracy and automation in scalable applications ranging from ecological monitoring to communications.

5 Comparative Analysis: ML Versus DL for Bioacoustic Data

ML and DL have been promising bioacoustic data in terms of increasing accuracy and efficiency while processing animal vocalizations, detecting species, and identifying behavioral indicators. But there are various methodologies and their approaches, computational requirements, and efficiency toward tasks that make a comparison between ML and DL which reveals some benefits and drawbacks for each that is given in Table 1.

5.1 Data Requirements and Feature Extraction

Machine Learning: In the classical approaches to machine learning, feature extraction is usually manual before the actual model prediction. In bioacoustics, it usually includes manual identification and quantification of attributes such as pitch, frequency, amplitude, and temporal patterns in animal vocalizations. All features need to be chosen based on the decisions of domain experts, with the quality of feature extraction being directly reflected in the quality of the model. Then, these machine learning algorithms are trained, including SVM, random forest, and k-NN set to carry out classification or clustering activities over these features.

Deep learning methodologies, particularly convolutional neural networks (CNNs) and recurrent neural networks (RNNs), possess the ability to independently discern relevant features from raw bioacoustic datasets, thereby obviating the need for human intervention. This makes deep learning especially suitable to processing huge and complex collections of data that, moreover, may involve hierarchical aspects which can learn from the data. Deep networks may search for patterns in audio or frequency, try to discern some sequences of time, and catch expressions that would go completely unnoticed under a straightforward application of machine learning.

Table 1 Comparative analysis of AI bioacoustics data

S. No.	Parameters	Machine learning (ML)	Deep learning (DL)
1	Data requirements and feature extraction	– Manual feature extraction is essential, relying on domain experts to identify attributes like pitch and frequency – Quality of feature extraction directly impacts model performance – Common algorithms include SVM, random forest, and k-NN	– Automatically extracts features from raw data, eliminating the need for manual intervention – Particularly effective for large, complex datasets with hierarchical aspects – Utilizes CNNs and RNNs for pattern recognition
2	Big data performance	– Performs well with small datasets, effectiveness declines with larger volumes due to plateauing performance – Requires advanced tools for processing high-dimensional bioacoustic data	– Excels with large, high-dimensional datasets; can learn minute patterns directly from raw data – Capable of training on extensive collections of animal vocalizations without predefined feature sets
3	Generalization and adaptability	– Struggles with generalization across different environments and species; often requires retraining with new datasets – Performance may degrade if training conditions differ significantly from application conditions	– Superior generalization capabilities; adapts well to new species and environmental changes – Learns discriminative representations directly from data, making it flexible in dynamic acoustic environments
4	Accuracy and precision	– High accuracy achievable if data is well-prepared; however, may struggle with complex acoustic structures in noisy conditions or recordings of multiple species	– Generally, outperforms traditional methods in accuracy for complex tasks like species identification and behavioral detection – Handles noisy data effectively, discovering advanced patterns without extensive feature engineering

5.2 Big Data Performance

Machine Learning: The standard machine learning models perform well with small-sized datasets, especially when the labeling is accurate and robust extraction techniques are applied. However, when this volume becomes more significant, traditional models become increasingly problematic or impractical and require an implementation of higher-order tools for processing larger amounts of information. For high-dimensional sets of complex data, particularly that typically associated with much bioacoustic research, machine learning model performance plateaus or worsens.

Deep Learning: Deep learning-based models work better with big and high-dimensional data. Because deep learning can pick very minute patterns from raw data, it performs much better in dealing with big amounts of bioacoustic data. For example, thousands of hours of animal vocalizations can be used for training deep neural networks; in turn, the network can learn all the features present in the spectrum of acoustic features without requiring a defined feature set from experts (Liu and Dhakal 2020).

5.3 Generalization and Adaptability

Machine Learning: The machine learning models suffer from generalization across different environments, acoustic contexts, and species. Traditional machine learning frameworks requiring explicit retraining with newer datasets may degrade in performance if the training acoustic conditions are significantly different than those experienced when using such a model. The task of using models in novel environments or for another species can be highly expensive in retraining with features that are salient to the target domain. The deep learning of superior generalization across a variety of datasets has been one significant advantage that deep learning possesses in the bioacoustics field.

Deep Learning: The biggest benefit of deep learning, other than anything done in bioacoustics, is that the models tend to generalize better on different sets of data. Because deep models can learn extremely discriminative representations directly from audio data without feature processing, they can even adapt to new species, environmental changes, or unusual patterns of acoustic signals with a much lesser degree of engineering upon features. Thus, deep learning models are much more flexible and robust, especially in dynamic and heterogeneous acoustic environments.

5.4 Accuracy and Precision

Machine Learning: If the data is adequately prepared with appropriate features, it is truly possible to get impressively high effectiveness in a traditional model of machine learning. Nevertheless, it is bound by the nature of complexity inherent in acoustic data. Such models might not be able to differentiate between subtle or complex acoustic structures, especially in noisy conditions or in recordings of multiple species.

Deep Learning: Deep learning models tend to outperform traditional machine learning methods regarding accuracy when dealing with complicated and vast data. Deep learning helps achieve better accuracy in activities like species identification, behavioral cue detection, and environmental change recognition by analyzing raw data and automatically discovering advanced patterns. In addition to this, deep learning performs impressively in bioacoustics with all this noisier data most of the time, almost very noisy. In bioacoustic data analysis, the specific demands of each study allow ML and DL to show its pros and cons. Machine learning algorithms are exceptionally suitable for a small set with a clearly defined acoustic feature wherein interpretability is an utmost necessity. However, for large-scale and complex data with little engineering of features, deep learning systems give more accuracy, agility, and accuracy but do this at a cost to higher computational complexity and poor interpretability. A mixed approach of machine learning using deep learning only for much more complex pattern recognition issues could prove the best of both worlds for many bioacoustic applications.

6 Technologies and Techniques

6.1 Sensors and Data Collection Systems

Sensor and data collection systems are critical for the interpretation of animal behavioral signs as precursors to natural disasters, providing the relevant information on environmental changes that occur before catastrophic events. Such systems are therefore designed to capture, monitor, and analyze the physical and behavioral responses of animals in changing environments, thus offering vital early warning signs to significantly improve disaster prediction and countermeasures. This objective requires that different categories of sensors are used, and each is supposed to identify specific alterations that might stimulate animals to respond. The most widespread acoustic sensors are microphones and hydrophones, used to identify animal vocalizations that may be modulated in frequency, amplitude, or pattern because of disturbances from environmental events, such as earthquakes, tsunamis, or storms. These sensors would be put in general areas of wildlife, such as forests, wetlands, or coastal areas, thereby offering direct observation of the vocalization of animals. Other detection technologies that use motion-sensitive capabilities, such as infrared cameras and GPS-enabled monitoring devices, can also measure the movement patterns of these animals since most species follow unusual patterns of movement before calamity strikes.

For instance, acceleration and motion detectors mounted to or placed at appropriate places on animals or objects may detect abnormal actions or movements of animals, say leaving a locality before an earthquake or volcanic eruption. Finally, temperature, humidity, and atmospheric pressure sensors play important roles because animals may respond to changes in those environmental conditions long before storm surges, flooding, and other meteorological changes manifest. For example, some species may be very sensitive to barometric pressure fluctuations leading to increased anxiety or abnormal behavior before the occurrence of a storm or tsunami. Some sensors in the environment installed in the field, including soil moisture sensors and water level monitoring devices, can be installed in flood-prone or landslide-prone areas giving immediate information regarding changes in the environment which will trigger a response from populations of animals. These sensors can be integrated into a network of IoT-compatible devices (Sergio et al. 2018; Mutanu et al. 2022).

The data can be transferred to central servers, which allow machine learning algorithms and AI systems to analyze data patterns that correspond to categories of natural disasters. In this manner, it is possible to monitor animals in real time, under normal conditions and in response to adverse environmental conditions that become valuable early warning indicators, leading to quick disaster preparedness. This is also possible through video and image sensors, such as drones and stationary cameras, to monitor large animal populations using visual means. Such approaches may then be used to collect behavioral sign data, like changed social relationships, unusual calls, or other stress responses. The integration of data from different sensor modalities will help scientists better understand the correlation between the behavior of animals and the occurrence of natural disasters. Thus, besides a more effective early warning system, the predictive capabilities may improve. All of this will be done through adequate assessment of environmental context and species in focus in tandem with technological infrastructure supporting smooth flow of data to be analyzed. In this regard, the systems would significantly improve early warnings in case of natural disasters; hence, lives and property lost would be curbed through early responses following indications of animals' behaviors (Tomanska et al. 2024).

6.2 AI Algorithms for Bioacoustic Signal Processing

Artificial intelligence algorithms have revolutionized bioacoustic signal processing, improving accuracy and efficiency and scalability in the analysis of animal vocalizations for the purposes of species identification, monitoring of behavior, and evaluation of ecological status. With modern ML and DL algorithms that allow pertinent features to be autonomously extracted from vast, complex acoustic datasets, a transformation occurs in the way these bioacoustic signals are processed and interpreted (Tomanska et al. 2024; Venkat Narayana Rao et al. 2024). Listed below is the outline of the main AI algorithms found in bioacoustic signal processing:

Convolutional Neural Networks (CNNs):

CNNs are largely applied in processing bioacoustic signals, as they can automatically derive hierarchical features from raw audio signals. In the domain of bioacoustics, CNNs is applied on images of spectrograms, which are visual illustrations of frequency of sound across time periods, for identifying and categorizing animal voices. The basic property of CNNs in an architectural sense is to apply attention to local patterns found on the spectrogram, and to which frequency bands and their correlations throughout time are essential for certain animal species-specific call recognition as well as changes in some animal behavior. Greatly successful in the identification of different species' vocalization in noisy environments, its great application in large-scale systems with continuous bioacoustic monitoring data analysis is done.

Recurrent Neural Networks (RNNs) and Long Short-Term Memory (LSTM) Networks:

It really takes highly efficient sequences to deal with sequential data as the acoustic signals come time-series. Bioacoustic applications will involve long-spanning vocalizations by animals which often exhibit temporal dependencies, thus well-suited to use RNNs and LSTMs to capture long-term patterns in the context of the data. Especially, LSTMs bypass the vanishing gradient problem with the memory of past sequences in an RNN to capture more complex and context-dependent acoustic features. Most commonly, these algorithms are applied in analyzing continuous audio streams, such as tracking of animal behavior or the detection of warning signals before a natural catastrophe occurs.

Support Vector Machines (SVMs):

Support vector machine is probably one of the commonest traditional machine learning classification approaches in bioacoustic signal processing. SVM finds, in this case, a best possible hyperplane with the greatest margin that allows separation between different classes and maybe species or behavioral cues being taken into account based on an appropriate extraction of features of an audio signal. Though SVMs demand feature extraction to be done manually, it is extremely efficient for usage in bioacoustic applications if the features are chosen wisely and the dataset is not too large or well-labeled. The most common use of SVMs is in species identification, call classification, or in distinguishing between normal and abnormal animal behavior (Chen and Wang 2020).

Random Forest (RF):

RF is a class of ensemble learning method, in which hundreds and thousands of decision trees are combined for enhancing the accuracy as well as robustness of the classification. Normally, in bioacoustic signal processing, the algorithm is used for animal vocalization classification based on already drawn characteristics, such as frequency, pitch, and amplitude. Models based on RF are highly powerful in the case of noisy data, and the feature importance scores provide interpretability to the academic researcher. Random forests are typically used for classification tasks, such as calls from various species, behavioral changes, or sounds in the environment.

Auto-encoders:

Auto-encoders are a category of unsupervised deep learning models used in the processing of bioacoustic signals for anomaly detection, noise removal, and feature extraction. In essence, an auto-encoder learns to compress input data into lower-dimensional representations that can be reconstructed, thereby enabling the detection of patterns and anomalies in data. Auto-encoders are one of the best tools within bioacoustics since they can clean datasets and making representations of audio signals that can then be denoised of extraneous background noises. This helps in predicting anomalous sounds from rare behavioral anomalies in animals and changes within the environmental condition.

K-Nearest Neighbors (K-NN):

K-NN algorithm is one of the simplest yet effective methods that are used in the process of handling bioacoustic signals, both classification and clustering. K-NN working relies on the analysis of the new point by its nearest neighbors in feature space and assigning the class that dominates. Due to high computational intensity and requiring careful choice of features, K-NN remains useful whenever data sizes are relatively small or fast classification which is not necessarily required. K-NN is a good classification useful application in bioacoustics. This might be used to classify sounds, for example, in animal species identification or behaviors (Karpakam et al. 2022).

Hidden Markov Models:

HMMs are probabilistic models. They are the most efficient models when working with data consisting of sequences which may have hidden states. Use of HMM in bioacoustic signal processing is much appreciated. Most of the tasks involve segmentation of vocalizations, behavior analysis, and recognition of patterns in the sounds created by animals. In bioacoustic research, HMMs can model the temporal structure of vocalizations or behavioral patterns, so will include probabilistic dependencies between states, for example, resting, feeding, or fleeing. HMMs are stronger when there exist long sequential or continuous patterns of behavior with precedence conditions influencing the state at hand.

Generative Adversarial Networks:

Recently, there has emerged a new generation of deep learning algorithms referred to as generative adversarial networks, or GANs. There have been very impressive applications in the processing of bioacoustic signals. Two types of neural networks are involved with GANs—the generator and the discriminator—combined for improved data quality (Chávez-Andrade et al. 2023). Bioacoustic applications of GANs are helpful in developing synthetic vocalizations of animals to serve as additional augmentations in the training datasets or even test processing of the signal. The applications involve enhancing the resolution of the noisy bioacoustic records, so that scientists would get the signal with a more accurate output to further analyze.

Clustering Algorithms:

The K-means, DBSCAN, and hierarchical are some examples of clustering algorithms that can be applied to unsupervised classification. They are highly efficient in finding patterns within very large datasets without any labels. These could be, for example, a group of similar animal calls that can be classified in terms of the vocalizations and help identify behavioral changes or sounds in the environment without an immediate perception.

Reinforcement Learning (RL):

Reinforcement learning (RL) is another sub-area in bioacoustic signal processing, which is a relatively emerging field; however, RL has immense potential in applying real-time decision-making based on the observation of animal behavior. The optimal actions for RL models are learned via environmental feedback and can be utilized for predicting or responding to behaviors of animals during natural disasters. For instance, an RL algorithm can be trained to determine early warning signals of disasters using the conduct of animals—for example, strange vocalizations or movements—and then institute appropriate disaster response measures.

Principal Component Analysis:

PCA is an efficient technique for extraction and dimensionality reduction relevant to the acoustic features obtained from the analysis of animal vocalizations, thus making the complex analysis and interpretation of bioacoustic data smoother. Usually, the characteristics attributed to the vocalizations of animals include pitch, frequency, amplitude, and temporal features of a bioacoustic signal. PCA reduces the features and condenses them into a lesser set of uncorrelated variables, called principal components, which captures most of the significant variations of the data. With transformation of the original high-dimensional feature space into a set of principal components, PCA diminishes the noise and redundant information, which focus on key patterns that differentiate between different species or behaviors (Mutanu et al. 2022). This not only enhances the classification and clustering algorithms themselves in use but also enriches the acoustic data representation of how to analyze relationships of many vocalizations. With PCA, animal scientists could easily analyze both their vocalizations and characteristics without having too much load during computations, yet still capture their crucial aspects for further study.

7 Drawbacks of Traditional Algorithm

The algorithms that are traditionally used in the analysis of animal behavior against natural disasters have many limitations. They do not support noisy and unstructured data handling and therefore cannot detect behavioral signals from environmental noise. Moreover, it has a manual feature extraction method which is time-consuming and error-prone in terms of missing important patterns. Temporal dependencies of animal behavior, which are critical for understanding slow responses to disasters, are also often neglected by traditional methods. They have poor generalization across species or environments, poor ability to process large datasets, and are not very adaptive to dynamic conditions; hence, they are less effective in real-time disaster monitoring (Stowell 2022; Gao et al. 2006).

Mel Frequency Cepstral Coefficients (MFCCs):

Of particular importance are MFCCs in animal behavior studies, particularly natural disasters. Since MFCCs are compact and efficient representations of the spectral properties of animal vocalizations, these features have broad applications in bioacoustic signal processing and may become the critical features for identifying changes in behavior relative to environmental disturbances (Coutant et al. 2024).

Regarding natural disasters, an animal can change its typical vocalization pattern due to perceiving changes in its environment, like increased fear or altered communication because of an impending storm, earthquake, or other ecological risks. This frequency content of such vocalizations can further be represented using MFCCs that can map these into a mel scale that is even more close to the human ear and somewhat sensitive to frequency ranges relevant for animal communication. This would be another way the researcher would know changes in voice patterns in recordings of animal sounds over time and hence subtle changes in pitch, tone, and even duration that reflect stress, disorientation, or other behavioral responses in front of a potential disaster (Venkat Narayana Rao et al. 2024).

The reason MFCCs are especially useful is because they filter out complexities in audio data but maintain a good amount of acoustic information, which means machine learning algorithms can better classify and differentiate among the patterns associated with animal behavior.

8 Integration of ML and DL

This introduces considerable opportunities for disaster management frameworks to be changed, in the form of the improvements that integration of ML and DL methodologies for the identification of animal behavioral irregularities present, such as precision and speed of preliminary alert notifications. Behavior of animals has long been considered one of the earliest warnings given by nature before the onset of earthquakes, tsunamis, and storms. Animals always show changes in their behavior just before the event happens; in other words, they sometimes make unusual noises, act restlessly, or change their migration routes. Machine learning algorithms can be trained on very large datasets regarding animal behavior where features could be related to vocalization patterns, trajectories of movements, and variables describing the environment for the eventual identification of disaster-related patterns (Wood et al. 2023; Şengöz 2024). In addition, their algorithms improve predictability with time. Even more advanced capability comes with more complex DL that is capable of processing raw data like audio and video without feature extraction by humans. The capability is even further advanced as neural networks can recognize complex patterns in bioacoustic signals, visual cues, and sensor data to thus detect subtle behavioral changes that may not be caught by human observers. When such systems are deployed for the discovery of anomalies in animal behavior, incorporating ML and DL techniques may give rise to more effective strategies for responding. Real-time recording and processing of responses by animals toward seismic activity, shift in pressure of the atmosphere, or any other change in conditions may present very valuable information for preparations against disaster situations. The integration is incorporated to improve the precision of early warning systems while offering scalability that will help deploy these systems across different ecosystems and species; this promises to improve disaster response worldwide. These advanced algorithms may make next-generation disaster management systems depend so much on animal behavior, potentially offering a unique and valuable insight into preventing loss of life during natural disasters and minimizing damage as well (Chávez-Andrade et al. 2023; Malini et al. 2022).

9 Conclusion and Future Trends in AI for Bioacoustic Analysis

The most important changes that are going to be observed within the field of artificial intelligence in terms of bioacoustic analysis are those specifically related to wildlife conservation, disaster response, and ecological oversight. Real-time processing capabilities in AI technologies have been improved with the developments in deep learning and edge computing. This means such systems will analyze large volumes of bioacoustic data in real time to detect unusual behaviors in species of animals. Critical aspects in disaster early warning systems that would be of great significance for quicker response times toward changes in animal behavior is another trend. Further, AI will become more focused on generalization across species by applying transfer learning and unsupervised learning techniques to generate models that adapt to various animal species and environmental conditions without requiring retraining on a large scale. With sensor technologies currently advancing and with an increased availability of data, the assertion is that with greater scalability, AI-based systems for bioacoustic biodetection or ecological disturbance indication with possible early disaster alerts will become characteristic around several parts of the globe (Teixeira et al. 2019). Features will continue being drawn automatically, while a person reduces processing data in the way that artificial intelligence could discern the subtle patterns of bioacoustic signals that a human will forget. These developments will eventually lead to an increase in precision, effectiveness, and scalability in bioacoustic analysis with new opportunities for investigations and practical realizations of different domains.

References

Chávez-Andrade M, de la Cueva H, Luévano-Esparza J, Lavín-Murcio PA (2023) Contributions of bioacoustics to the scientific knowledge of marine mammals in Latin America. Lat Am J Aquat Mammals 18(1):96–113

Chen L, Wang S (2020) Advances in machine learning for early warning systems: a systematic review. Nat Hazards 102(2):345–362. https://doi.org/10.1007/s11069-020-04346-0

Coutant M, Villain AS, Briefer EF (2024) A scoping review of the use of bioacoustics to assess various components of farm animal welfare. Appl Anim Behav Sci 106286

Gao Y, Chen YX, Ding YS, Tang BY (2006) Immune genetic algorithm based on network model for flood disaster evaluation. J Nat Disaster 15

Gorain S, Barik S, Patra M, Pal J, Pramanik P, Mahato M et al (2023) Use of biotic and abiotic factors to predict natural calamities: a case study concerning the Santali tribe in four districts of West Bengal, India. In: Indigenous knowledge and disaster risk reduction: insight towards perception, response, adaptation and sustainability, pp 187–207

Hayakawa M, Yamauchi H (2024) Unusual animal behavior as a possible candidate of earthquake prediction. Appl Sci 14(10):4317. https://doi.org/10.3390/app14104317

Karpakam S, Malini P, Kumar S, Vijayakumari G, Thirukkumaran R, Hanne LS (2022) Deep learning and fuzzy decision support system for visually impaired persons. IEEE xplore 7

Kvsn RR, Montgomery J, Garg S, Charleston M (2020) Bioacoustics data analysis—a taxonomy, survey and open challenges. IEEE Access 8:57684–57708

Liu Y, Dhakal S (2020) Animal behavior abnormality and earthquake disaster early warning. Rev Cient Fac Cien Vet 30(3):1371–1381

Malini P, Vijayakumari G, Ganesh C, Femina V, Ramani U, Basha SI (2022, October) A deep learning based framework implementation and artificial intelligence using optical communication. In: 3rd international conference on smart electronics and communication (ICOSEC). IEEE, pp 1077–1080

Mutanu L, Gohil J, Gupta K, Wagio P, Kotonya G (2022) A review of automated bioacoustics and general acoustics classification research. Sensors 22(21):8361. https://doi.org/10.3390/s22218361

Nandan A, Tiwari SN, Yadav AL (2024) AI-driven wildlife monitoring and protection system. Available at SSRN

Oestreich WK, Oliver RY, Chapman MS, Go M, McKenna MF (2024) Listening to animal behavior to understand changing ecosystems. Trends Ecol Evol

Şengöz M (2024) Harnessing artificial intelligence and big data for proactive disaster management: strategies, challenges, and future directions. Haliç Üniversitesi Fen Bilimleri Dergisi 7(2):57–91

Sergio F, Blas J, Hiraldo F (2018) Animal responses to natural disturbance and climate extremes: a review. Global Planet Change 161:28–40

Sharma RK (2023) A study of animal behavior and natural disasters. In: International handbook of disaster research. Springer Nature Singapore, Singapore, pp 197–209

Sharma S, Sato K, Gautam BP (2023) A methodological literature review of acoustic wildlife monitoring using artificial intelligence tools and techniques. Sustainability 15(9):7128

Stowell D (2022) Computational bioacoustics with deep learning: a review and roadmap. PeerJ 10:e13152

Teixeira D, Maron M, van Rensburg BJ (2019) Bioacoustic monitoring of animal vocal behavior for conservation. Conserv Sci Practi 1(8):e72

Tomanska A, Chorbinski P, Klimowicz-Bodys M, Mill P (2024) Communication among animals and bioacoustics studies on bees. Med Weterynaryjna-Vet Med-Sci Pract 80(5):200–211

Venkat Narayana Rao T, Jakkam P, Medipally S (2024) Future trends and innovations in natural disaster detection using AI and ML. In: Predicting natural disasters with AI and machine learning. https://doi.org/10.4018/979-8-3693-2280-2.ch005

Wood CM, Champion J, Brown C, Brommelsiek W, Laredo I, Rogers R, Chaopricha P (2023) Challenges and opportunities for bioacoustics in the study of rare species in remote environments. Conserv Sci Practi 5(6):e12941

Zschau J, Küppers AN (eds) (2013) Early warning systems for natural disaster reduction. Springer Science & Business Media

Bioacoustics Feature Engineering: Techniques and Applications for Ecological Insight

G. Geetha, J. Godwin Ponsam, K. Nimala, D. Saveetha, and R. Sujatha

Abstract

Bioacoustics has rapidly grown into a key branch of science that investigates all types of acoustic signals from biological systems, enabling novel research concerning the monitoring of wildlife and environmental health for the purpose of species conservation. Transformation of raw acoustic data into meaningful information requires robust techniques of feature engineering, which bridge signal acquisition to actionable insights. This chapter reviews principles, methods, and applications of bioacoustic feature engineering under an emphasis on unlocking the mysteries of biological communication and soundscapes. We review time-domain, frequency-domain, and time–frequency features of extraction, focusing on their biological relevance and computational efficiency. Concrete techniques like MFCC, spectral centroid, and wavelet analysis will be studied along with the state-of-the-art techniques that embody deep learning frameworks for the automatic extraction of features. Other challenges are related to noise interference, scalability for large datasets, and cross-species generalization. This chapter goes on to describe case studies in various applications-avian song analysis, marine mammal monitoring, insect sound identification, among others-to illustrate the transformative potential of bioacoustic data within ecological research and conservation. This chapter synthesizes current methodologies and emerging trends to provide a comprehensive guide for researchers and practitioners in the advancement of bioacoustic feature engineering.

Keywords

Bioacoustics · Feature engineering · Mel-frequency cepstral coefficients (MFCCs) · Frequency-domain · Time–frequency features

1 Introduction to Bioacoustics and Feature Engineering

1.1 Definition and Importance of Bioacoustics

Bioacoustics is an interdisciplinary study that deals with the production, transmission, and reception of sound in biological systems. It encompasses knowledge of biology, ecology, and acoustics to study the vocal behaviors of various organisms, ranging from terrestrial animals to marine species. It has been instrumental in comprehending the mechanism of communication, behavioral patterns, and ecological interactions. The importance of bioacoustics is because, within its scope, bioacoustic monitoring and its interpretations have been noninvasively able to monitor a wider range of biological phenomena. For instance, one can study seldom-out-sighted species in nature with the help of bioacoustics by detecting their emissions, studying environmental health through soundscapes analysis, and anthropogenic noise impacts on ecosystem function (Kvsn et al. 2020).

G. Geetha · J. Godwin Ponsam (✉) · K. Nimala · D. Saveetha · R. Sujatha
Department of Networking and Communications,
School of Computing, SRM Institute of Science and Technology,
Kattankulathur, India
e-mail: godwinj@srmist.edu.in

G. Geetha
e-mail: geethag@srmist.edu.in

K. Nimala
e-mail: nimalak@srmist.edu.in

D. Saveetha
e-mail: saveethd@srmist.edu.in

R. Sujatha
e-mail: sujathar2@srmist.edu.in

K. Subbaraj et al. (eds.), *Computational Bioacoustic Artificial Intelligence*, Sustainable Artificial Intelligence-Powered Applications, https://doi.org/10.1007/978-3-032-06088-4_9

1.2 Role of Feature Engineering in Bioacoustic Data Analysis

Feature engineering is a very important step, acting as a bridge between raw bioacoustic data and insights. Intrinsically, acoustic signals are complex and need to be transformed into structured features to be analyzed effectively. This transformation basically means extracting relevant characteristics representing the biological and physical properties of sounds. The most commonly used features include time-domain attributes, like amplitude and zero-crossing rate; frequency-domain parameters, like spectral centroid and bandwidth; and time–frequency representations, such as spectrograms and Mel-frequency cepstral coefficients (Ganchev et al. 2017). Feature engineering will enable tasks of species classification, behavior identification, and soundscape analysis by providing a way to identify patterns and anomalies. Machine learning models can also be more accurate and efficient to ensure robust ecological monitoring (Mcloughlin et al. 2019; McEwen 2024).

1.3 Applications in Ecology, Wildlife Monitoring, and Conservation

The applications of bioacoustic research are immense, helping in ecological understanding and conservation efforts:

- **Ecology**: Bioacoustics contributes to the study of species interactions, biodiversity assessments, and ecosystem dynamics. For example, the study of environmental soundscapes gives information on the presence and activity of organisms.
- **Monitoring Wildlife**: Acoustic monitoring will enable the researcher to trace the species distribution, population density, and behavioral pattern. Passive acoustic recorders will be deployed at the sites for continuous data gathering without disturbing the wildlife.
- **Conservation**: Bioacoustics identifies species reaching critical lists, studies habitat health, and determines effects of human activities like deforestation and industrial noise on animal populations. This would provide very important information toward the policymaking relevant for effective conservation (Napier et al. 2024).

Analysis of Fig. 1 emphasizes important details on bioacoustic feature engineering and applications. The time-domain features have been excellent in computational efficiency and low cost, making them very ideal for resource-constrained applications; however, they fail to achieve biological relevance since they are unable to provide the required spectral details. While frequency-domain features are costlier computationally, they imply higher biological relevance, which turns out to be indispensable with tasks of species identification or behavioral analysis. This constitutes a trade-off: features engineering techniques are chosen as per the requirements of application, such as precision and computational feasibility.

Applications, on the other hand, have focused on wildlife monitoring, especially because of its immediate practical rele-

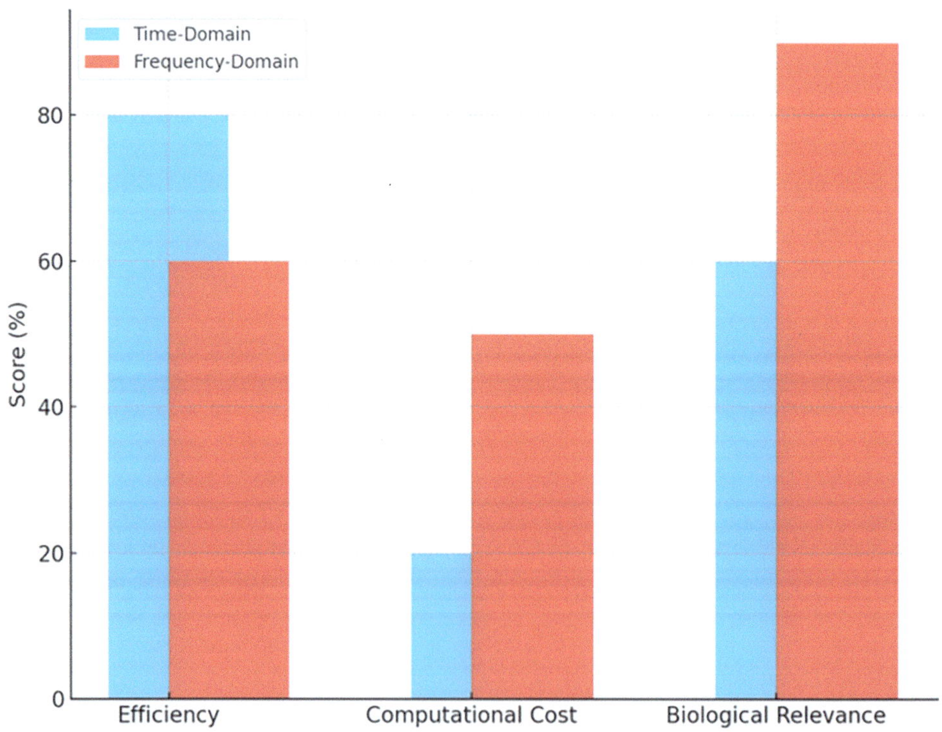

Fig. 1 Comparison of time-domain versus frequency-domain features

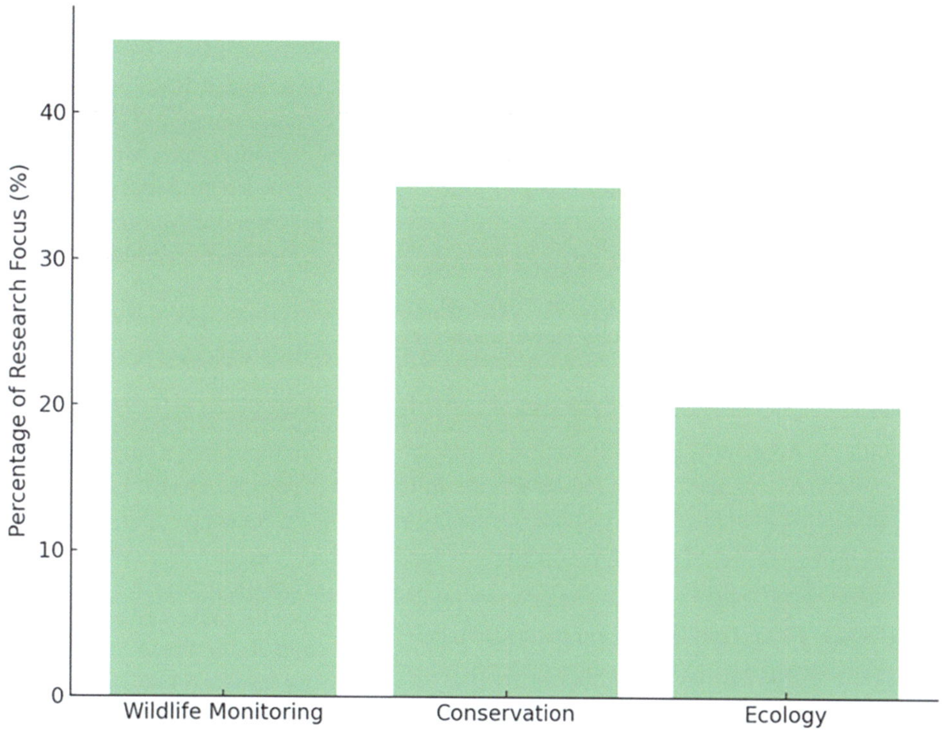

Fig. 2 Applications of bioacoustics across fields

vance in tracking the distribution of species and the study of their behaviors, whereas conservation, in turn, comes second, and that has to do with how urgent the application of bioacoustic tools is in the protection of biodiversity and assessment of human activities with respect to their impact on ecosystems; see Fig. 2. Ecological studies, though important, have received comparatively less attention probably due to the more complex issues related to the integration of bioacoustics into broader ecosystem-level analyses. These findings suggest that the bioacoustic research currently might be driven, perhaps disproportionately, by its utility for immediate conservation and monitoring needs, with further growth in its ecological applications. Further feature engineering techniques and interdisciplinary approaches will continue to increase the role of bioacoustics in promoting biodiversity and ecological sustainability.

2 Overview of Bioacoustic Data Types and Sources

Biological or ecoacoustic data spans, by name, a multitude of sound recording types associated with living biological systems, including animal vocalizations: that is to say, from birdsong to the calls of whales; environment soundscapes, more commonly known as the ambiance or background sounds of ecology: that is, either terrestrial, forest, or underwater environments; and anthropogenic noise levels impinging into natural habitats. Each format corresponds, appropriately, to the heterogeneous diversity in the sources of collection-things range from handheld audio recorders and hydrophones all the way up to advanced sensor networks and ARUs. Animal vocalizations often provide information on species identity, behavior, and habitat use. Environmental soundscapes capture the overall acoustic activity within an ecosystem, reflecting both biological and non-biological elements. Growing availability of open-access bioacoustic datasets has further facilitated research and enabled large-scale analyses (Nieto-Mora et al. 2023).

2.1 Basics of Signal Acquisition and Preprocessing

Bioacoustic data acquisition involves the recording of signals using specific equipment, appropriate to the environment and species in question. Hydrophones would be used to capture underwater sounds, for example, while parabolic microphones would be ideal for terrestrial wildlife. Proper placement and calibration of the recording devices are very important for high-quality data collection. Preprocessing is a significant step in the preparation of raw audio data for analysis. It involves the process of noise reduction, filtering, normalization, and segmentation. Noise reduction techniques, such as spectral subtraction, eliminate the unwanted background sounds. The filtering removes frequencies outside the range of interest, while normalization standardizes the amplitude levels across recordings. Segmentation divides continuous recordings into smaller, analyzable units, such as individual calls or vocalizations (Brandes 2008).

2.2 Challenges in Bioacoustic Signal Processing

Noise Interference: Bioacoustic recordings are often contaminated with environmental noise, such as wind, rain, or human activity. This interference complicates the extraction of meaningful features and requires sophisticated noise reduction techniques.

Variability in Acoustic Signals: Biological sounds are inherently variable due to, but not limited to, specific differences between species, individuals, and context. Variability in pitch, duration, and intensity opens issues in the development of generally applicable methods.

Large-Scale Data Handling: Continuous acoustic monitoring generates a large volume of data, which in turn requires efficient storage, processing, and analysis pipelines. A trade-off between computational resources and analytical accuracy remains one of the biggest challenges (Kaluri 2022).

Overlapping of the Sounds: Several sounds may run over each other in complex vocalizations, making it hard to isolate an individual signal for analysis. Advanced source separation algorithms are often required to address this issue.

These challenges were partly overcome by the interaction among strict methods of recording, advanced approaches to signal processing, and new computational tools. The confrontation of these challenges in the future areas of bioacoustic research carries the keys that will open completely new avenues into the investigation of ecological and conservation questions (Jayanthi et al. 2025; Stowell 2022).

Figure 3 presents the distribution of the bioacoustic data, where 50% is from an animal vocalization that has predominant usage in species identification and, to a lesser extent, behavioral study; environmental soundscapes take up a formidable contribution of 35%, indicating remarkable applications in ecosystem-level analysis relevant to biodiversity monitoring; a small contribution of 15% comes from anthropogenic noises, but it nonetheless holds critical importance in understanding how people's activities are affecting life in natural habitats. These ratios reflect the focus of bioacoustic research on biological and ecological phenomena, while anthropogenic influences are increasingly receiving attention. Figure 4 presents comparative levels of the impact concerning key challenges in the processing of bioacoustic signals. Thus, the interference of noise is indicated as the most important (40%) that refers to the difficulty to isolate the biological sounds in presence of environmental and anthropogenic noise. Variability in the signal follows at 30%, while large-scale data management is at 20%, with sound overlap at 10%, being other challenges, mainly in long-term monitoring and in complex soundscapes. It would, therefore, imply from

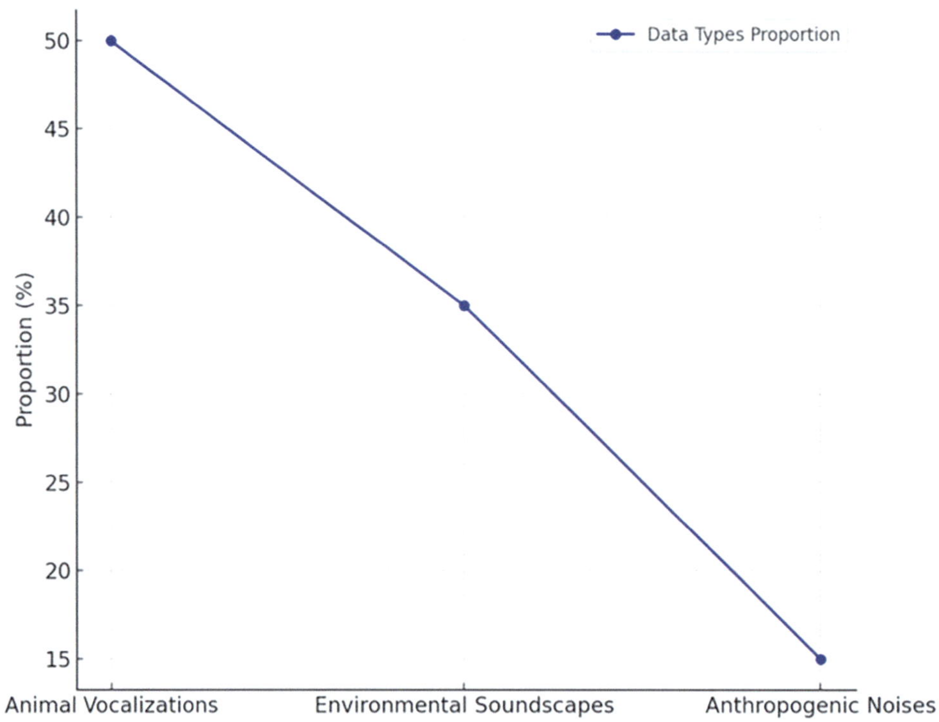

Fig. 3 Proportions of bioacoustic data types

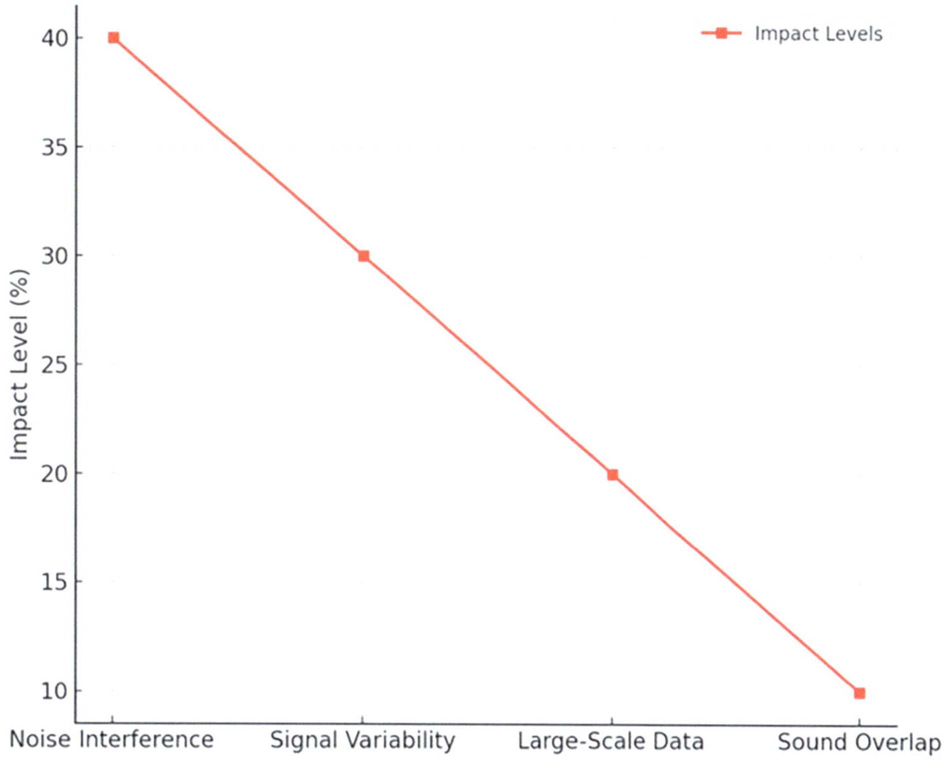

Fig. 4 Challenges in bioacoustic signal processing

these findings that noise reduction and standardization of the signal are the two factors, if improved that would immediately benefit bioacoustic research.

3 Bioacoustic Feature Extraction Techniques

3.1 Time-Domain Features

Time-domain features are directly obtained from the amplitude variations of the acoustic signals with respect to time. The two most important commonly used features are zero-crossing rate and root mean square energy (Scarpelli et al. 2021):

- **Zero-Crossing Rate (ZCR)**: This refers to the rate at which the waveform of the signal crosses the zero-amplitude axis. The ZCR proves helpful in discriminating between voiced and unvoiced sounds, and hence, it is a basic feature in animal call analysis for identifying the type of sound. It is, however, sensitive to noise and might not capture detailed frequency characteristics (Bonet-Solà and Alsina-Pagès 2021).
- **Root Mean Square (RMS) Energy**: RMS energy defines the average power of a signal and gives an idea of its intensity. This feature turns out to be very useful for detecting vocalizations in noisy recordings. The drawback is that it cannot represent spectral details, which are quite relevant for distinguishing between species whose call intensities may be similar (Chhaya et al. 2021).

3.2 Frequency-Domain Features

Frequency-domain features are extracted by the analysis of spectral properties of acoustic signals by methods such as Fourier transform. Some important features include:

- **Spectral Centroid**: This is the "center of mass" of a sound spectrum, showing the dominant frequency. It is useful for classifying sounds based on their tonal characteristics (Dangwal 2024).
- **Bandwidth**: Bandwidth is the range of frequencies present in the signal. It helps to characterize species that have unique frequency ranges in their calls.
- **Spectral Roll-Off**: It designates the frequency below which a certain percentage, usually 85–90%, of the signal energy is concentrated. It is usually used to make a distinction between tonal and broadband sounds (Gul and Khan 2023).

These features are quite important in species identification and sound classification tasks but usually require high-quality recordings and preprocessing in order to reduce the impact of noise as much as possible.

3.3 Time–Frequency Features

Time–frequency features represent a combination of temporal and spectral information, hence providing a complete representation of acoustic signals. The most salient techniques include:

- **Mel-Frequency Cepstral Coefficients (MFCCs)**: Extracted from the mel scale, MFCCs have features of human auditory perception. Because of this fact, they are highly effective for analyzing animal vocalizations, finding wide application in most modern automated classification systems.
- **Wavelet Analysis**: It decomposes the wavelets into its time–frequency components to perform transient event analyses for more complex sounds. Wavelets would, therefore, be particularly ideal in studies focused on either short-duration vocalizations or rapid changes in frequency.
- **Spectrogram-Based Feature Extraction**: A spectrogram represents the time course of the spectral content of signals. It, therefore, may provide a very rich source of features. Using techniques in image processing, it is possible to extract patterns from the output which will represent the unique signature of species.

Figure 5 presents the relative applications of time-domain, frequency-domain, and time–frequency features to species classification and behavior analysis. Time–frequency features, such as MFCCs and spectrogram-based techniques, dominate both applications due to their ability in combining temporal and spectral information to achieve superior accuracy for complex tasks. Frequency-domain features are also widely used, especially for sound classification, because of their ability to capture spectral properties. While time-domain features are simpler to compute, they reveal limited applications due to a lack of spectral detail. This comparative study underlines the growing preference of time–frequency features in modern bioacoustic analysis. Figure 6 compares the noisy sensitivity and computational complexity of using time-domain, frequency-domain, and time–frequency features. Time-domain features are most sensitive to noise due to their reliance on raw signal characteristics. Frequency-domain features offer moderate noise resilience but demand more computational resources. Time–frequency features, despite being the least sensitive to noise and the most accurate, exhibit the highest computational complexity, necessitating powerful processing systems. This trade-off underscores the need for tailored feature selection based on application goals and available resources.

4 Advanced Techniques in Feature Engineering

Feature engineering in bioacoustics has undergone rapid evolution with the introduction of advanced methodologies to incorporate artificial intelligence and machine learning, overcoming traditional limitations. Some of the key techniques which are revolutionizing this field include the following.

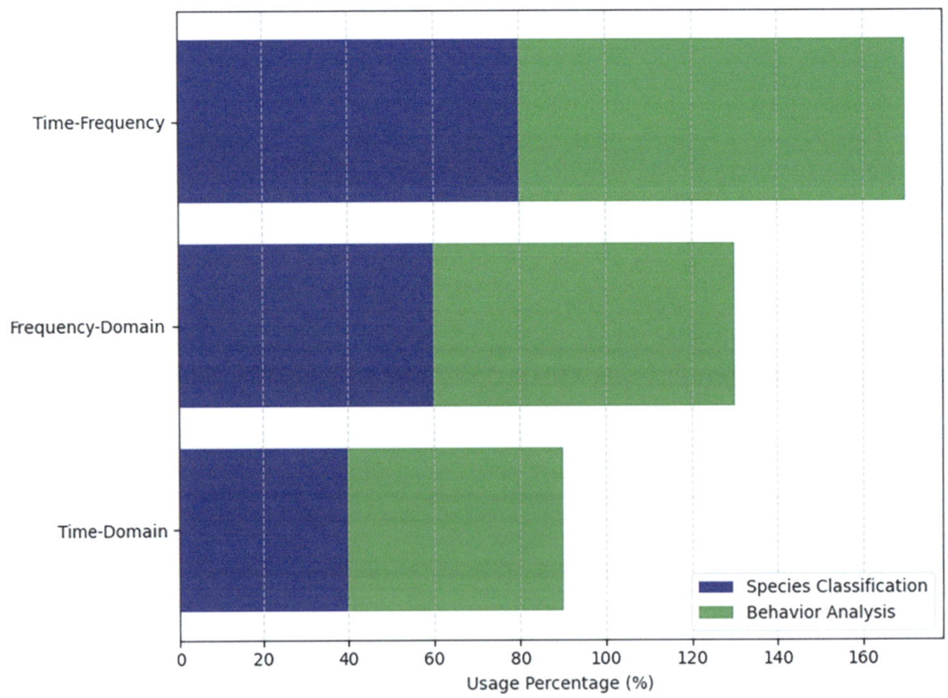

Fig. 5 Applications of feature types

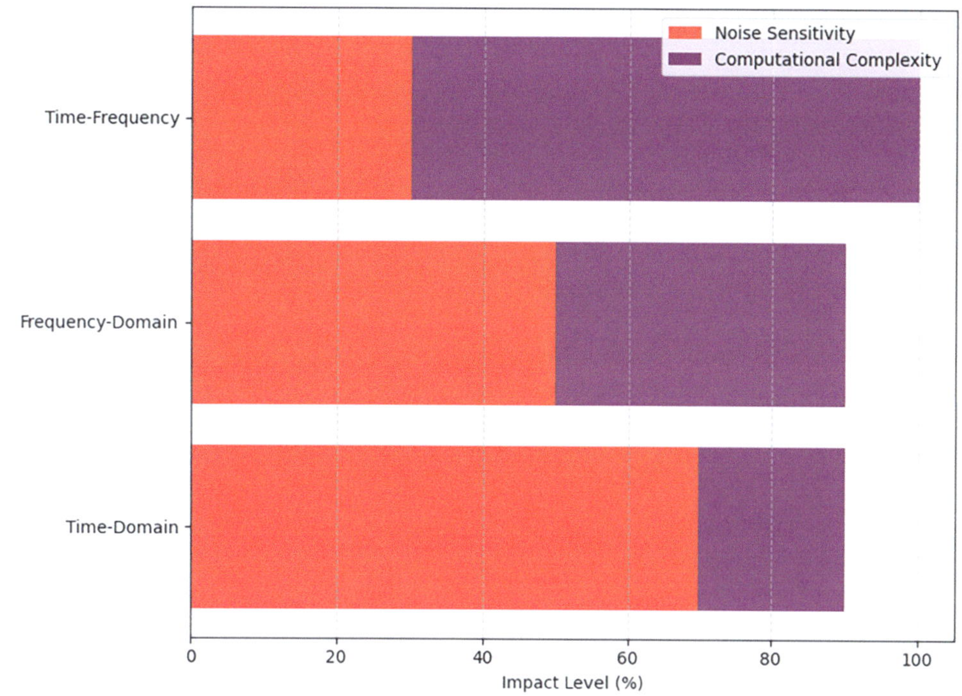

Fig. 6 Challenges in using features

4.1 Use of Deep Learning for Automated Feature Extraction

It is now done automatically in a shift from raw acoustic data by deep learning, a technique that has revolutionized feature engineering through complex and high-dimensional features. In particular, convolutional neural networks and recurrent neural networks, especially long short-term memory networks, are very useful in bioacoustics and are used for performing a variety of tasks like species classification and activity recognition (Odom et al. 2021).

- **Advantages**: These models learn a hierarchical feature representation that models complex temporal and spectral patterns without the need for any manual feature selection. For example, spectrogram-based input to CNNs allows subtle bioacoustic signatures—such as individual call variation or faint signals—to be captured.
- **Limitations**: Admittedly, despite their enormous potential, deep learning models require huge annotated datasets and computational powers for training, which is mostly challenging in bioacoustics due to a lack of labeled data.

4.2 Transfer Learning for Cross-Species Analysis

The process of pretraining a deep learning model on one dataset and fine-tuning it for another, different but related task, is known as transfer learning. This makes the approach quite useful in bioacoustics, where data are limited both in diversity and amount (Devi et al. 2025).

- **Applications**: A model that is pretrained using general sound classification can be fine-tuned for the detection of particular species or behaviors. For example, a model that was pretrained using a large set of labeled data including bird calls can then, with little additional training data, adapt to the same task for marine mammal vocalizations.
- **Benefit**: Transfer learning helps in greatly reducing large data collections and computational burdens. Enhanced cross-species comparison/generalization.

4.3 Hybrid Approaches Combining Traditional and AI-Driven Methods

Hybrid methods combine classical signal processing methods with AI-driven models in such a way that each can leverage the strengths of the other (Kershenbaum et al. 2024).

- **Methodology**: Noise reduction, time–frequency transformations (like generating spectrograms), along with manual feature extraction steps like MFCCs result in input features which may be fed into machine learning models for enhanced accuracy. Example: Studies show that the combination of wavelet analysis for feature extraction and deep learning models demonstrate much better performance in the automatic detection and classification of noisy and complex bioacoustic signals, such as humpback whales' over-running calls.

- **Advantages**: These methods balance computational efficiency with model performance and are hence suitable for real-world applications where resources may be limited.

5 Addressing Challenges in Bioacoustic Feature Engineering

5.1 Noise Interference and Signal Variability

Noise interference and signal variability have always been the major concerns in bioacoustic feature engineering. Most bioacoustic data is acquired outdoors, where external factors such as wind, rain, and anthropogenic noise mask the signals of interest. Overlapping calls from different species contribute more complications to meaningful feature extraction. These interferences reduce the quality of the acoustic signal and subsequently introduce errors into the processes of feature extraction and classification (Ghosh et al. 2024).

Advanced denoising methods including spectral subtraction, adaptive filtering, and deep learning-based noise reduction models have recently been explored to deal with noisy interference. Such techniques reduce the background noise while preserving acoustic features that are of biological importance. Signal variability driven by a variety of recording equipment and environmental conditions, along with the individual variation within a species, can be minimized using techniques like normalization along with appropriate and robust preprocessing workflows. For example, amplitude scaling and frequency alignment are some of the methods applied on the features extracted from different datasets aiming to be consistent. Probably, just by combining these methods is one way to reduce the disturbing effects of noise and the variability to obtain an appropriate bioacoustic analysis.

5.2 Scalability and Computational Efficiency for Large Datasets

The growing availability of long-term acoustic recordings from multiple sites gives rise to very large bioacoustic databases. Processing and analysis of such unprecedented datasets pose significant challenges in relation to scalability and computational complexity: traditional feature extraction means cannot be applied in efficient, sometimes even feasible, manner in the case of collections composed of thousands of recording hours.

In an effort to meet these challenges, scalable algorithms have been developed that take advantage of parallel processing and distributed computing environments. Scalability of processing large datasets is facilitated through cloud-based platforms, GPU acceleration, lightweight models, and optimized feature extraction pipelines. Dimensionality reduction techniques, including PCA and t-SNE, are also applied in order to reduce computational complexity while retaining the most important features. Automated pipelines that integrate feature extraction, classification, and result validation further streamline the workflow and make bioacoustic analyses more efficient and accessible (Manikandan and Jayashree 2024).

5.3 Generalization Across Species and Environments

Probably the biggest challenge in this domain is generalizing the current bioacoustic feature engineering methods across diverse species and habitats. Features effective enough to analyze bird songs in a temperate forest may not perform well when applied to marine mammal vocalizations in underwater environments. It follows that the lack of generalization in feature engineering techniques and models is limited in their application to different ecological contexts. It has lately emerged as a powerful tool to enhance generalization: transfer learning. The models pretrained on one species or habitat can be fine-tuned on the small data from another domain and hence need not develop large datasets all over again. On the other hand, feature alignment across environments is obtained with the help of domain adaptation techniques such as adversarial training that enhance the robustness of models. While single models are rigid, ensemble learning approaches—that combine the predictions of several models—can adapt more flexibly to datasets of diverse origins. Also, incorporating a wide range of acoustic data during training makes the models learn features from multiple species and different environmental conditions, thus improving the generalization capability (Williams et al. 2022).

Figure 7 compares traditional methods with advanced techniques in handling three key challenges: noise interference, scalability, and generalization, which have been encountered or are yet to be encountered in bioacoustic feature engineering. From the graph, advanced techniques have consistently outperformed traditional approaches for each of these three challenges. Noise interference is addressed better with 90% effectiveness through deep learning-based denoising compared to traditional noise reduction techniques, which only achieve 70%. This further pinpoint the enhanced capability of state-of-the-art approaches in handling complex and dynamic noise patterns common in bioacoustic data. In terms of scalability, the advanced computational methods of cloud-based processing and GPU acceleration are much more effective, at 85%, than traditional sequential processing methods, which are at 60%. These modern approaches have allowed large-scale datasets, which are becoming increasingly common in bioacoustic studies, to be handled effi-

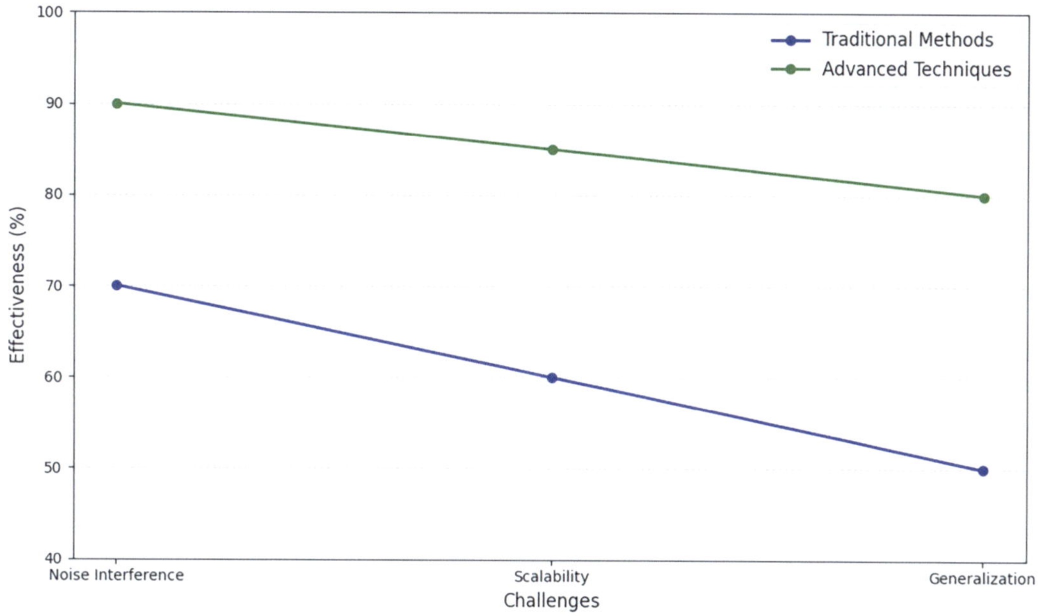

Fig. 7 Comparative analysis of the effectiveness of traditional methods and advanced techniques

ciently. Finally, generalization, which is one of the most important techniques contributing to the effectiveness of a number of state-of-the-art methods, such as transfer learning in general, achieves effectiveness results as high as 80%, far higher than when traditional methods are used-the maximum effectiveness of which stops at 50%. Feature adaptability among diverse species and variable environments is of great essences to ensure that recently advanced methods offer robust bioacoustic analysis. In a nutshell, the comparison graph outlines how sophisticated techniques transformed those main challenges into opportunities toward bettering more accurate, scalable, and adaptable methodologies of conducting bioacoustic research.

6 Case Studies in Bioacoustic Feature Engineering

Applications in bioacoustic research have been numerous and quite successful in showcasing feature engineering as a transformational power. Advanced computational methods have been able to empower researchers to achieve unparalleled insight into avian, marine, and terrestrial ecosystems.

6.1 Bird Ecosystems

Feature engineering has been an important part of avian bioacoustics, be it species identification or behavioral analysis. For example, MFCCs and spectral roll-off have been used in the classification of bird calls across diverse habitats. One such example is the monitoring of endangered bird species in temperate forests using ARUs. By extracting frequency-domain features, researchers were able to distinguish closely related species with over 90% accuracy. This approach not only allows for real-time monitoring of birds but also provides detailed insight into their seasonal variations in vocal activities. Such findings are bound to impact the bottom line in the carving out of appropriate conservation strategies regarding the identification of key breeding grounds, for instance (Gibb et al. 2024).

6.2 Marine Ecosystems

Bioacoustic studies concerning marine animals mostly relate to cetaceans, including whales and dolphins, whose communication occurs with very low-frequency calls. This was one of the milestone studies that implemented time–frequency features in order to conduct monitoring on the migration of whales across ocean basins. Information from the acoustic data collected by underwater hydrophones revealed clear migrations and feeding behaviors of whales. Another example is the use of wavelet analysis to detect echolocation clicks of dolphins, allowing population density estimates in areas where visual observation may be limited. These findings have spurred international conservation to address major threats such as ship strikes and underwater noise pollution (Vasconcelos et al. 2024).

6.3 Terrestrial Ecosystems

Bioacoustic techniques have presently been used in monitoring insects and mammalian communication in terrestrial ecosystems. For instance, in studying the insect sounds in agricultural landscapes, the investigators used zero-crossing rate and root mean square energy. Data from these manifested that insect activities correlated well with environmental factors that in turn helped in strategizing the management of pests. Similar methods for feature extraction have also been used to study terrestrial mammals, for example, elephants, who use their low-frequency rumbles to be able to communicate over extremely long distances. These types of studies have helped anti-poaching efforts and habitat management.

7 Future Trends and Emerging Research Areas in Bioacoustic Feature Engineering

Bioacoustics is a rapidly developing field, driven by rapid technological development and a growing appreciation of its role in ecological monitoring. In the coming years, a number of emerging research areas are likely to further strengthen the role of bioacoustics in environmental science. Herein, three critical trends of the future in bioacoustic research are presented, each dealing with integration into other ecological datasets, real-time monitoring using edge computing/IoT, and ethics concerning bioacoustic research.

7.1 Integration of Bioacoustics with Other Ecological Datasets

With the maturation of the field of bioacoustics, there is a growing emphasis on the integration of acoustic data with other ecological data types, including environmental, climatic, and geographical. This does indeed promise a more holistic understanding of ecosystems and deeper insights into species behavior, distribution, and environmental health. In that regard, combining acoustic data with the use of satellite imagery or sensor data, for example, will enable researchers to find various patterns in the migration, habitat use of a species, and the effects of climate change on biodiversity. Another exciting application of this integration is in the development of multi-modal systems for environmental monitoring. By fusing acoustic data with data from other sensors, including temperature, humidity, and atmospheric pressure, changes in soundscapes can be tracked that can be correlated with shifts in the environment. Such an integrated approach would perhaps serve to make very effective predictions of ecological events-migratory species arrivals, onsets of seasonal behaviors, shifts in habitat that would result from climate change (Sethi et al. 2020).

The integration of bioacoustics with other data will also further enhance the scalability of research efforts, allowing large-scale, long-term monitoring projects. By integrating these datasets, researchers can build comprehensive models that can make more accurate predictions about ecosystem status, inform conservation strategies, and enhance wildlife management.

7.2 Real-Time Bioacoustic Monitoring Using Edge Computing and IoT

The employment of edge computing and the Internet of Things for the real-time monitoring of biological soundscapes perhaps stands out as one of the most promising future trends in bioacoustic research. Generally, data collection and analysis have been delayed in many bioacoustic studies, with most data needing to be moved for processing at central processing systems. With the recent development of edge computing, acoustic sensors are now capable of processing data locally. This reduces latency and thus enables near-instant analysis of bioacoustic data. Large-scale autonomous monitoring systems enabled by edge computing can be deployed onto the most remote and uninviting environments through a network of IoTs. Sensor networks that, for instance, install recurrence records in conservation areas, forests, or oceans do this through continuous recording and sound analysis of wildlife. That means, by using this technology today, we are in a position to monitor real-time changes within the environment, perhaps vulnerable species or illegal poaching activity, affecting the biodiversity. In this sense, real-time analytics applied to bioacoustic monitoring can make conservation much more responsive by orders of magnitude and more dynamic in terms of decision-making regarding ecological management (Hilasaca et al. 2021).

Another critical benefit of edge computing combined with IoT is substantial cost efficiency and storage of data. By processing data locally, sending only the relevant information or the preprocessed one to central systems reduces the bandwidth and storage burden, thereby making large-scale continuous bioacoustic monitoring systems feasible.

7.3 Ethical and Ecological Considerations in Bioacoustic Research

The time has come to give further consideration to the ethical and ecological issues raised by collection and use of acoustic data in the ongoing growth of both academic and applied bioacoustics. Acoustic monitoring in remote or protected ecosystems faces potential issues with privacy, disturbance, and human impact on wildlife. From an ethical perspective, researchers will have to consider the impacts of monitoring activities on the studied species. In some instances, the presence of acoustic sensors or recording of the sounds of certain species may cause stress or disrupt natural behaviors. Ethical guidelines and best practices will have to be developed to ensure that monitoring activities are noninvasive and the well-being of the species and ecosystems being studied is considered first. Ecological concerns, such as a general understanding by bioacoustic research of its impact on the whole spectrum of soundscapes and biodiversity, also show that human-made noise, from industrials, transport, and urbanization, interferes with the acoustical environment and could negatively affect the possibility of species communicating, reproducing, and surviving. Bioacoustics might provide a way to monitor and assess those impacts, but researchers need to be very aware of their own monitoring activities, which again may contribute to sound pollution or alter the ecological balance of the environments under study. There is also a growing concern about the access and sharing of bioacoustic data. While open data sharing may accelerate global collaboration in the conservation process, it does raise serious questions with regard to data ownership, privacy, and the potential for misuse. Clear protocols regarding data management, sharing, and usage need to be laid down to ensure that bioacoustic research will remain responsible, equitable, and beneficial to scientific communities and local stakeholders alike.

8 Conclusion

Feature engineering in bioacoustics lies at the heart of advancing deep knowledge of biological soundscapes and ecological systems. The present chapter has reviewed key techniques currently in use within bioacoustic data extraction and analytic methods, which range from the time–frequency domain feature to the state-of-the-art deep learning for automatic features extraction. In view of this, addressing such thorny issues as those imposed by noise interference, computational efficiency, and cross-species generalization, refinement after refinement is continuously worked out to enable the translation of raw acoustic data into actionable insight. The applications of bioacoustic research are huge in number and transformational-including case studies in analysis of avian song to marine mammal monitoring through identification of insect sounds. These examples, while underlining the versatility of bioacoustics, bring out its essentiality in ecological research and conservation efforts. As this field is in continuous development, the integration of bioacoustic data with other ecological datasets will surely enrich our comprehension of complex ecosystems and afford a holistic approach toward environmental monitoring. Looking ahead, real-time bioacoustic monitoring, at the hands of edge and IoT, will let researchers collect and process data in ways nobody has imagined, enabling researchers to promptly react to variation in ecology. Like all the scientific gains, however, bioacoustic research draws on important ethical and ecological lessons that should be accounted for as well. It is important not to disturb the wildlife or the ecosystem while carrying out monitoring activities, and also to make sure that data sharing is responsibly done to support conservation efforts globally.

Eventually, it will be the crossings among technological innovation, ecological insight, and ethical responsibility where a bioacoustics future will be made. By embracing the emerging trends and challenges ahead, both researchers and practitioners alike can further unlock the vast potential that bioacoustic feature engineering has in store for the furtherance of biodiversity conservation, environmental health, and wildlife management. This chapter provides a critical framework for understanding the present position and future directions that bioacoustic research may take, equipping those working within the field with the necessary tool kit and knowledge to take forward this exciting and rapidly developing discipline.

References

Bonet-Solà D, Alsina-Pagès RM (2021) A comparative survey of feature extraction and machine learning methods in diverse acoustic environments. Sensors 21(4):1274

Brandes TS (2008) Feature vector selection and use with hidden Markov models to identify frequency-modulated bioacoustic signals amidst noise. IEEE Trans Audio Speech Lang Process 16(6):1173–1180

Chhaya V, Lahiri S, Abhinava Jagan M, Mohan R, Pathaw NA, Krishnan A (2021) Community bioacoustics: studying acoustic community structure for ecological and conservation insights. Front Ecol Evol 9:706445

Dangwal K (2024) AI-enhanced bee bio-acoustics with data augmentation and applications in early swarming detection

Devi GD, Kumar N, Manikandan J, Rekha V (2025) Innovative brain tumor detection: stacked random support vector-based hybrid gazelle coati algorithm. Biomed Sig Process Control 101:107156

Ganchev T (2017) Computational bioacoustics: biodiversity monitoring and assessment, vol 4. Walter de Gruyter GmbH & Co KG

Ghosh F, Nair N, Elayadam P, Xavier Savarimuthu SJ (2024) Application of machine learning algorithms in the field of bioacoustics. In: Artificial intelligence for multimedia information processing. CRC Press, pp 233–249

Gibb KA, Eldridge A, Sandom CJ, Simpson IJA (2024) Towards interpretable learned representations for Ecoacoustics using variational auto-encoding. Ecol Inf 80:102449

Gul S, Salman Khan M (2023) A survey of audio enhancement algorithms for music, speech, bioacoustics, biomedical, industrial and environmental sounds by image U-Net. IEEE Access

Hilasaca LMH, Gaspar LP, Ribeiro MC, Minghim R (2021) Visualization and categorization of ecological acoustic events based on discriminant features. Ecol Ind 126:107316

Jayanthi K, Chitradevi D, Muthaiyah S (2025) Early prediction of harmful environmental changes based on bioacoustics analysis. In:

Technological advancements for deep sea ecosystem conservation and exploration. IGI Global Scientific Publishing, pp 129–148
Kaluri VSNRR (2022) Real time biodiversity measurement in large bioacoustic datasets. University Of Tasmania, PhD diss.
Kershenbaum A, Akçay C, Babu-Saheer L, Barnhill A, Best P, Cauzinille J, Clink D et al (2024) Automatic detection for bioacoustic research: a practical guide from and for biologists and computer scientists. Biol Rev
Kvsn RR, Montgomery J, Garg S, Charleston M (2020) Bioacoustics data analysis–a taxonomy, survey and open challenges. IEEE Access 8:57684–57708
Manikandan J, Jayashree K (2024) Enhancing lung nodule classification: a novel CViEBi-CBGWO approach with integrated image preprocessing. J Imag Inf Med 37(5):2108
McEwen B (2024) Computational bioacoustics for the detection of rare acoustic events
McGinn K, Kahl S, Zachariah Peery M, Klinck H, Wood CM (2023) Feature embeddings from the BirdNET algorithm provide insights into avian ecology. Ecol Inf 74:101995
Mcloughlin MP, Stewart R, McElligott AG (2019) Automated bioacoustics: methods in ecology and conservation and their potential for animal welfare monitoring. J R Soc Interface 16(155):20190225
Napier T, Ahn E, Allen-Ankins S, Schwarzkopf L, Lee I (2024) Advancements in preprocessing, detection and classification techniques for ecoacoustic data: a comprehensive review for large-scale passive acoustic monitoring. Expert Syst Appl 124220
Nieto-Mora DA, Rodríguez-Buritica S, Rodríguez-Marín P, Martínez-Vargaz JD, Isaza-Narváez C (2023) Systematic review of machine learning methods applied to ecoacoustics and soundscape monitoring. Heliyon
Nieto-Mora DA, de Oliveira MCF, Sanchez-Giraldo C, Duque-Muñoz L, Isaza-Narváez C, Martínez-Vargas JD (2024) Soundscape characterization using autoencoders and unsupervised learning. Sensors 24(8):2597
Odom KJ, Araya-Salas M, Morano JL, Ligon RA, Leighton GM, Taff CC, Dalziell AH et al (2021) Comparative bioacoustics: a roadmap for quantifying and comparing animal sounds across diverse taxa. Biol Rev 96(4):1135–1159
Scarpelli MDA, Liquet B, Tucker D, Fuller S, Roe P (2021) Multi-index ecoacoustics analysis for terrestrial soundscapes: a new semi-automated approach using time-series motif discovery and random forest classification. Front Ecol Evol 9:738537
Sethi SS, Jones NS, Fulcher BD, Picinali L, Clink DJ, Klinck H, Orme CDL, Wrege PH, Ewers RM (2020) Characterizing soundscapes across diverse ecosystems using a universal acoustic feature set. Proceed Nat Acad Sci 117(29):17049–17055
Stowell D (2022) Computational bioacoustics with deep learning: a review and roadmap. PeerJ 10:e13152
Vasconcelos D, Nunes NJ, Förster A, Gomes JP (2024) Optimal 2D audio features estimation for a lightweight application in mosquitoes' species: ecoacoustics detection and classification purposes. Comput Biol Med 168:107787
Williams B, Lamont TAC, Chapuis L, Harding HR, May EB, Prasetya ME, Seraphim MJ et al (2022) Enhancing automated analysis of marine soundscapes using ecoacoustic indices and machine learning. Ecol Ind 140:108986

Detection of Ecosystem Changes Using Bioacoustics Techniques

C. Fancy, S. Metilda Florence, M. Anand, and V. Joseph Raymond

Abstract

Bioacoustics has emerged as a powerful tool for monitoring and understanding ecological and environmental changes. This chapter reviews some applications of bioacoustic monitoring techniques in concert with advanced change detection methodologies for the assessment of biodiversity and ecosystem dynamics. The acoustic environment or soundscape is a reflection of the ecological interactions within an ecosystem and serves as a reliable indicator of environmental health. Change detection in soundscapes due to natural or anthropogenic factors informs changes in species composition, habitat quality, and climate impacts. This chapter also deals with state-of-the-art tools and technologies, from automated recording units and machine learning algorithms to data visualization techniques that have transformed the gathering and processing of acoustic data. Such innovation has now allowed for real-time and long-term monitoring across diverse habitats, from tropical rainforests to urban environments. Additionally, bioacoustic monitoring has proven extremely helpful for studying elusive or nocturnal species and detecting subtle ecosystem changes that might go unnoticed in more traditional methods. Case studies illustrate how bioacoustic monitoring serves to track the impacts of both deforestation and urbanization on animal behavior, as well as the impacts of climate change on behavioral and ecosystem health. Discussed also are challenges associated with data management, technology limitations, and ethical issues, along with strategies overcoming them. Finally, this chapter contemplates a future in which bioacoustics, integrated with interdisciplinary approaches, will be at the epicenter of global conservation efforts and environmentally sustainable management. This comprehensive review on detecting and monitoring bioacoustic change should inspire researchers, conservationists, and policymakers in realizing the potential of sound for understanding and conserving the natural world.

Keywords

Bioacoustics · Ecological interactions · Deforestation · Urbanization · Sustainable management

1 Introduction to Bioacoustics and Change Detection

1.1 Basics of Bioacoustics and Its Role in Ecological Monitoring

Bioacoustics, as determined, is the interdisciplinary science that melds biology and acoustics in the study of ways in which living organisms are able to produce, sense, and use sound. This science has now turned out to be of even greater importance in the twenty-first century, especially turning out to be one of effective noninvasive tools needed in ecological monitoring. Besides, sound is a remarkably important component that characterizes such ecosystems because it can give information on the existence, ways of behavior, interactions among those species, and environmental background. Bioacoustic signals span, from the songs of birds and calls of amphibians down to rustling leaves and humming insects, in datasets rich for the comprehension of the dynamics of biodiversity.

C. Fancy · S. Metilda Florence (✉) · M. Anand · V. Joseph Raymond
Department of Networking and Communications, School of Computing, SRM Institute of Science and Technology, Kattankulathur, India
e-mail: metildam@srmist.edu.in

C. Fancy
e-mail: fancyc@srmist.edu.in

K. Subbaraj et al. (eds.), *Computational Bioacoustic Artificial Intelligence*, Sustainable Artificial Intelligence-Powered Applications, https://doi.org/10.1007/978-3-032-06088-4_10

Ecological monitoring through the use of bioacoustics first deals with capturing soundscapes, composite recordings of biophony, geophony, and anthropophony. These soundscapes are important in identifying trends in species abundance, invasive species detection, and monitoring ecosystem health. Changes in the soundscapes over time can be analyzed to determine the effects of environmental stressors such as habitat loss, climate change, and urbanization. Bioacoustic monitoring boasts several advantages over traditional ecological methods. It allows for continuous, large-scale data collection with minimal disturbance to wildlife. Automated recording devices can also allow for long-term monitoring of remote or physically difficult-to-reach areas. This technology has been especially successful for studying secretive or nocturnal species, which are often difficult to detect visually (Jayanthi et al. 2025; Oestreich et al. 2024).

1.2 Overview of Change Detection Methodologies

In the field of bioacoustics, change detection would refer to change over time or space to acoustic patterns indicative of ecological fluctuations. Techniques for the detection of changes in biological sounds could be broadly ranged from manual analysis to different automated methods.

Manual Analysis: It relies on the work of human experts in interpreting spectrograms, which are visual representations of sound frequencies over time. While this method is highly accurate, it is also labor-intensive and cannot be done at scale.

Automated Methods: Contrasting this, the automated methods are harnessed from state-of-the-art developments in signal processing, machine learning, and artificial intelligence. Such methods leverage extensive data by recognizing patterns, anomaly detection, and classification into predefined sound categories. Among these, CNNs and random forests have become popular approaches for identifying species-specific calls or abrupt changes in soundscapes (Bobba 2410; Mcloughlin et al. 2019).

Key metrics for detecting bioacoustic change include acoustic diversity, acoustic evenness, and acoustic entropy, which, respectively, quantify the richness and complexity of soundscapes. A temporal shift in these metrics often correlates with an ecological disturbance or shift in biodiversity. Integration of bioacoustics with remote sensing, environmental data, and predictive modeling further enhances its utility in change detection. For instance, merging acoustic data with satellite imagery would develop a multidimensional perspective on changes to ecosystems by relating acoustic signals produced to physical habitat alteration. The role of bioacoustics and change detection methodologies in enabling precise, scalable, cost-effective monitoring becomes very important for advancing the cause of conservation science, timely intervention, and sustainable environmental management.

Comparing the manual and automated techniques of bioacoustic analysis yields advantages in leveraging technology for effective ecological monitoring, as depicted in Fig. 1. Indeed, automated techniques show high improvement in time efficiency at 90%, in contrast to 40% obtained for manual analysis. This indicates that they are able to process large volumes of data within minimal time, a major prerequisite in long-term and large-scale monitoring projects. Besides, the accuracy of automated methods is higher than that of manual techniques by about 7% (92% vs. 85%), which again speaks to their reliability in the detection and classification of acoustic signals. These results show that automatic analysis saves not only time but also enhances the precision of bioacoustic monitoring, which is indispensable in modern ecological studies.

Figure 2 explores the analysis of acoustic metrics before and after an ecological disturbance has shown a definite decline in ecosystem health through soundscape parameters. Diversity, evenness, and entropy of acoustics are all significantly lower, with scores dropping from 85, 80, and 75 to 60, 55, and 50, respectively. These changes represent the losses in species richness and sound complexity due to reduced species presence or disrupted ecological interaction. Such sensitivity of acoustic metrics to environmental changes adds to their value as reliable indicators in assessing ecosystem disturbances. Their monitoring over time thus allows for effective tracking of the impacts of habitat degradation, climate change, or anthropogenic activities, enabling timely conservation interventions.

2 Technological Advances in Acoustic Monitoring

2.1 New Tools and Technologies for Bioacoustic Data Collection

Recent advances in technology have greatly improved bioacoustic monitoring to the point that researchers are now capable of recording, processing, and analyzing acoustic data with far greater accuracy and speed. One of the most common tools in bioacoustics is usually an automated recording unit. These can run for days, weeks, or even months without much human interaction, recording soundscapes from the densest forests to the urban environment. Added to these factors are their robustness and the capability to operate in harsh conditions, which make them quite suitable for long-term ecological studies. Another important development involves underwater hydrophones, which are special microphones designed for recording sounds produced in aquatic

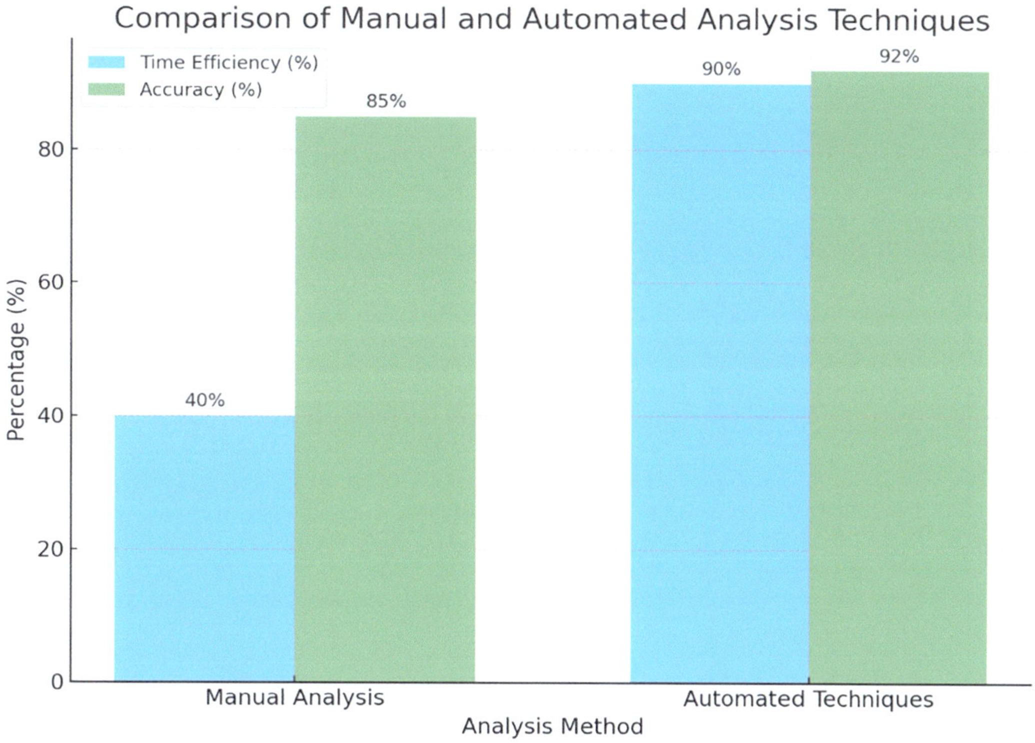

Fig. 1 Comparison of manual and automated analysis techniques

ecosystems. Hydrophones have contributed much to marine biodiversity monitoring, the detection of aquatic species, and research into underwater acoustic environments. Drones equipped with acoustic sensors have also emerged as game-changing tools. They allow researchers to access remote or challenging terrains and capture real-time acoustic data across large spatial scales. This capability is particularly valuable in the case of species inhabiting hard-to-reach areas, such as cliffs, wetlands, or dense canopies (Ross et al. 2023). Recent cloud storage and data-sharing platform developments further enhance acoustic monitoring with the facility for storage, organization, and global sharing of large datasets. Open-access bioacoustic libraries such as the Macaulay Library and Xeno-Canto provide reference databases that can be used to identify species and perform comparative studies (Nel et al. 2024).

2.2 AI and Machine Learning in Audio Analysis

The integration of AI and ML is changing the whole approach to the analysis of acoustic data in bioacoustics. Conventionally, the means for analyzing sounds depended on human interpretation; hence, it was slow and full of potential mistakes. In contrast, AI-driven approaches open an avenue for fast, accurate, and automated processing of large datasets. Algorithms created by machine learning were supposed to recognize patterns within the acoustical data. For instance, these supervised learning models learn species-specific calls from training datasets with labeled examples, and in classification of sounds into their respective classes with very high accuracy, even in noisy conditions. Techniques like convolutional neural networks will work most effectively for the analysis of spectrograms, that are visualizations displaying the frequency of sound as it changes over time (Metcalf 2021; Zilli 2015).

Other unsupervised learning methods include clustering algorithms, which could also be useful in the exploration of unlabeled datasets. It will detect anomalies or group similar sounds, hence aiding in the discovery of unknown species or environmental changes. AI tools are further enhanced through techniques related to NLP, enabling analysis of text-based annotations associated with acoustic data. This allows streamlining metadata processing and improving the efficiency of bioacoustic databases. Another critical development is the real-time monitoring systems powered by AI. These can classify sounds at the time of recording and thus allow for immediate responses to ecological events, such as the presence of invasive species or illegal activities like logging and poaching. Integration of AI with IoT devices creates smart monitoring networks capable of operating autonomously in remote areas. Such networks are able to continuously collect and analyze data, thus sending alerts or reports in real time to the researchers. Overall, synergy between emergent tools and AI-driven sound analysis forms a transformative era in

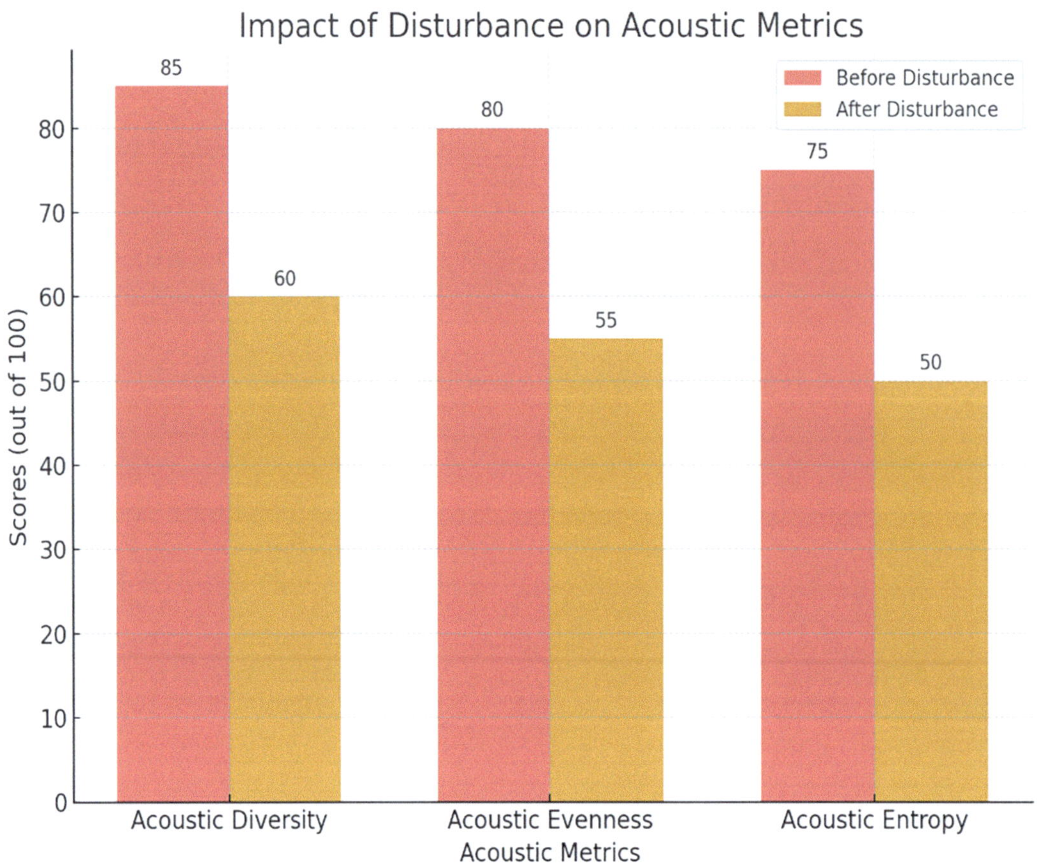

Fig. 2 Impact of disturbance on acoustic metrics

bioacoustics, allowing an extension of ecological research scope and giving a deeper look into ecosystems with proactive conservation measures (Müller et al. 2023).

Figure 3 compares the efficiency between manual and automated methods of data collection from 2015 through 2022. While the graph indicates a gradual rise in efficiency for manual methods that of automated methods is rather steep: 92% efficiency in automated methods in 2022 against 68% in manual methods. The difference in efficiency between the manual and automated methods of collecting data grew wider with time, with automated methods way ahead in a landslide. Such a trend could be explained by rapid developments and improvements of technologies, especially in the frames of ARU development and drones, which simplify bioacoustic data gathering and processing. The efficiency increase of the automated methods rises so steeply that it is underlined how important these are for modern ecological monitoring.

Figure 4 illustrates the accuracy of the AI-based analysis compared with that using traditional methods for an increased dataset size. Traditional methods show marginal increases in accuracy, peaking at 76%, with larger datasets. In contrast, the AI-based methods have shown a gradual increase and attained 92% with 100,000 samples. AI-based analysis is much more scalable and accurate than traditional methods, especially for larger datasets. While traditional approaches cannot improve much, AI-driven systems go on to excel by learning and adapting to complex patterns in data. That's why AI is going to be such a key component of scaling bioacoustic monitoring applications in general: data sets are ever increasing.

3 Applications in Ecosystem Monitoring

3.1 Detecting Changes in Biodiversity Through Soundscapes

Soundscapes are the composite recordings of biological, environmental, and human-made sounds that serve as strong indicators of biodiversity in ecosystems. It further enables the researchers to analyze the changes in the species composition, abundance, and behavior in time through the analysis of soundscapes. Changes in biodiversity often manifest as alterations in acoustic metrics such as acoustic diversity, acoustic evenness, and entropy. These metrics quantify the richness and complexity of the sounds emanating from an ecosystem. For instance, a reduction in acoustic diversity should imply degradation in habitat or key species loss, while an increase in anthropophony could indicate higher environmental quality

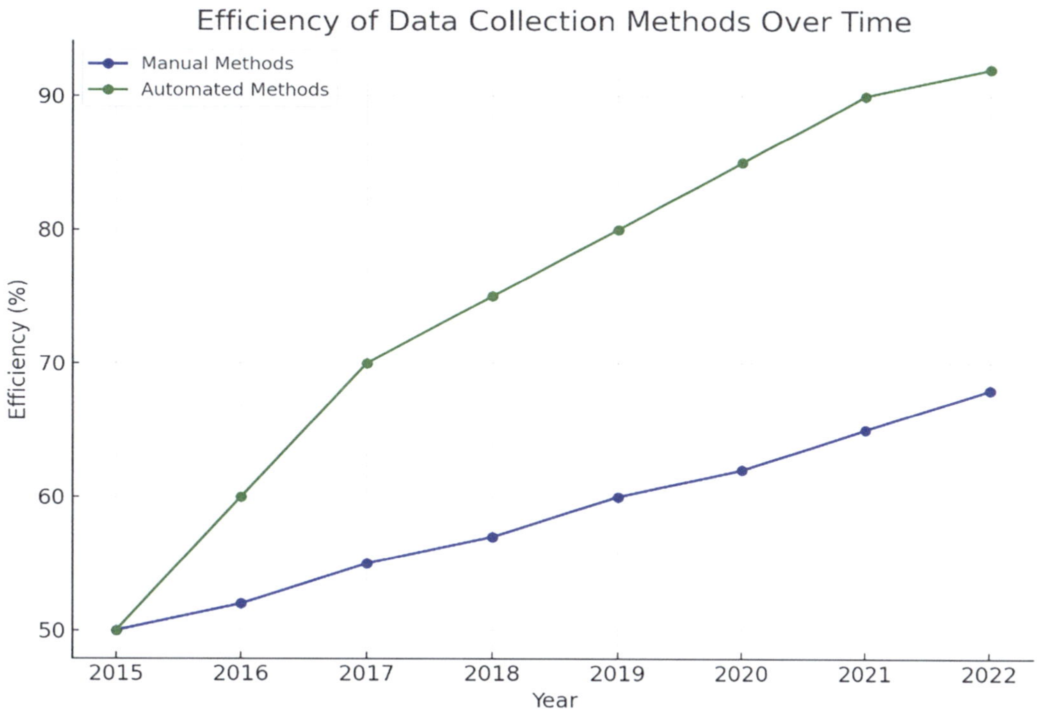

Fig. 3 Efficiency of data collection methods over time

or increased human activity into previously undisturbed areas. It is also possible that soundscapes can indicate ecological thresholds by highlighting sudden shifts in the acoustic environment. This analysis of soundscapes will prove to be particularly effective for extensive and remote areas, as it will provide a low-cost and noninvasive means of studying biodiversity without the physical capture or observation of species. It is also very useful in regions where traditional surveys cannot be done for various reasons, including logistics and safety (Borker 2018). Additionally, soundscapes can provide researchers with the opportunity to monitor temporal patterns, such as seasonal migrations, breeding calls, and diurnal activity cycles, as a means of insight into the health and dynamics of ecosystems. Long-term monitoring of soundscapes enables early detection of environmental stressors and thus assists in the timely implementation of conservation strategies (Growcott et al. 2024).

3.2 Monitoring Specific Species and Habitats

It is an important tool in studying certain species and their habitats, mainly those that are cryptic, nocturnal, or otherwise difficult to observe visually. Acoustic recordings can capture unique vocalizations, like the songs of birds, the calls of frogs, or even bat echolocation signals, that will permit species identification and allow assessments of their populations. For example, bioacoustics has been used in the monitoring of amphibian populations because their calls are species specific and vary with environmental conditions. Similarly, passive acoustic monitoring is commonplace in marine mammals, such as whales and dolphins, whose vocalizations carry information on their distribution, behavior, and population trends. In habitat monitoring through bioacoustics, the goal is to find changes in the acoustic environment indicative of habitat quality. For example, a decrease in biophony may indicate habitat loss or fragmentation, while alterations to geophony may suggest changes in the physical environment, such as river flow or wind patterns. Bioacoustic monitoring also offers an effective method for invasive species detection. Acoustic signals show the presence of non-native species before visual detection can, enabling early intervention. Additionally, bioacoustics helps identify critical habitats by mapping regions that possess high acoustic activity and have usually been biodiversity hotspots (Desai and Sharma 2024).

Figure 5 compares the diversity, evenness, and entropy of three acoustic metrics between forest and urban habitats. For all these metrics, the score in the forest habitat is remarkably higher, indicating a far richer and more complex soundscape compared to the one in the urban habitat dominated by human activities. These acoustic metrics indicate the ecological health of forest habitats over their urban counterparts. On the other hand, forests show higher values of diversity, evenness, and entropy, indicating a more balanced and biologically rich environment. Urban habitats reduce the acoustic metrics due to reduced species presence and reduced sound complexity because of anthropogenic noise and habitat

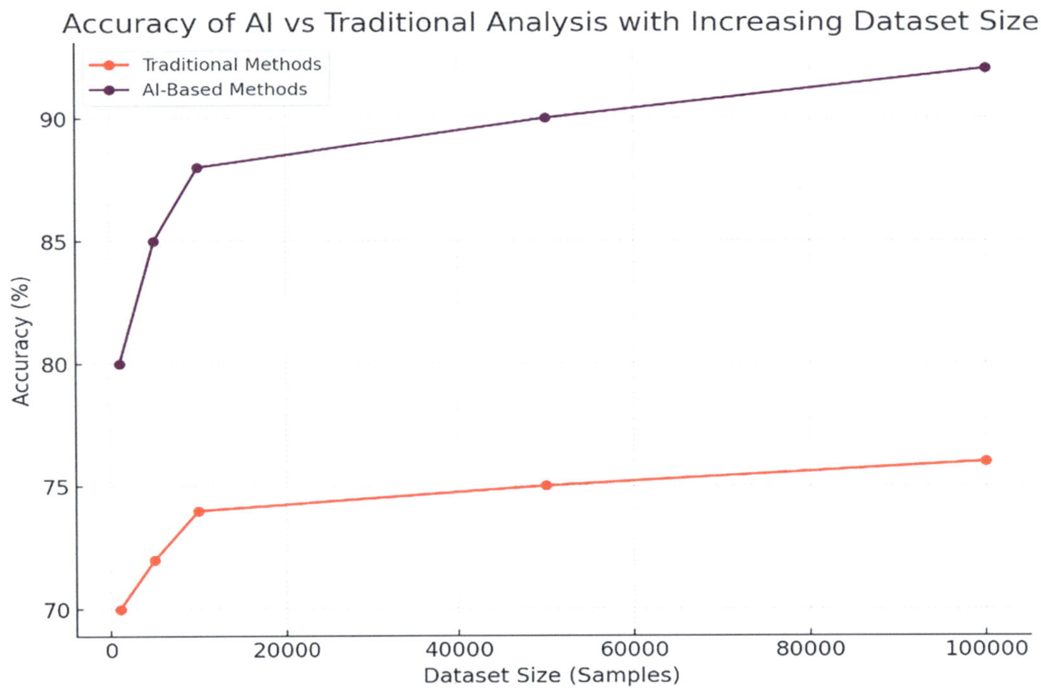

Fig. 4 Accuracy of AI versus traditional analysis with increasing dataset size

Fig. 5 Comparison of acoustic metrics across two habitats

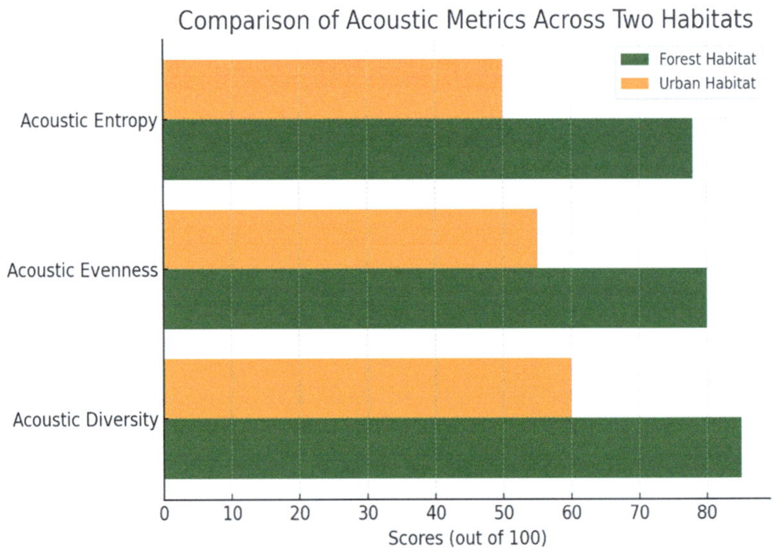

disruption. The comparison thus calls for a need for conservation efforts toward the preservation and restoration of natural soundscapes.

Figure 6 shows the day and night detection percentage of different species, like birds, frogs, insects, and mammals. These results clearly indicate that compared to daytime, frogs and mammals are more detected at nighttime, whereas birds are mostly detected during the daytime. There is a clear temporal difference in activity, with high daytime activities from birds, while at night, the activities for frogs and mammals are higher. Insects were more or less uniformly detected across both periods. The findings highlight the importance of considering round-the-clock monitoring when assessing species' presence and behavior, given that nighttime records of nocturnal species may be underestimated if observations are strictly conducted during daytime.

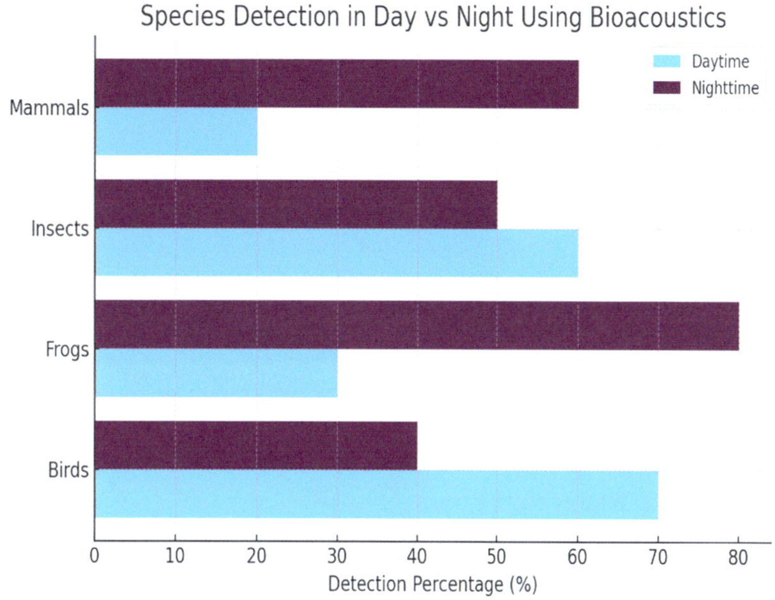

Fig. 6 Species detection in day versus night using bioacoustics

4 Bioacoustics in Climate Change Studies

4.1 Linking Acoustic Changes to Environmental Stressors

In recent times, bioacoustics has emerged as one of the critical tools that help in understanding the changes of climate on ecosystems. This is because climate change due to environmental stressors such as temperature fluctuations, altered precipitation patterns, and extreme weather events influences species' behavior and distribution. The characteristics of these changes are then recorded in the acoustics of the ecosystems. For instance, temperature increases can alter the seasonal phenology of species calls and songs, thus modulating acoustic activity. Birds will most probably change the date of breeding calls as per earlier springs in temperate regions while amphibians may relate to drought periods of tropical seasons to adjust calling events. Similarly, the sound of waterfalls, flowing rivers, and aquatic species will also be controlled by changes in precipitation (Gouvêa et al. 2023).

These various compositions of soundscapes have changed, however, due to human-induced environmental stressors in the form of habitat losses brought about by sea-level rise and deforestation, among other causes. In this manner, changes might be quantified through certain acoustic indices representing a measure of richness or complexity of the sounds, namely acoustic diversity and entropy. With these, by correlating such indices to environmental variables, researchers are able to establish specific stressors that can cause disturbance on biodiversity. Moreover, it can be useful in the monitoring of invasive species that might be related to changing climatic conditions. The detection of novel acoustic signals in an ecosystem often points to the arrival of non-native species that may disrupt local ecological dynamics (Garapati et al. 2024).

4.2 Long-Term Monitoring of Climate Impact on Ecosystems

Long-term bioacoustic monitoring forms the basis for understanding how ecosystems respond to climate change over a long period. Continuous acoustic data collection allows researchers to track trends in species activity, migration, and population dynamics. For instance, the gradual decline in acoustic diversity over decades may indicate the loss of species due to habitat degradation or changing climate conditions. Seasonal variations in soundscapes also act as indicators of ecosystem health. Comparing acoustic data between successive years will reveal the shift of the phenological events: dates of bird migrations or dates of the commencement of breeding seasons of frogs. These shifts are in most instances due to changing climate, having cascading effects through food webs and ecosystem stability. Bioacoustics is used uniquely to monitor sensitive and remote systems—such as polar regions—critical to monitor directly by observation. By showing changes across areas, where seals especially rely on vocals directly with the availability of sea ice, bioacoustic data capture subtle, less recognized effects of this global phenomenon. In these particular applications, such integration with a complement of satellite imagery and environmentally deployable sensors allows for true holism in examining the impacts caused by changes in climate. This multidisciplinary framework enables the researcher to relate acoustic changes with physical and biological processes and therefore provide a comprehensive understanding of ecosystem dynamics (Manikandan and Jayashree 2024).

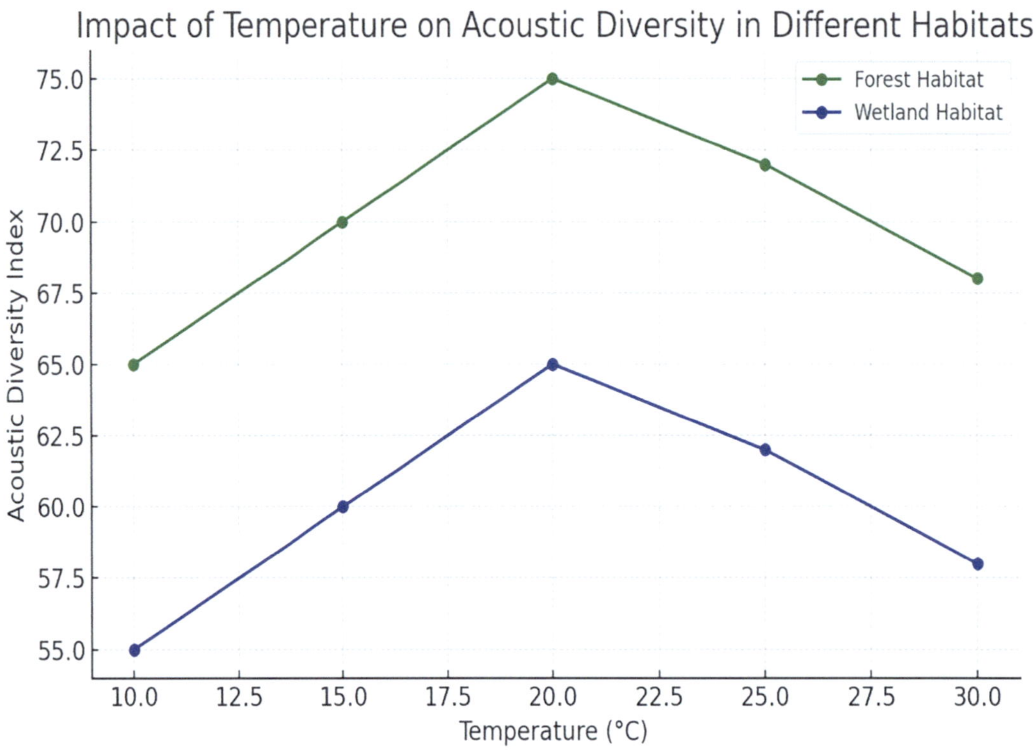

Fig. 7 Impact of temperature on acoustic diversity in different habitats

Figure 7 shows how acoustic diversity varies with temperature in both forest and wetland habitats. The acoustic diversity increases with a rise in temperature up to a certain threshold, beyond which it decreases, at 25 °C for the forest and 20 °C for the wetland. This pattern is indicative of the optimal temperatures for species activity and adverse effects of extreme temperatures. The relationship between temperature and acoustic diversity highlights the sensitivity of ecosystems to climatic conditions. Indeed, forests across the temperature gradient are acoustically more diverse than wetlands, showing a wider range of species and interactions. Diversity in both ecosystems declines beyond their optimum temperature, reflecting a heat stress effect. This goes to suggest that rising temperatures around the world may push many ecosystems past their ecological thresholds beyond which biodiversity and functional integrity would be compromised.

Figure 8 presents the decline in acoustic activity across forest and wetland habitats over two decades. This decline represents the cumulative impact of climate change, including habitat degradation, species loss, and behavioral pattern changes. The continuous decline in acoustic activity for two decades in both forest and wetland habitats shows the long-term effects of climate change. Reductions in acoustic activity indicate a decline in the populations of the species, disruption of phonological cycles, and degraded habitats. Although the forest, which was initially performing better in terms of soundscapes, shows a steeper decline, this could be because it is much more sensitive to longer-term variations in environmental conditions. Under such a scenario, continuous and informed ecological monitoring is required for advancing mitigative conservation measures through strategic interventions against climate variability and its negative impacts on biodiversity.

5 Case Studies in Bioacoustic Monitoring

5.1 Examples of Successful Bioacoustic Monitoring Projects

Applications of bioacoustic monitoring have already been put to work in several ecological projects across the world, showing the potentiality of the technology in tracking biodiversity and environmental health. A case example involves the use of acoustic recorders in tropical rainforests to monitor avian populations. Researchers deployed automated recording units—ARUs—which recorded bird songs across very large and inaccessible areas. From these recordings, they were able to pick up changes in species composition due to deforestation, providing vital data that guides the conservation of the ecosystems (Wrege et al. 2017). The second successful project targeted marine ecosystems. Acoustic sensors were installed to track the vocalizations of

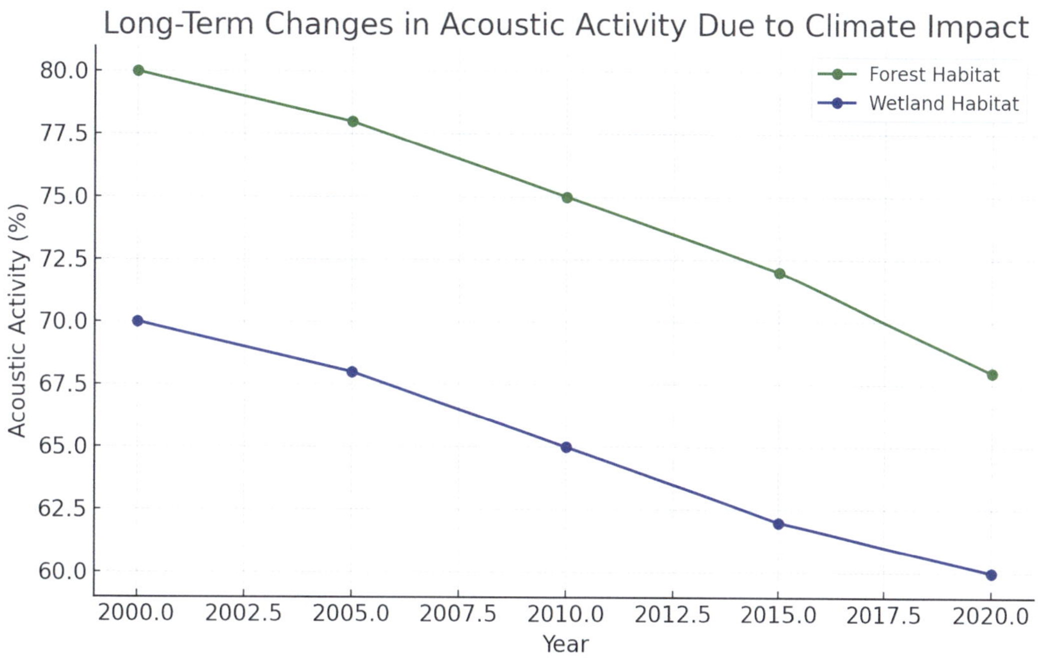

Fig. 8 Long-term changes in acoustic activity due to climate impact

marine mammals, such as whales and dolphins, to understand their migration patterns and habitat use. This method proved especially useful for areas affected by shipping and offshore industrial activities. The results indicated the effects of noise pollution on marine life, which developed noise mitigation measures. Bioacoustics has also been used to track the population of pests in agro-ecosystems. Recordings of insects' sounds show characteristic frequencies when outbreaks of pests occur and hence early intervention with harvest protection is possible. This not only increased agricultural output but also minimized the number of chemical pesticides used. A very good example for urban areas is the assessment of city noise effects on bird populations by bioacoustics. Researchers have compared urban and rural soundscapes, revealing behavioral adaptations among urban birds, including frequency changes to overcome interference with their calls (Kotera and Phillott 2022).

5.2 Lessons Learned and Practical Applications

These case studies offer a number of important lessons in the process of bioacoustic monitoring. First, technology ranging from ARUs and hydrophones enables researchers to capture substantial amounts of data non-invasively and for very extended periods of time. The functionality is important for studying some of the most remote regions on Earth or sensitive ecosystems unaltered by the species being observed. This, therefore, means that interdisciplinary collaboration becomes important. This blend of skills in ecology, acoustics, and data science empowered the development of new solutions, including AI algorithms for species identification. The tools developed considerably reduce both time and effort otherwise used by a human analyst and thus enabled large-scale bioacoustic monitoring. Besides purely scientific applications, there are quite a lot of practical usages for such projects. Bioacoustic monitoring now helps policy decisions on the identification of protected areas and regulation of human activities with implications on ecosystems. Besides, bioacoustic methods are important parts of citizen science projects whereby communities help gather data on different initiatives for conservation (Devi et al. 2025).

Figure 9 provides the value proposition of four bioacoustic monitoring projects on biodiversity and technology utilization. The marine ecosystem projects are ranked highest in both biodiversity and technology utilization, showing advanced uses of technology and substantial contributions to biodiversity conservation. Agricultural and urban bird projects rank a little lower, reflecting the potential for further improvement in these domains. Analysis: This shows the efficiency of bioacoustic monitoring in various ecosystems. The project on marine ecosystems stands out for its advanced deployment of hydrophones and noise mitigation strategies; thus, it is great for biodiversity preservation. In contrast, the projects regarding urban bird monitoring, even though effective, have their effectiveness hindered by high levels of anthropogenic noise.

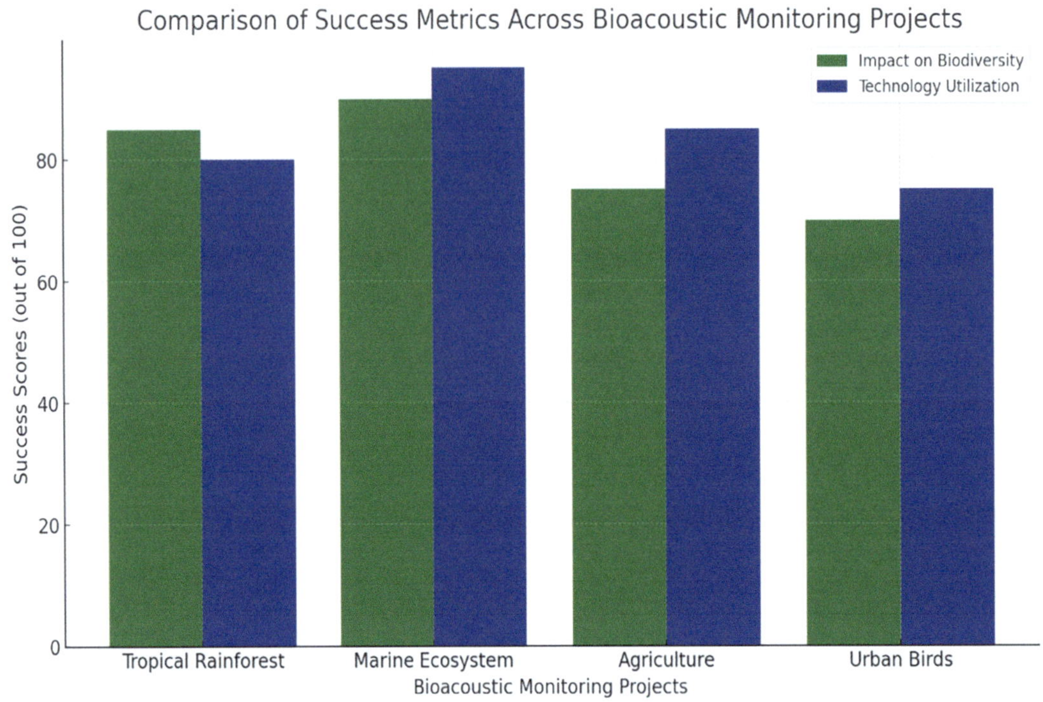

Fig. 9 Comparison of success metrics across bioacoustic monitoring projects

6 Challenges and Future Directions in Bioacoustic Monitoring

6.1 Technical, Ethical, and Logistical Challenges in Bioacoustics

While bioacoustic monitoring is a very promising technique, there are several challenges that may be limiting its effectiveness and scalability.

Technical Issues/Challenges

One major technical issue is the high volume of data produced by acoustic monitoring devices. Long-term recording can easily result in terabytes of audio data, which is extremely resource-intensive to store, process, and analyze. Efficient algorithms for automatic sound identification remain one of the main challenges in the presence of overlapping sounds or noisy conditions. Besides, equipment robustness when exposed to harsh environmental conditions—for example, tropical rainforests or marine ecosystems—can reduce the reliability of data gathering (Hoyos Cardona 2024).

Ethical Challenges

However, bioacoustic monitoring raises several ethical concerns, especially regarding its application in human-impacted areas. For instance, acoustic devices record human conversations even when it is not the intention of the study, hence a violation of privacy. Additionally, with the increase in such studies, there will be a need for guidelines on ethical issues that ensure these collections do not harm wildlife or alter their natural behaviors. There has to be transparency in the use and sharing of the bioacoustic data as a means of helping gain the trust of the local people and other stakeholders (Johnson et al. 2023).

Logistical Issues

Deployment and maintenance of recorders in remote or inaccessible areas pose a number of logistical challenges, including transport of equipment, reliable power supplies, and retrieval of data without device damage. Besides, access to quality recording equipment and analysis tools can be expensive, especially in projects from developing countries or with limited funding (Goodwin et al. 2022).

6.2 Future Trends and Innovations in Bioacoustic Monitoring

The field of bioacoustics is rapidly changing—innovations are trying to reach most of the fields which, at the time, present some kind of limitations.

AI and Machine Learning

Improvement in artificial intelligence and machine learning is rapidly changing the analysis of bioacoustic data. It is possible for the current machine learning models to identify species based on vocalizations with very high precision, even from complex soundscapes. The analytics from these new tools are really fast to process, even with large volumes of data for almost near real-time monitoring. This limits manual analysis; therefore, a large amount of field data to be recorded and analyzed over a pretty extensive area for longer (McGinn et al. 2023).

Edge Computing and Cloud Integration

This is made possible by the adoption of edge computing, which will, therefore, minimize the need to transmit and store data; the processing of acoustic data can be done directly on recording devices. This enables the seamless sharing of data with the cloud for collaboration in research across geographic boundaries.

Miniaturization and Reduction of Cost

New sensor technologies have provided the opportunity for smaller and cheaper devices to record, making this discipline more deployable, feasible, and accessible. Results: Increasingly compact sensor technology makes bioacoustic monitoring easier and increasingly used by both researchers and by citizen scientists (Ghani et al. 2307).

Integration with Other Technologies

Coupling bioacoustics with other monitoring technologies, like satellite imagery and environmental sensors, holistic approaches are achieved in ecosystem studies. For example, the combination of bioacoustic data with climate models may show more precisely how changing conditions are impacting biodiversity.

Community-Based Monitoring

Other future directions for bioacoustic monitoring involve the engagement of local communities. This can be achieved through citizen science approaches that allow people to take ownership and responsibility in the conservation of nature by collecting and analyzing acoustic data themselves (Munger et al. 2022).

Expanding Applications

Besides monitoring biodiversity, bioacoustics now finds its applications in fields as far-reaching as agriculture for pest outbreak detection and in urban planning for dealing with noise pollution. Applications of bioacoustics to health and well-being research, including studies of the effects of natural soundscapes on mental health, are also emerging.

7 Role of Bioacoustics in Urban Ecology

7.1 Monitoring the Impact of Urbanization on Wildlife

This has taken its toll on the natural world-urbanization transformed landscapes into fragmented habitats that put several constraints on wildlife. The construction of roads and buildings into natural habitats creates fragmentation in the habitats of animals, introducing anthropogenic noise to their environments and disrupting behaviors and communications of animals. Bioacoustics have become an essential tool to monitor these impacts, thus providing insight into how species adapt or decline in these ever-changing urban environments. For example, many urban bird species change their song by increasing frequency to cut through the constant low-frequency background noise generated by traffic and machinery. This may help in improving communication, but could eventually reduce the diversity and complexity of calls, further altering mating and territorial behaviors. Similarly, other animals such as bats may adjust the pattern of echolocation with a view to navigate the urban structure, influencing hunting efficiencies. Bioacoustics can also help in identifying species that are urban exploiters—those that thrive in cities—and urban avoiders, those that do not. Recordings from green spaces, parks, and urban fringes can tell researchers which species are resilient to urbanization and which need conservation attention (Goodwin et al. 2022).

7.2 Ecological Health Assessment in Cities Using Soundscapes

Soundscapes, therefore, give a holistic approach to evaluating ecological health within urban areas. A soundscape consists of all the sounds in any given environment, including biophony, or animal sounds; geophony, consisting of natural phenomena such as wind and water; and anthrophony, comprising human-made sounds. Changes in these sound types can indicate an ecological imbalance. For example, for an urban park, high biophony, like birds singing and insects calling, would indicate a rather good and healthy ecosystem with acceptable biodiversity. On the other hand, a soundscape composed of anthrophony mainly without biophony may denote poor ecological health. Bioacoustic monitoring over time can detect changes in soundscapes, offering a nonintrusive means of assessing and monitoring the effectiveness of measures taken in urban greening project implementation and biodiversity. Acoustic properties, such as the use of bioacoustic indices [ABI], Acoustic Complexity Index, and Normalized Difference Soundscapes Index, reflect the diversity of sounds making up the soundscape within a given habitat. Analysis of these indices allows monitoring of "acoustic hotspots" that support vast species diversity and provides real value to urban planning informed by conservation imperatives at both local and landscape planning levels (Lavner et al. 2024).

7.3 Application and Implication

In addition, there is practical application of urban ecology on issues that go beyond research alone; for instance, analyses done on soundscapes give guidance on proper designs of noise barriers and spacing of green areas, in the end enhancing biodiversity. Bioacoustics also assists policy makers by providing evidence that is credible to support regulations put forth concerning noise pollution.

8 Policy and Conservation Implications of Bioacoustic Monitoring

8.1 Translating Bioacoustic Data into Actionable Policies

Bioacoustic monitoring thus gives out immense data on status and trends, forming a vital basis for policy decisions toward the protection of ecosystems and biodiversity. In general, bioacoustics provides concrete evidence of changes in ecology and human impact by capturing sounds from wildlife and environmental conditions. Such information can be used to shape policy decisions across sectors, ranging from conservation to urban planning. For example, bioacoustic monitoring creates the potential to locate critical habitats through specific calls of endangered species, on which policymakers can take further action by setting aside these places as protected areas or bringing in regulations that limit certain human activities within those places, such as logging or mining. In a similar way, acoustic data regarding the impacts of noise pollution on wildlife would also result in noise abatement policies, including quiet zones in urban areas or industrial operations that employ noise-reducing measures. The noninvasiveness of bioacoustic monitoring also makes it apt in sensitive areas where more conventionally invasive surveying may otherwise be disruptive. This gives politicians the ability to gain material from locations where human access can be minimal or a task not feasible. Furthermore, the use of bioacoustic data within geographical information systems enables fuller achievement of spatial planning to coalign with sustainable development programs (Sievers et al. 2024).

8.2 Contributions to Global Biodiversity Conservation Goals

It directly contributes to global goals of biodiversity conservation, such as those specified within the Convention on Biological Diversity and the SDGs of the United Nations. Bioacoustic monitoring is a means of detailed insights into ecosystem dynamics for targeted conservation strategies. The monitoring of biodiversity at large scale and long-time frame is one of the essential contributions. Acoustic surveys, in turn, offer a valid tool for monitoring population change, changes in community composition as an early warning of loss, and reach goals of halting extinctions and restoration of degraded ecosystems. It is also a part of adaptation and mitigation strategies involving climate change. For example, monitoring of climate-induced habitat changes on species distribution provides valuable information toward adaptive management practices. Similarly, data on acoustic activity in forests and wetlands help to assess their roles as carbon sinks and support efforts in combating climate change. Another important contribution is to international collaboration. These bioacoustic networks will allow the sharing of data on a global scale and, hence, facilitate concerted actions by practitioners in tackling transboundary conservation issues. These networks also contribute to capacity building through providing equipment and training in regions where such resources are scarce.

9 Conclusion

Bioacoustic monitoring has emerged as a game-changing tool in the study and conservation of biodiversity. This new approach offers unparalleled insight into ecosystem dynamics, species behavior, and environmental changes by capturing and analyzing the sounds of nature. Bioacoustics has proven its versatility in addressing pressing ecological challenges—from monitoring biodiversity loss to understanding the effects of climate change and human activities—on terrestrial, aquatic, and urban landscapes. In this journey of discovery, the role and potential of emerging technologies, especially artificial intelligence and machine learning in enhancing efficiency and precision in bioacoustic data analysis, have been indicated. These innovations are expanding the applicability of bioacoustics to a far larger range of ecological and conservation contexts, aside from making the processing of enormously large datasets more efficient. The uses of bioacoustic devices range from the detection of endangered species in remote habitat areas to the assessment of the health of urban soundscapes, driving insights into action and empowering the making of informed decisions. Bioacoustics, besides its huge potential, involves technical, ethical, and logistical challenges to be surmounted for full capacity. Data management, equipment access, and the ethics involved in data collection are a few of the issues arising, which raise the need for continuous research, interdisciplinary collaboration, and development of standard practices. In the future, bioacoustics is likely to play an even greater role in global conservation. It provides real-time, non-invasive, and scalable monitoring solutions that meet or closely align with most international goals on biodiversity targets and sustainable development initiatives. Coupled with integrating bioacoustic monitoring with other rapidly developing technologies and community-based engagement, this can provide an integrated, inclusive approach toward conservation.

References

Bobba KS, Krishna Sai V, Bugga D, Surendra Bolla VM (2024) AI-enhanced acoustic analysis for comprehensive biodiversity monitoring and assessment. arXiv preprint arXiv:2410.12897

Borker AL (2018) Applying ecoacoustics to bird conservation and monitoring. PhD diss., UC Santa Cruz

Desai K, Sharma N (2024) Harnessing the power of machine learning to predict and mitigate the impacts of climate change on global ecosystems. In: 2024 international conference on advances in computing research on science engineering and technology (ACROSET). IEEE, pp 1–6

Devi GD, Kumar N, Manikandan J, Rekha V (2025) Innovative brain tumor detection: stacked random support vector-based hybrid gazelle coati algorithm. Biomed Signal Process Control 101:107156

Goodwin M, Halvorsen KT, Jiao L, Knausgård KM, Martin AH, Moyano M, Oomen RA, Rasmussen JH, Sørdalen TK, Thorbjørnsen SH (2022) Unlocking the potential of deep learning for marine ecology: overview, applications, and outlook. ICES J Marine Sci 79(2):319–336

Garapati D, Burla N, Shaik N, Praveen Kumar K, Yamarthi NR (2024) Integrating deep learning techniques for wildlife species identification using vocalization analysis. Lib Progr Int 44(3):28784–28794

Ghani B, Denton T, Kahl S, Klinck H (2023) Feature embeddings from large-scale acoustic bird classifiers enable few-shot transfer learning. arXiv preprint arXiv:2307.06292

Goodwin M, Halvorsen KT, Jiao L, Knausgård KM, Martin AH, Moyano M, Oomen RA, Rasmussen JH, Sørdalen TK, Thorbjørnsen SH, Vadimus Q (2022) Unlocking the potential of deep learning for marine ecology: overview, applications, and outlook

Gouvêa TS, Kath H, Troshani I, Luers B, Serafini PP, Campos IB, Afonso AS et al (2023) Interactive machine learning solutions for acoustic monitoring of animal wildlife in biosphere reserves

Growcott J, Lobora A, Markham A, Searle CE, Wahlström J, Wijers M, Simmons BI (2024) The secret acoustic world of leopards: a paired camera trap and bioacoustics survey facilitates the individual identification of leopards via their roars. Remote Sens Ecol Conserv

Hoyos Cardona LA (2024) Addressing the nocturnal problem: the ecology of cryptic birds through bioacoustics in the neotropics. PhD diss., Universidad Nacional de Colombia

Jayanthi K, Chitradevi D, Muthaiyah S (2025) Early prediction of harmful environmental changes based on bioacoustics analysis. In: Technological advancements for deep sea ecosystem conservation and exploration. IGI Global Scientific Publishing, pp 129–148

Johnson E, Campos-Cerqueira M, Jumail A, Yusni ASA, Salgado-Lynn M, Fornace K (2023) Applications and advances in acoustic monitoring for infectious disease epidemiology. Trends Parasitol 39(5):386–399

Kotera MM, Phillott AD (2022) Calls for conservation: a review of bioacoustics monitoring with case studies from India. Asian J Environ Ecol 19(4):142–150

Lavner Y, Melamed R, Bashan M, Vortman Y (2024) The bioacoustic soundscape of a pandemic: continuous annual monitoring using a deep learning system in Agmon Hula Lake Park. Eco Inform 80:102528

Manikandan J, Jayashree K (2024) Enhancing lung nodule classification: a novel CViEBi-CBGWO approach with integrated image preprocessing. J Imag Inf Med 37(5):2108

McGinn K, Kahl S, Zachariah Peery M, Klinck H, Wood CM (2023) Feature embeddings from the BirdNET algorithm provide insights into avian ecology. Ecol Inf 74:101995

Mcloughlin MP, Stewart R, McElligott AG (2019) Automated bioacoustics: methods in ecology and conservation and their potential for animal welfare monitoring. J R Soc Interface 16(155):20190225

Metcalf OC (2021) Acoustic monitoring of Amazonian wildlife in human-modified landscapes. PhD diss., Manchester Metropolitan University

Müller J, Mitesser O, Martin Schaefer H, Seibold S, Busse A, Kriegel P, Rabl D et al (2023) Soundscapes and deep learning enable tracking biodiversity recovery in tropical forests. Nat Commun 14(1):6191

Munger JE, Herrera DP, Haver SM, Waterhouse L, McKenna MF, Dziak RP, Gedamke J, Heppell SA, Haxel JH (2022) Machine learning analysis reveals relationship between pomacentrid calls and environmental cues. Mar Ecol Prog Ser 681:197–210

Nel H, Mishra AK, Schonken F (2024) EcoSonicML: harnessing machine learning for biodiversity monitoring in South African wetlands. SN Comput Sci 5(5):1–9

Oestreich WK, Oliver RY, Chapman MS, Go M, McKenna MF (2024) Listening to animal behavior to understand changing ecosystems. Trends Ecol Evol

Ross SRP-J, O'Connell DP, Deichmann JL, Desjonquères C, Gasc A, Phillips JN, Sethi SS, Wood CM, Burivalova Z (2023) Passive acoustic monitoring provides a fresh perspective on fundamental ecological questions. Funct Ecol 37(4):959–975

Sievers BL, Siegers JY, Cadènes JM, Hyder S, Sparaciari FE, Claes F, Firth C, Horwood PF, Karlsson EA (2024) "Smart markets": harnessing the potential of new technologies for endemic and emerging infectious disease surveillance in traditional food markets. J Virol 98(2):e01683-e1723

Wrege PH, Rowland ED, Keen S, Shiu Y (2017) Acoustic monitoring for conservation in tropical forests: examples from forest elephants. Methods Ecol Evol 8(10):1292–1301

Zilli D (2015) Smartphone-powered citizen science for bioacoustic monitoring. University of Southampton, PhD diss.

Human-Computer Interaction and Multimodal Analysis with Bioacoustics

Integrating Bioacoustic Sensors for Improved Human–Computer Interaction

M. Shanmugham, G. Sekar, S. Kavitha, S. D. Vijayakumar, and Chithirai Pon Selvan M.

Abstract

The integration of these bioacoustic sensors in AAL environments into multimodal human–machine interaction systems provides nonintrusive real-time monitoring of vital signs and enhances the detection of falls. The biological patterns of sounds include respiratory patterns, coughing, and heartbeat, which are captured by the bioacoustic sensor and presented with information regarding vital health status of a person. This characteristic is very beneficial for older adults or patients with chronic medical conditions, as it allows continuous health monitoring without having to physically interact with intrusive medical devices. Bioacoustic sensors take advantage of acoustic irregularities that could represent early manifestations of respiratory and cardiovascular illnesses, such as unusual patterns of breathing or aberrant sounds in the heart. Furthermore, the auxiliary modalities of motion sensors and visual data significantly enhance the performance of fall-detection systems. The acoustic features typically characteristic of a fall can include sudden impact or voice changes, which may be captured better by integrating the data from acoustic sensors with that from motion and visual data. This helps minimize the occurrence of false positives and ensures that responses to real emergencies are immediate and appropriate. The falling can be assessed how major it is by getting its vital signs after the incident through sensing breathing rates and the like. In that regard, the system can now differentiate minor falls from emergency health cases, hence making more intervention occur promptly by notifying caregivers or health responders. This is the way resources are rationed; unnecessary interventions that had to take place ensured but one got attention soon as possible when the fall had been severe. This creates longitudinal health data for the healthcare providers to identify trends and monitor early signs of chronic diseases, besides allowing for preventative healthcare interventions that can lead to improvement in the general well-being of individuals in AAL environments. In conclusion, bioacoustic sensors integrated into multimodal HMI systems will ensure a complete and non-invasive method of continuous health monitoring and emergency response in AL environments. This will considerably improve the quality of care for most elderly patients and patients suffering from chronic conditions since fall detection is enhanced to prompt swift intervention, above all, health data that could be gained over time.

Keywords

Bioacoustic sensors · Multimodal human–machine interaction (HMI) · Ambient-assisted living (AAL) · False-positives · Longitudinal health data

M. Shanmugham (✉) · S. Kavitha · S. D. Vijayakumar
Department of ECE, Nandha Engineering College, Erode, India
e-mail: shanmuecengg@gmail.com

S. Kavitha
e-mail: deanece@nandhaengg.org

S. D. Vijayakumar
e-mail: vijayakumar@nandhaengg.org

G. Sekar
Department of ECE, V.S.B. College of Engineering Technical Campus, Coimbatore, India
e-mail: gsekar@vsbcetc.com

M. Chithirai Pon Selvan
Department of AI&DS, Nandha Engineering College, Erode, India

Head of School, Science and Engineering, Curtin University, Dubai, UAE

M. Chithirai Pon Selvan
e-mail: pon.selvan@curtindubai.ac.ae

K. Subbaraj et al. (eds.), *Computational Bioacoustic Artificial Intelligence*, Sustainable Artificial Intelligence-Powered Applications, https://doi.org/10.1007/978-3-032-06088-4_11

1 Introduction

The creation of ambient-assisted living (AAL) environments is a big leap in healthcare and life support in the overall aim of making life easier for people, especially older people and those with disabilities. Advanced technologies that will be integrated into the residential environment will enable developing intelligent and flexible user-centered environments supporting autonomy and security and thus enabling optimal holistic well-being via AAL systems. Ambient-assisted living addresses an infusion of developments to everyday living environments with which a constant monitoring, support and intervening is possible. Organic structure of sensors, actuators, and communication technologies could be networked over Internet of Things (IoT) (Dey et al. 2019). Altogether, when combining these instruments, it leads to capturing information, processes, and generates responses concerning appropriate issues pertaining to user requirements or situational conditions. A remarkable feature of AAL systems is that they can work completely non-intrusive, thus reducing the user's effort while increasing overall utility. For example, smart sensors in a home can monitor vital signs, analyze movement behaviors, or detect fall incidents. These data are then analyzed by machine learning algorithms to check for any patterns, predict potential health risks, or recommend lifestyle changes. These are quite important to use in the health industry due to offering a remedy to chronic disease management, postoperative care, and even psychoneurosis management. This makes reduction possible at health facilities, enabling the use of early interventions with reduced workload; they help patients get the desired outcome from health. Also, these AAL systems make living better due to automated day-to-day life processes like lightings, temperatures, medication, or home security. Though there is tremendous transformative potential, its realization is not free from numerous challenges. Ensuring that privacy and security are given precedence is paramount, for this system is essentially managing the sensitive personal information of citizens. Adaptive systems require to be user-centered designs if these systems are going to support diverse cultural and individual preferences. This technology will help scale further AAL environments by the introduction of innovations through personalization with AI, wearable sensors, and robotics. The advancement promises to uniquely change the way people live assisted living to make room for more livings in a healthy, safe, and independent life in their environment.

1.1 Bioacoustic Sensors in Ambient-Assisted Living

Environments designed for ambient-assisted living (AAL) allow the elderly and other disabled people to live independently by permanent monitoring and intervention carried out by high-tech instruments. The biological sounds gathered and processed by Bioacoustic sensors play an important role in improving these environments as they provide ample information about the status of a person's health and behavioral habits (Kavitha et al. 2014).

Such sensors are able to sense the sound of breathing, coughing, and even heartbeat, making them very viable non-invasive monitoring equipment to monitor physiological conditions in real time. However, applying bioacoustic sensors comes with significant opportunities as well as challenges in AAL environments.

1.2 Opportunities and Challenges of Bioacoustic Sensors in AAL

The bioacoustic sensors can monitor heart rates and respiratory frequency. These are of extreme value to the AAL system as they are silent sensors operating in the background, gathering data without actually involving the user. This should sense any changes in normal breath sounds or new unfamiliar auditory presentation of physiological sounds; therefore, triggering automatic alerts to a caregiver or emergency response systems. For example, bioacoustic sensors can monitor a person's breathing patterns or distinguish anomalies by respiratory depth or frequency, which is suggestive of respiratory distress. It is one of the significant health concerns for the elderly, and the use of bioacoustic sensors will improve the timeliness of detection. This should sense any changes in normal breath sounds or new unfamiliar auditory presentation of physiological sounds; therefore, triggering automatic alerts to a caregiver or emergency response systems. It also assesses the heart function of the patient postfall to determine the magnitude of the fall and thus sets the right emergency intervention (Das et al. 2015). Bioacoustic sensors will allow the caregiver and the healthcare staff to obtain direct information regarding the health status of the user. Then, this can lead to personalized care. For example, in respiratory disease like asthma or COPD, bioacoustic sensors track the frequency and amplitude of coughing. Thus, early treatment becomes apparent upon the onset of the said symptoms. Currently most sensor technologies are highly invasive on several counts. Bioacoustic sensors make location inconspicuously possible at inconspicuous points in the house. In this respect, therefore, it is highly relevant to user comfort and dignity, not least among the elderly potential end-users, who tend to be sensitive about invasive surveillance and its implications for social stigma. Such sensors being implanted in devices that are practiced daily, like the smart speakers and lighting systems, makes the device to be as effective to individuals.

Another well-known limitation of bioacoustic sensor function is the interference of ambient noise. It usually means external auditory sounds: a background conversation, TV music, traffic noise, which might dgrade the functionality and efficacy of biological sounds by sensors. Other interfering factors could be an inappropriate design of the habitat in regard to sensors; more mobile than average users whose activity could compromise the integrity of data acquisition. Bioacoustic sensors would collect sensitive health information so the privacy and security problem is to be solved first in terms of it, that the user data transmitted into networks in AAL would pose a potential threat or flaw where those hackers' opportunities. All precautions needed regarding data encryption and restricted entry based upon permissions are necessary for respect toward users' privacy besides complying with healthcare rules in society. Although bioacoustic sensors provide a wealth of health-related information, their use often requires them to be integrated with other sensor types, such as motion sensors, temperature sensors, or cameras, to form multimodal ambient-assisted living systems. A great deal of technical effort and the development of suitable algorithms are also necessary to ensure that integration is seamless and that different forms of data are synthesized in real time. Multimodal systems are also designed to offer user interfaces that improve ease of their use in everyday living. For the system to work effectively in AAL settings, the user should feel comfortable with the system. Thus, users and caregivers should be sensitized on the advantages and operations of the technology. Further, users are reluctant to use such systems as they are viewed as surveillance and the feeling of always being watched (Singh et al. 2012).

2 Fundamental Concepts

2.1 Pattern Identification in Biological Acoustics for Health Monitoring

Bioacoustic sensors record diverse biological auditory patterns, which provide significant insights into the health status of an individual. Acoustic signals associated with respiration, coughing, and cardiac functions can be used as potential markers of physiological conditions and can greatly help in early detection, continuous monitoring, and urgent interventions within the healthcare domain. Continuous, nonintrusive and noninvasive health monitoring opportunity, because the introduction of biological sound into an AAL environment. A very precious biological signal is respiration that varies easily with the health status; dyspnoe or stridory with an exaggerated motion might relate to respiratory diseases such as asthma, pneumonia, and others. Bioacoustic sensors developed for frequency and rate of breathing along with an amplitude can monitor slight variations that may well predict acute conditions, thus providing very useful time for doctors to react. Coughing is a biological sound that has great diagnostic value. A constant or nonproductive cough may be a manifestation of an underlying medical condition; this can be viral infections, asthma, or pulmonary diseases such as tuberculosis. The bioacoustic sensors can monitor the patterns of coughing and differentiate between productive and dry types of coughs, making it possible to detect respiratory illnesses early. Sensors are deployable in an AAL framework monitoring the number of cough episodes as well as the duration of cough, therefore alerting caregivers or health professionals if such variables move beyond thresholds set. Bioacoustic sensors can be used in wearable devices as well as ambient systems in assisted living to ensure uninterrupted monitoring of the heart functions. This will lead to sudden peaks and troughs, including tachycardia or bradycardia as well as other arrhythmias with abnormal sounds that can be fast tracked to reduce the likelihood of any serious medical complications. The sensor offers a noninvasive method to track cardiac health, particularly among the elderly population who would not have had the means for constant clinical follow-up (Yashpalsing Chavhan and Dhore 2010; Cano et al. 2020).

2.2 Integration of Biological Acoustic Patterns into Multimodal Systems

The integration of respiratory, coughing, and cardiac acoustics into a holistic multimodal health monitoring framework provides an effective means for individualized healthcare provision in ambient-assisted living applications. Information from bioacoustic features can be integrated with other sensors-such as fall detection systems, thermal sensors, or voice recognition equipment-for holistic AAL application-providing an overall view of the health condition of the user in question. Integration of various sources of data will eventually culminate into more advanced diagnostic accuracy, timely intervention, and more enhanced support systems for caregivers. Biological auditory patterns can be analyzed for anomalies using advanced signal processing technologies in conjunction with machine learning algorithms.

3 Multimodal Systems for Human–Computer Interaction

Multimodal HCI systems attempt to enhance user interaction through multiple modes of input and output like voice, gestures, touch, bioacoustic signals, to name but a few. In

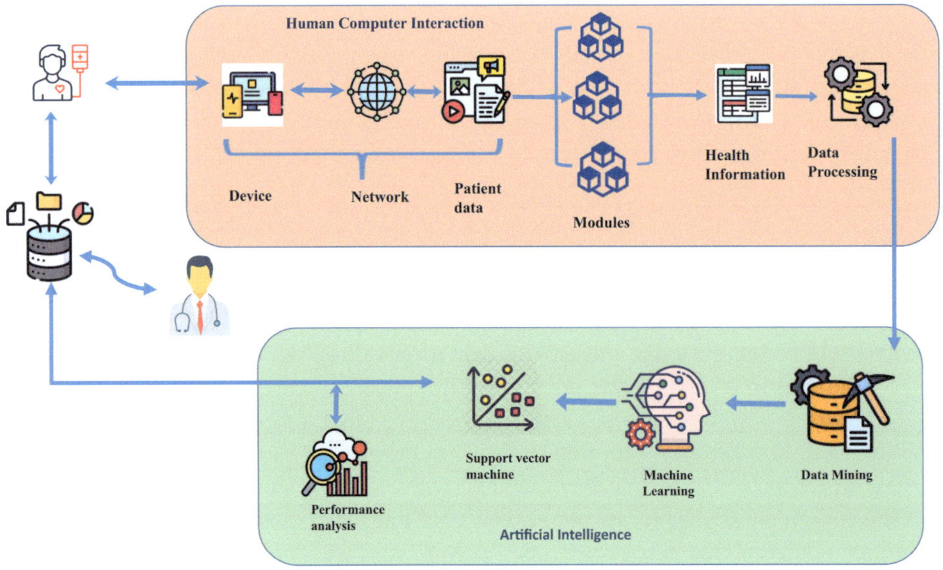

Fig. 1 Multimodal systems for human–computer interaction

ambient-assisted living, multimodal systems are extremely important in enhancing the quality of life for the elderly or disabled since it makes users interact smoothly and intuitively with the technology. Then, health monitoring and support services in terms of technology accessibility and efficiency would increase. The integration of bioacoustic sensors with conventional interaction technologies, such as voice recognition or gesture recognition creates a richer experience while involving the user.

These bioacoustic sensors can monitor very subtle physiological markers, like changes in respiratory and heart rates, which are necessary for monitoring a person's health status in real time. Once these signals are aggregated with other data inputs, it can enable a much more complete and contextually aware interaction because the system will have a better sense of what the user actually needs (Fig. 1).

In the case of AALs, the promise is rather one of support without intrusion; and in this regard, multimodal systems play a very important role in achieving the balance between responsiveness and subtlety. For instance, a multimodal HCI can recognize whether the user's speech shows some sudden change in their respiratory pattern or coughing, pointing to some health problem. This would include alerting a caregiver, beginning to initiate a health assessment, or nothing if the person was perfectly healthy—all of this could be done automatically without needing the user's input. And, in the case of a fall, unusual sounds detected by the bioacoustic sensors like the inability to breathe normally or irregular heartbeat would prompt and interact with other sensor types including accelerometers and pressure sensors, creating a clear image of what's happening so that proper procedures would start. This integration of the modalities within a framework will bring greater functionalities to personalized interaction. The user will differ in preferences and needs. For example, some users may prefer voice commands because they become easier to use while others may prefer gesture-based inputs when their hands are occupied in other things. Such sensors can continuously monitor vital signs; thus, the system can identify trends in an individual's health and respond accordingly with suitable assistance in a timely manner. This comes with cautions in the design of multimodal HCI systems for AAL due to their nature regarding user experience. This intuitively should be an easy user interface, accessible to different technological literacy levels. The system needs to react promptly and accurately in respect of inputs as it tries to keep trust and reliability; else, managing sensitive health data may turn out quite dicey. False-positives must be kept at bay lest unnecessary alarms or reactions result, more so in a health context. Bioacoustic sensors would bring cohesion and adaptability into the environment. Traditional human–computer interaction techniques are used that include speech, gesture, or touch inputs. Improving quality of care comes with better user satisfaction and, more significantly, with ever-changing technologies developed to meet user needs (Golestan et al. 2019).

4 Components of the Sensor Integration Framework

Multimodal human–computer interaction systems in ambient-assisted living (AAL) settings will feature the integration of bioacoustic sensors. Systematically designing the above framework would result in integrating hundreds of data streams. There is a necessity to process and synchronize these inputs to ensure information derived could be responded to in time about health and emergency situations of the users. It would then seek to develop an innovative adaptive system in order to help an individual feel secure but also address his medical requirements through the assimilation of studying biological auditory patterns with various other sensory inputs.

4.1 Data Collection and Sensor Integration

Bioacoustic sensors capture all key biological sounds that represent key health indicators, such as heartbeats, breathing patterns, episodes of coughing, or fall incidents. These sensors can be in the form of microphones or wearable devices or embedded into smart furniture or flooring systems in any environment. The other type is the integration with others; this includes motion detectors, temperature sensors, or cameras; therefore, holistic health condition assessment and evaluation of their environmental setting are facilitated. This part of the framework is crucial in developing the integrated interfaces between different sensor types and ensures that all data are available in real time.

4.2 Noise Cancelation and Signal Enhancement

Bioacoustic signals are susceptible to interference from environmental noise or artifacts due to physical movement or other exterior influences. Hence, preprocessing techniques are essential to optimize and enhance the quality of bioacoustic recordings. Filtering algorithms are applied in order to filter out unwanted frequencies and feature extraction focuses on critical biological sounds such as fast heart rate or breathing.

4.3 Data Integration and Synchronization

All the data in the process are taken as inputs from different sensors. Notably, these range from bioacoustic data to visual and motion data. In theory, it aggregates all those inputs into one model to encompass this condition of the user on a comprehensive basis. For instance, bioacoustic signals can display abnormal breathing or heart rate patterns that could be more easily interpreted with motion data, like the user falling. This way, real-time synchronization ensures that data from sensor groups are processed and temporally aligned so meaningful conclusions can be drawn from these data (Banda and Robinson 2011; Golestan et al. 2019).

4.4 Data Analysis and Conclusion

Perhaps the most basic characteristic of the multimodal system is its ability to analyze integrated data in search for potential health risks or events. It can utilize machine learning algorithms that are able to recognize patterns in bioacoustic signals, such as those associated with irregular heartbeat or labored breathing, and then cross-reference them with other variables like changes in body position or jerky movements.

4.5 User Interface and Feedback

Once the analysis is done, the research findings have to be communicated in a manner that is understandable to the targeted users. For caregivers, health care professionals, or even family members of the patient, the alerts received must be based on the health anomalies identified by the system. These could be sent directly from their phones into their desktops or may even trigger alerts in the local environment. Ambient feedback to the user—from visual signals to auditory cues—may be offered for confirmations of awareness of the existence of the system's observations and any important actions taken by it.

4.6 Security and Privacy Concerns

The collection and analysis of sensitive biological data require robust compliance with best privacy standards and regulatory frameworks such as HIPAA in the health sector. This means, therefore, that the established framework must protect the confidentiality and integrity of data transmissions and storage through authentication mechanisms of users with regard to their personal health information. Moreover, encryption and anonymization can prove to be good strategies to further enhance the privacy of the user especially in a public or collective setting.

5 Ambient-Assisted Living System Designing Using Bioacoustic Sensors

An AAL system with bioacoustic sensors would, therefore, be a great opportunity for elderly or disabled people to increase their independence and quality of life. Such systems are constructed to create an environment where technology is seamlessly integrated to become part and parcel of daily activities but still provides both physical and emotional support. Bioacoustic sensors incorporated in these systems can heavily complement real-time health monitoring systems to be able to detect health-related incidents like respiratory complications, falls, and cardiovascular events early on, with prompt feedback given to caregivers or family members (Elboushaki et al. 2020) (Fig. 2).

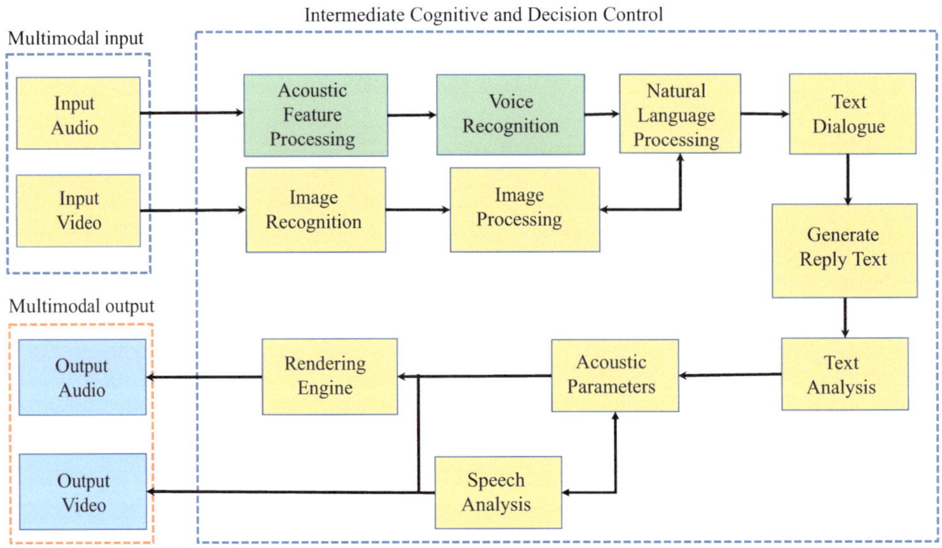

Fig. 2 Design of ambient-assisted living by multimodal data processing

5.1 System Architecture

Design the system in a living environment and user need must be modular and scalable. Sensor networks provide the core for the system integrating bioacoustic sensors, which can pick up on biological patterns, such as breathing, coughing, and cardiac activity; therefore, they need to be installed in the environment in a nonintrusive and continuous fashion. For instance, it is also possible to fit sensors within a user's clothing, for example, bed, chair, and even in the walls. Sensor data is collected and processed in real time with a central processing unit or edge computing device. Edge computing is one example of processing unit which will distinguish normal signals from warning signs, such as wheezing or abnormal heartbeats. These signals therefore require complex signal-processing algorithms to clear noise and retrieve meaningful health indicators. The use of machine learning methods will improve the system's ability to give anomaly detection through knowledge learned from data gathered over time. Integration with multimodal interfaces one characteristic of AAL systems would be multimodal interaction, making the system intuitive and user-friendly to different levels of users. This is because other modes like gesture recognition, speech input, and facial expression may facilitate user interactions with the system in more natural ways. For example, if an anomaly has been detected in the sounds of the user's respiratory or cardiac system, the system shall alert the user with a voice command or via a gesture-based interface.

5.2 User-Centered Design and Accessibility

In designing this AAL system, software has to user-friendly in adaptability to differ for each person. Given those ideas, implementing principles about user-centered design guarantees that an interface of the system developed must be sufficiently accessible by an older user with deficient skills in handling technology, having large buttons, peculiar visual indicators, and clear voice instructions provided. Furthermore, the system should adapt to the changes in the health status of the user through offering appropriate health information that would be derived from bioacoustic data. The interface should also feature aspects that would help the remote monitoring of the users by their caregivers. In this manner, there could be reasurance given and intervention conducted fast whenever critical cases arise.

5.3 Security and Privacy Issues

Since the bioacoustic sensors collect sensitive health-related information, privacy and security must be the primary concern in the core design of the system. The data need to be encrypted very strongly and must maintain secure communication channels. Mechanisms for getting user consent need to be developed as well to minimize the risk associated with violation of privacy. All regulations for handling personal health data like HIPAA or GDPR will be strictly followed while managing the data. Bioacoustic sensors in an AAL system should be able to introduce seamless integration that promotes the existence of monitoring and interpretation of health data that bioacoustic sensing and multimodal interfaces generate such opportunities for gaining very invaluable knowledge of the health of a user in a manner such as its being accessible and secure at once.

6 Case Study: Bioacoustic Sensors for Fall and Accident Event Detection

Falls are the second most common cause of injury in elderly subjects or patients with certain medical conditions. Thus, falls detection is counted as one of the fundamental technologies within AAL. Bioacoustic sensors form a new innovative way in monitoring and assessment of falls or other forms of injuries with the aid of specific patterns of biological sounds that arise during such an incident. Such noises can be altered breathing and heartbeat. They may even feature coughing or gasping, which are the best indicators of the seriousness of the fall and quick medical care (Jagnade et al. 2023).

6.1 Bioacoustic Patterns in Falls and Accidents

The other indicators are restlessness or irritability, and patterns of changes in breathing, showing potential chest or abdominal injuries. A classic indicator is that of having symptoms of breathing short or breathing unevenly. Pain often occurs along with elevated pulse, shock, or injury (Shaikh et al. 2017). Another such indicator includes a coughing or gurgling noise resulting from sudden movement or injuries within the body. The real-time captured bioacoustic signals would then be processed further to determine the characteristics and magnitude of the fall.

Bioacoustic sensors are embedded in the physical environment or on the human body. The sensor may pick up various acoustic signals like breathing sounds and heart rhythms. Such signals present different patterns when a person in pain is concerned. Sensors can be set up to identify the situation of falling down for a person. Such messages can send to the care giver or team of emergency medical doctors to immediately respond to such a person.

6.2 Other Modalities of Bioacoustic Sensors Integrations

Bioacoustic sensors often appear in multimodal human–computer interaction systems because of their ability to contribute to the enhancement of detection and assessment of falls. For example, the incorporation of bioacoustic data with motion sensors, accelerometers, or video surveillance will make the system analyze incidents even more accurately. The integration of sound-based data, including the abrupt movement, with motion-related data will enable the system to differentiate between different classes of accidents, such as falls, fainting fits, or seizures.

If any of these sensing types sense a fall, then it is possible to use an algorithm to process the collected bioacoustic information on how serious the fall is. For example, if there are sounds indicating heavy breathing or pulse changes, it will be possible to activate alarm calls that alert for emergent medical care. On the other hand, if things have only slightly changed, a carer will be notified to ensure that undue medical interventions are kept to the minimum (Lew et al. 2007).

6.3 Applied Effects and Advantages

The practical deployment indicates that the bioacoustic sensors can be applied for fall detection. Elderly persons and patients suffering from chronic diseases may be monitored continuously without requiring obtrusive wearable devices. The sensors act as passive, nonintrusive health parameter monitoring systems in domestic environments. The integration of bioacoustic sensors with other health-monitoring systems provides a holistic view of the health being measured. This integration allows for instant detection of falls and their related severity. The synergistic approach improves reaction time and reduces the time taken to reach medical help, thus providing better health results for users in an ambient-assisted living environment.

7 Data Analysis and Processing in Assisted Living Systems Using Bioacoustic Sensors

Bioacoustic sensors are one of the significant tools used in the health and welfare monitoring systems developed by AAL for elderly and medical conditions patients. Such systems' performance relies on complex processing and analytical techniques, which transform raw acoustic signals into meaningful information derived from the sensor signals. Quality data processing is, therefore, indispensable to guarantee accuracy, reliability, and prompt response.

7.1 Signal-Processing Techniques

Bioacoustic sensors can detect a large number of biological noises-inhalation, coughing, heartbeats, and fall-associated noises among others. These raw signals are mostly noise and interference induced by ambient noise during data processing.

The raw signal is purified and filtered to achieve bioacoustic features that will help to monitor health.

7.1.1 Noise Cancelation

In fact, most acoustic signals encountered are accompanied by background noise. There is a wide variety of schemes used to eliminate unwanted frequencies, and the signal-to-noise ratios are improved using band-pass filters, spectral subtraction, and adaptive filtering. For example, one may focus on the human biological sounds between 100 Hz and 4 kHz for heartbeats or sounds from breathing by using a band-pass filter.

7.1.2 Feature Extraction

Once the signal has been preprocessed, it is then forwarded to the feature extraction part. Bioacoustic signals convey key attributes that describe a biological phenomenon. Such critical features include time-domain-based attributes like mean and variance and frequency-domain-based elements like spectral density or harmonic analysis. For example, arrhythmia can be indicated if the frequency of heartbeat has changed, whereas loud coughing sounds can imply respiratory disorders.

7.1.3 Pattern Recognition

Following extraction, a number of pattern recognition algorithms will be applied to data to attempt to recognize specific health events, including unusual patterns of breathing, fall indications, or abnormal cardiac rhythms. Most of the machine learning algorithms are used for classification, but the most popular ones among them are SVM, decision trees, and neural networks since they help the system classify the various biological sounds and whether or not they are normal (Fig. 3).

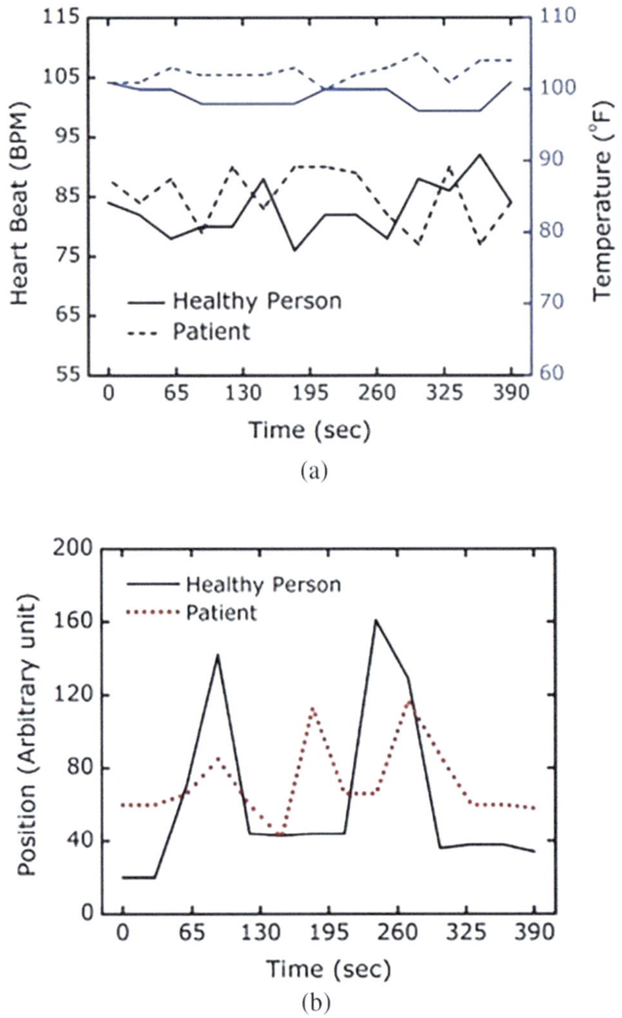

Fig. 3 Graphical depiction of **a** heart rate and body temperature and **b** patient's mobility in real time

7.2 Real-Time Data Processing

The ultimate requirement for AAL systems, in most cases of emergencies, is real-time processing of data. For instance, while sensing the fall, it should immediately send an alarm signal and classify the degree according to patterns of bioacoustic signals such as heart or breathing rates. So, keeping the viewpoint of real-time processing, the system has to be optimized according to low latency with high throughput in order to process data without delays.

Cloud and edge computing architectures can be very supportive in this context. Data preprocessing can be carried out locally at the edge using wearables or IoT-enabled sensors, with only the relevant data being sent to the central server for further analysis in a cloud environment. It reduces the burden of computation on the central server and reduces the response times, thus making quicker decisions in emergencies possible.

7.3 Bioacoustic Data Analysis—AI and ML

AI significantly upgrades the functioning of bioacoustic sensors due to facilitating predictive analytics with adaptive responses, which includes possible integration with machine learning algorithms. In the sense of using ML algorithms in making patterns relating to respiratory distress for prediction before its actual manifestation, thus allowing suitable intervention well before actual distress, this could help for instance. Deep learning techniques CNN and RNN works surprisingly in identifying complex sequential patterns from the acoustic data (Cano et al. 2020).

7.4 Data Intercombination and Integration

Most multimodal HCI systems integrate bioacoustic data with other sensor inputs-for example, motion sensors, temperature, or visual inputs. The techniques for data fusion consequently combine diverse streams of data, and as such, more accurate and robust system functioning is obtained, and this is demonstrated in the fused bioacoustic data with the motion sensor data, which enhances the prospects of fall detection accuracies by characterizing movement patterns associated with inordinate sounds.

8 Ambient-Assisted Living Systems Using Bioacoustic Sensors: Challenges for User Experience and Accessibility

This great opportunity for the enrichment of older adults' lives and people with disabilities is given by integrating bioacoustic sensors into AAL systems combined with multimodal human–computer interaction. The design of user experience and accessibility of interfaces are among the major factors that will determine the success of such systems. It should be functional, intuitive, and responsive.

8.1 User Interface Design

For AAL systems, an intuitive interface is very crucial. Elderly and people with body and cognitive impairments will fail to interact with complex systems. So, it's important to simplify the interface so that it can be used by the user with easy navigation. Large buttons, high contrast vision and voice prompts may enable users with limited vision or dexterity in navigating comfortably. This requires voice recognition as an input. Bioacoustic sensors also fall into the category, including devices such as microphones and heart monitors that could potentially work with voice commands. The device might be controlled to modify surroundings or summon help without physically touching the device.

8.2 Facilitating Flexibility to Accommodate Variations in User Requirements

Flexible AAL systems that accommodate a wide range of users with diverse abilities, preferences, and needs. An elderly designed system, for instance, demands readjustable interfaces to compensate for hearing loss or cognitive decline. Bioacoustic sensors can monitor continuously the health condition of the user during the interaction process (Wright et al. 2013). For example, if signals from surveillance show that respiratory patterns are abnormal or at a rate higher than the normal rate of beats, the system usually changes into a more intuitive way of communicating or provides further instructions that tell the user what to do.

Furthermore, AAL systems possess adaptive algorithms that are assumed to collect gradually knowledge regarding lifestyle and preferences. The system can observe patterns of interaction in order to make more refined and therefore more fluent and natural responses toward the increase in interaction fluency. For example, knowing beforehand when a user generally needs support, the system can thus be proactive about providing that kind of support, from reminding a user to take his or her medication to altering environmental conditions and lighting and temperature.

8.3 Inclusive Design for Diversely Abled Populations

Because access to these AAL systems should be offered to not only the elderly but other disability forms like motor, visual, auditory, and cognitive impairments, inclusive design will ensure that such systems are accessible for use by all. For instance, mechanically disabled individuals could fully control a system purely through non-contact interfaces like eye-tracking or head-tracking devices in combination with bioacoustic sensors.

9 Security and Privacy Issues in Bioacoustic-Based Assisted Living Systems

Bioacoustic sensors are included within AAL frameworks, opening up exciting new possibilities for monitoring elderly or disabled people in real time for the information about their health and welfare. However, like all healthcare systems handling sensitive personal information, there are massive security and privacy concerns.

It is, therefore, very important that these concerns are addressed adequately to allow such systems to be implemented in a manner that safeguards user data while preserving confidence in their efficacy (Pantic et al. 2008).

9.1 Security Issues of Bioacoustic-Based Systems

Bioacoustic sensors can monitor a wide range of subtle biological signals such as heartbeats, respiratory frequencies, coughs, and other sounds from a human body. Such analysis of signals can provide very important information about the health status of a person. Abnormal breathing patterns might indicate respiratory distress, and abnormal heartbeats may indicate possible cardiovascular problems. Inadequate protection of this health information may lead to serious breaches of privacy, identity theft, or other forms of abuse (Young 2010).

In essence, the threats to security of bioacoustic-based systems are:

Data Break-in: The dependence of bioacoustic data on others renders it vulnerable to interception by cyber criminals. With or without security measures such as encryption for data and a secure transmission protocol, potential hackers may obtain confidential healthcare information.

Unauthorized access: Bioacoustic data usually are stored in cloud-based servers or local databases. This fact poses a problem when unauthorized parties gain access to the information because of weak authentication practices. The breaches might arise from the systemic flaws existing in the system or by means of flawed access policies.

Data integrity: It should be assured that the data collected must be maintained in an undiluted form so that the AAL system would be able to provide proper diagnoses and health evaluations. Violation of integrity might result in incorrect recordings or tampered information, hence damaging the AAL system.

9.2 Privacy Issue from Bioacoustic Systems

User privacy is yet another critical aspect in AAL systems. Quite much information about a person's health and activities can be identified from the bioacoustic data, hence capable of being used to trace people's movement activities, emotional states, or even predict future health outcomes. In certain instances, the bioacoustic data may reveal personal habits or conditions in the health of users, which would be considered secret.

Therefore, strategies should include all the essential elements while resolving the privacy issues.

Data Anonymization: This technique will help in maintaining anonymity to ensure privacy. Data within sets of bioacoustic data containing details such as names or addresses can be deleted so that the data can then be provided for the analysis of health trends without compromising any confidential information.

Consent Management: It has to ensure that the consent procedures are clear and very transparent, which implies information on what type of data will be collected, how it will be applied, and who will be allowed access. Users ought to be free to opt-in or to opt-out at any time.

Data Minimization: The practice of collecting only data deemed relevant and necessary for health monitoring, deliberately excluding those unrelated and unnecessary data, to thereby reduce exposure to sensitive data of individuals and therefore minimize the opportunity to commit the violation of privacy.

10 Analysis of Ambient-Assisted Living Framework

Testing and evaluating the AAL system, especially for systems that deploy bioacoustic sensors in combination with multimodal HCI, is vital to ensure the functionality and reliability of the systems. Mainly, such systems are meant to aid people in performing daily life activities, managing health-related conditions, and reaction to emergencies. Systems like this require proper assessment for justifying their effective operation in real-world environments.

10.1 Performance and Reliability

The main aim of AAL systems is to improve the quality of life for the aged persons or individuals who suffer from any health-related problem. Therefore, it would make sense to measure its performance and reliability. In case of bioacoustic sensors, that would mean evaluation based on the extent of its performance in capturing physiological sound patterns accurately and converting them. This would include capturing heartbeats, breath rates, and coughing pattern. The sensors should give readings under various conditions, including motion, noise, or other interferences in the environment (Pantic et al. 2007). This can be tested by repeated testing in various settings, ensuring that the system gives real-time and accurate measurements, especially in health indicators. Bioacoustic sensors, for example, monitoring cardiac or respiratory functions are considered examples that require considerate design and intended function, even when speaking or walking. In this case, it is likely that an effectiveness study will focus on the device's ability to distinguish between pertinent signal events and irrelevant background noise to avoid false-positives and false-negatives.

10.2 Usability and User Experience (UX) Usability

It is, at the same time a prerequisite for any assistive technology, especially when the end-user belongs to the elderly or otherwise with disabilities. AAL systems will be designed to be accessible, while non-technical users ought not to face significant hardship while operating them. Actual users interacting with the system in controlled environments will constitute the basis for conducting user-centered usability testing by gathering information about usability, understandability, and responsiveness. In this context, it is aimed at enhancing the user interface to support the cognitive and physical differences of individuals.

10.3 Interoperability and Integration Ambient-Assisted Living

It consists of a wide array of technologies and systems working parallelly. Thus, interoperability comes forward as an essential part of the assessment. For example, bioacoustic sensors, multimodal interfaces, and other types of assistive technologies such as smart home devices and medical monitoring apparatus must integrate into an integral system. It combines other types of sensor data like the environment that includes temperature, humidity, light and physiological metrics of wearable devices such as a smart watch or a fitness band and the result is then tested.

It should be integrated with external databases of healthcare or emergency services, more so when there is a more serious incident such as fall or any medical emergency case. Strong testing protocol about integration ensures that the entire integrated system will work and data exchange will be with no delays or lost data.

10.4 Security, Privacy, and Compliance

Because health information and personal data are sensitive, one of the biggest areas AAL systems need to be able to provide adequate security and privacy. More specific to bioacoustic data, it is for the very first-person view of your state of health. Therefore, system evaluation needs to be tailored by how it could help protect users' data in terms of encryption and secure storage and communication protocols.

Other key elements of testing include legislation that is HIPAA or GDPR. Health information should not be in breach of any legal requirements surrounding its gathering, processing, and use. Secondly, the penetration test should identify vulnerability assessments to remove possible threats related to the breach of integrity or breaches of data.

10.5 Empirical Test and Pre-experimental Research Investigations

Although controlled testing environments are very informative, real-world assessments are also necessary to understand how the system performs in natural contexts. Therefore, pilot studies or field trials are included in the evaluation process (Alvarez-Cortes et al. 2007). Pilot studies refer to the experimental use of the AAL system in the actual homes or caregiving environments of users, where the system can be observed in real-time situations.

This implementation of practical testing is vital in determining whether a problem is really known to be realized by its applications in normal everyday lives or in response to real changes of behaviors of users, noise and/or light, and also usability in expanded periods. Besides that, this type of evaluation will enhance performance or accuracy and its impact toward the experience of users in using a system.

11 Conclusion

It would thus offer effective integration into AAL systems, with which it aims to bring multimodal human–computer interaction into the scenario. Ambient-assisted living is described as that vision of using technology toward aiding living for people by providing safe, comfortable living environments. Introduction of bioacoustic monitoring together with other input modalities, such as voice, gesture, and touch, may increase the quality of life, better health conditions, and facilitate emergency response to be less complicated and quicker. The merging would result in living environments that are intelligent, responsive, and adaptive.

Perhaps the greatest contribution of bioacoustic sensors in AAL systems is continuous monitoring of vital signs and identification of health anomalies. Traditional health monitoring through checking pulse or respiratory rates through manual means is clumsy and bound to have human errors. Bioacoustic sensors can detect changes in biological sound patterns, such as breathing, coughing, and heartbeats. These elements can be integrated into the products of everyday life, such as mattresses, furniture, and wearable technologies. These operate unobtrusively, without spoiling the real-time health information of any user.

For instance, bioacoustic sensors may be able to sense anomalies in a person's breathing pattern that may indicate respiratory distress or even an impending cardiac event. In the case of a fall, such sensors would be able to measure heart rate and respiratory rate and determine the gravity of the accident. This is extremely beneficial for AAL settings where there must be quick response to health conditions. The initial sign of the acoustic signal from the body should be proactively discovered by the health care provider or emergency responder and therefore allow for immediate, effective interventions.

Bioacoustic sensors in multimodal human–computer interaction for ambient-assisted living mark very profound steps on how we can potentially begin to rethink support systems concerning aging and other disabilities. It forms an entirely holistic way of managing health, emergencies, but also the user interface-in a way that maintains better independence and safety on behalf of the users. The future outlook of the AAL system promises much with the advancement that is being made in the technology of AI, ML, and sensors. Bioacoustic sensors and multimodal HCI are going to play a critical role in the improvement of life for vulnerable individuals in creating environments that will be adaptive, responsive, and personalized. Therefore, ethics on considerations on privacy and who rights give all to be well understood to develop and implement it effectively and responsibly.

References

Alvarez-Cortes V, Zayas-Perez BE, Zarate-Silva VH, Uresti JAR (2007) Current trends in adaptive user interfaces: challenges and applications. In: Proceedings of the electronics, robotics and automotive mechanics conference (CERMA 2007), Cuernavaca, Mexico, 25–28 Sep 2007, pp 312–317

Banda N, Robinson P (2011) Multimodal affect recognition in intelligent tutoring systems. In: D'Mello S et al (eds) ACII, Part II, LCNS 6975, pp 200–207

Cano S, Araujo N, Guzman C, Rusu C, Albiol-Perez S (2020) Low-cost assessment of user eXperience through EEG signals. IEEE Access 8:158475–158487

Das P, Acharjee K, Das P, Prasad V (2015) Voice recognition system: speech to text. J Appl Fundam Sci 1:2395–5562

Dey S et al (2019) Real time driver fatigue detection based on facial behaviour along with machine learning approaches. IEEE Int Conf Sig Process Inf Commun Syst (SPICSCON) 135–140

Elboushaki A, Hannane R, Afdel K, Koutti L (2020) MultiD-CNN: a multi-dimensional feature learning approach based on deep convolutional networks for gesture recognition in RGB-D image sequences. Expert Syst Appl 139:112829

Golestan S, Mahmoudi-Nejad A, Moradi H (2019) A framework for easier designs: augmented intelligence in serious games for cognitive development. IEEE Consum Electron Mag 8:19–24

Jagnade G, Ikar M, Chaudhari N, Chaware M (2023) Hand gesture-based virtual mouse using open CV. In: 2023 international conference on intelligent data communication technologies and internet of things (IDCIoT), Bengaluru, India, pp 820–825. https://doi.org/10.1109/IDCIoT56793.2023.10053488

Kavitha R, Nachammai N, Ranjani R, Shifali J (2024) Speech-based voice recognition system for natural language processing. Int J Comput Sci Inf Technol 5(4)

Lew M, Bakker EM, Sebe N, Huang TS (2007) Human computer intelligent interaction: a survey. In: Proceedings of international workshop on human-computer interaction (HCI 2007), pp 1–5

Pantic M, Nijholt A, Pentland A, Huanag TS (2008) Human-centred intelligent human computer interaction (HCI2): how far are we from attaining it? Int J Auton Adapt Commun Syst 1:168

Pantic M, Pentland A, Nijholt A, Huang TS (2007) Human computing and machine understanding of human behavior: a survey. In: Huang TS, Nijholt A, Pantic M, Pentland A (eds) Human computing of the series LNAI 4451, pp 47–71

Shaikh S, Ajmal M, Ahmed N, Badar F (2017) Comprehensive understanding of intelligent user interfaces. Int J Adv Comput Sci Appl 8:395–401

Singh B, Kapur N, Kaur P (2012) Speech recognition with hidden Markov model: a review. Int J Adv Res Comput Softw Eng 2(3)

Wright SJ, Kanevsky D, Deng L, He X, Heigold G, Li H (2013) Optimization algorithms and applications for speech and language processing. IEEE Trans Audio Speech Lang Process 1558–7916

Yashpalsing Chavhan ML, Dhore PY (2010) Speech emotion recognition using support vector machine. Int J Comput Appl 1:6–9

Young S (2010) Cognitive user interfaces. IEEE Signal Process Mag 27:128–140

Recent Advances in Bioacoustic Data Fusion and Multimodal Analysis

Amj Muthukumaran, K. M. Umamaheswari, K. Senthil Kumar, J. Shobana, and Vijayalakshmi Kaliyaperumal

Abstract

Bioacoustics, or the study of sound in biological contexts, has emerged as a point of great interest in helping further understand species behavior, biodiversity, and ecosystem dynamics. Traditional bioacoustic analysis relies on single-modality features, focusing mostly on audio. However, ecology is a complex process that requires more systems thinking. The integration of bioacoustic data with multiple other modalities-such as environmental variables, visual imaging, and behavioral metrics-carries transformative potential. This chapter explores bioacoustic data fusion and multimodal analysis as a comprehensive approach to ecological research. Recent developments in sensor technology, machine learning, and big data analytics have been able to make possible the collection, integration, and interpretation of diverse datasets. Bioacoustic data fusion takes advantage of this synergy, adding complementary modalities to acoustic signals in order to enhance the resolution and context of ecological insights. Applications range from monitoring animal populations to the detection of endangered species and evaluation of the impacts of climate change on ecosystems. Some of the important challenges that have arisen relate to data heterogeneity, synchronization, and computational scalability. Recent advances using techniques such as deep learning models, graph-based methods, and cloud-based platforms have overcome these challenges to pave the way toward real-time, scalable systems. This chapter further demonstrates case studies that reveal the possibility of integrating multiple modes of bioacoustics into practical applications: from mapping sounds to habitat to linking calls with physiological states. Bioacoustic data fusion and its multivariate analyses have great potential to bring large changes in ecological science and in conservation through promoting collaborations within and outside disciplines using leading-edge analytics frameworks. This chapter serves to provide a guide on state-of-the-art techniques, applications, and future directions in this dynamic field.

Keywords

Bioacoustics · Single-modality · Multimodal analysis · Ecological science · Dynamic field

1 Introduction to Bioacoustic Data Fusion and Multimodal Analysis

The aim of bioacoustics has been defined as the basic or pure scientific investigation into the generation, transmission, and reception of sound in biological organisms. While having ecological monitoring, identifying species, behavioral studies, and environmental concerns in mind, most recently developed sensor technology along with sophisticated and rapidly evolving analytic methods has more or less expanded the scopes of the research being made within this field. By this very reasoning, tradition has it that there used to be more dependence upon the analysis of sound alone as a singular modality in bioacoustics. But with the rising complexity in

A. Muthukumaran · K. M. Umamaheswari (✉) · K. Senthil Kumar
Department of Computing Technologies, School of Computing, SRM Institute of Science and Technology, Kattankulathur, India
e-mail: umamahek@srmist.edu.in

A. Muthukumaran
e-mail: muthua@srmist.edu.in

K. Senthil Kumar
e-mail: senthilk3@srmist.edu.in

J. Shobana
Department of Data Science and Business Systems, SRM Institute of Science and Technology, Kattankulathur, India
e-mail: shobanaj1@srmist.edu.in

V. Kaliyaperumal
Department of CINTEL, SRM Institute of Science and Technology, Kattankulathur, India
e-mail: vijaylak@srmist.edu.in

K. Subbaraj et al. (eds.), *Computational Bioacoustic Artificial Intelligence*, Sustainable Artificial Intelligence-Powered Applications, https://doi.org/10.1007/978-3-032-06088-4_12

the ecological system, and a move toward holistic insights, this has resulted in the inclusion of many information sources into one. Bioacoustic data fusion and multimodal analysis constitute such a paradigm shift that has recently taken place within the framework (Pandiyan et al. 2024).

1.1 The Need for Multimodal Integration

Biologically, systems are bound to be intricate, with integration at the level of organisms, behaviors, and environmental interactions. Purely acoustic information is very frequently rich but can lack vital contextual details. For example, animal calls may be interpreted only by relating the sounds with concurrent visual display, parameters of the environment, or physiological measurements. Data fusion bridges this gap and offers holistic understanding of ecological phenomena by amalgamation of such datasets of dissimilar natures (Nieri et al. 2021).

1.2 Technologies Driving Data Fusion

Recent technological innovations have enabled the efficient collection and integration of multimodal data. Advanced microphones, environmental sensors, camera traps, and GPS devices are now commonly used to gather data in bioacoustic studies. Cloud computing and IoT networks enable real-time transmission and storage of data. Furthermore, machine learning algorithms and deep learning techniques drive the synthesis of these data streams in the identification of patterns across modalities (Winter et al. 2023).

1.3 Benefits of Bioacoustic Data Fusion

- **Better Data Context**: Fusion can integrate audio data with ecological and behavioral metrics for better ecologically relevant insights. For example, the integration of audio data with temperature or humidity will be able to develop relationships between animal activities with climatic conditions.
- **Higher Accuracy**: Multimodal analysis reduces ambiguity stemming from any single-modality data alone. Cross-referencing acoustic signals with visual observations may prove useful in verifying species identity or behavioral events.
- **Wider applications**: The applications of bioacoustic data fusion range from basic monitoring of wildlife and biodiversity assessments to understanding anthropogenic noise and climate change impacts (Pearson 2024).

1.4 Challenges and Solutions

The implementation of bioacoustic data fusion is not without challenges. The primary hurdles include:

- **Data Heterogeneity**: Different modalities of data should be combined and standardized for harmonization. Diversity in data format and/or sampling rate may create difficulties in data integration. Solution: Application of data preprocessing techniques such as resampling and normalization may align datasets for analysis.
- **Synchronization**: Real-time alignment of data streams bears paramount importance. For instance, temporally correlating the oral expressions with the video of expressions at the same time requires accurate time stamping. Solution: Sensor arrays that are synchronized on every level and the institution of robust time-stamping protocol will ensure that the data sets correspond accurately.
- **Computational Complexity**: Handling large-scale multimodal data really requires huge computational resources. Solution: Cloud computing, high-performance computing systems can help to improve scalability issues (Neethirajan 2024).

Figure 1 presents the results of multimodal fusion compared with audio-only approaches to bioacoustic research. The accuracy and context resolution for multimodal fusion are 20% higher, with 90% and 95%, respectively, as compared to the audio-only method, which stands at 70% and 60%, respectively. This discrepancy underlines the limitation of solely relying on acoustic data, which is often starved of environmental or behavioral context. Multimodal systems integrate various data streams, such as visual, environmental, and acoustic data, to provide a richer and more nuanced understanding of ecological phenomena, enabling better species identification, behavioral analysis, and habitat monitoring.

Figure 2 gives evidence of the added complexity introduced by multimodal data fusion. While a single-modality system faces difficulty in approximately 30% regarding data heterogeneity, 20% in the case of synchronization, and 10% on computation complexity issues, each of these may be faced up to as much as 70, 80, and 85% levels in a multimodal system. Especially challenging in the proposed multimodal integration include: (a) synchronizing various datasets in space and time, (b) computation-intensive analysis tasks resulting from the integration of multimodality datasets. The result points out that for tackling the management-related challenges associated with multimodal analysis, the adoption of advanced solutions must be considered related to timesstamped sensor arrays, data preprocessing techniques, and cloud computing frameworks.

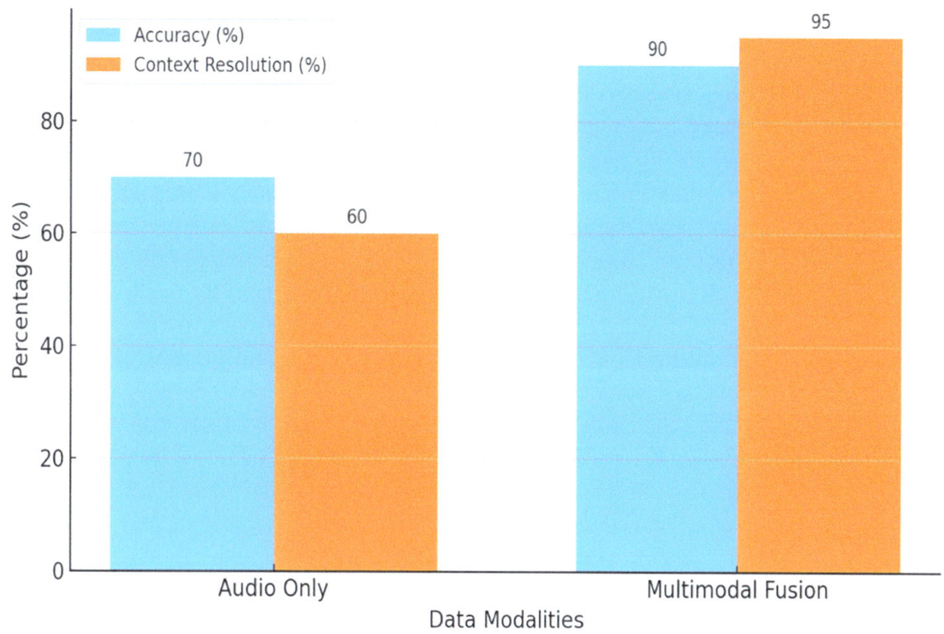

Fig. 1 Comparison of accuracy and context resolution by data modality

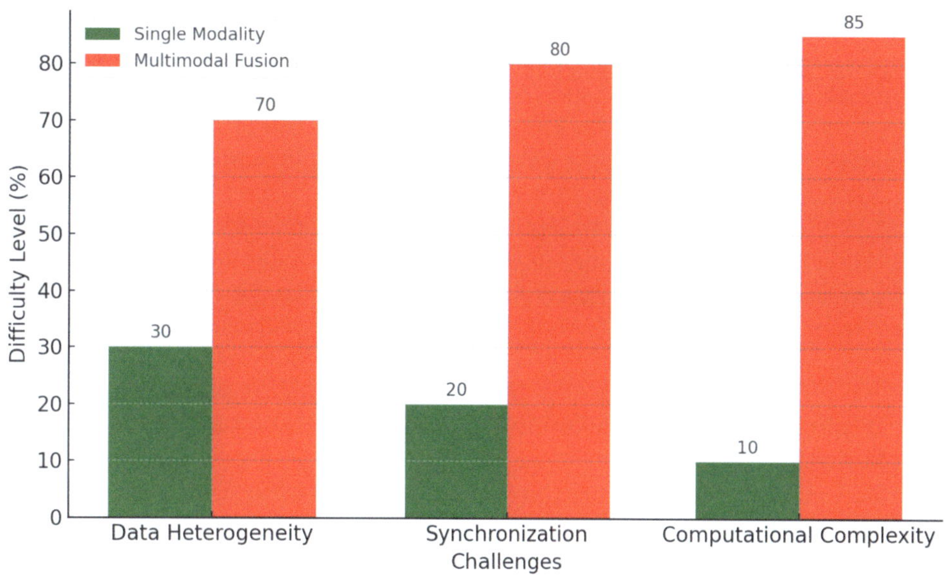

Fig. 2 Comparison of challenges in data processing

2 Sensor Technologies for Multimodal Data Collection in Bioacoustics

The advancement of sensor technologies has revolutionized bioacoustic research, enabling the collection of multimodal datasets that integrate sound with other environmental and biological parameters. These innovations provide researchers with comprehensive tools to study complex ecological systems, monitor biodiversity, and inform conservation efforts (Neethirajan 2024).

2.1 Key Types of Sensors in Multimodal Bioacoustic Data Collection

The sensors that are used in bioacoustic studies are acoustic, ranging from hydrophones and microphones, which are important. They form the backbone in the capture of detailed soundscapes. These sensors record a wide range of frequencies, from ultrasonic bat calls to infrasonic elephant communications. With advancements in features such as directional recording, noise reduction, and high sampling rates, they enhance the detection and classification of species-specific vocalizations, thereby becoming indispensable in species monitoring and ecological research. Complementary to the acoustic data, visual sensors such as camera traps and video recorders provide contextually visual information on animal behavior and environmental surroundings (Bain et al. 2021). Infrared and thermal imaging cameras have proven especially useful in observing nocturnal species or environments with low light conditions. By synchronizing these visual tools with acoustic recordings, researchers can draw meaningful correlations between vocalizations and physical activities. Then the environmental sensors add another level of depth by capturing temperature, humidity, air pressure, and light intensity. This type of data is crucial for understanding how changes in the environment affect animal behavior and vocal patterns, combining information toward an overall ecological view (Sathi 2024). GPS and location tracking devices enable bioacoustic studies to take a further leap into precise geolocation for recorded data. GPS-tagged animals provide information on movement and habitat use to which spatial behaviors can be matched with vocal activity. Physiological sensors, such as heart rate monitors and accelerometers, deliver information about internal states—like levels of stress or physical exertion by animals—that can be directly linked to vocalization patterns. Finally, multifunctional sensor arrays, such as ARUs, integrate acoustic, visual, and environmental sensors in one device. These systems allow for an integrated approach to data collection, especially over long periods or in areas inaccessible to humans, and thus guarantee the widest possible scope of monitoring (Latif et al. 2023).

2.2 Deployment Strategies for Multimodal Sensors

It includes such systems typically involve the deployment of sensors into relatively fixed locations, like a forest canopy, underwater, or open fields. Hence, stationary systems record long-term acoustics and environmental records much better while recording the spatial patterns across areas under observation. By opposition, mobile platforms—including but not restricted to drones, autonomous vehicles, or robotic systems—enable wide-scale data collection (Rappaport et al. 2020). These platforms help immensely in surveying species that have large habitat ranges, besides facilitating access to highly remote and inaccessible environmental sites. Wearable technologies such as sensors worn by the animals directly provide regular data about the behavior of animals, which include their vocalizations and physiological states. The use of wearables has immense significance regarding rare or migratory species because it can give one close, detailed, real-time insight into the activity of species and its internal condition (Stefano et al. 2022).

2.3 Challenges for Sensor Technologies

The integration of data from several sensors implies significant challenges regarding synchronization, as for meaningful analysis, temporal alignment should be very accurate. This task is increasingly difficult with large datasets and a variety of sensor types, and robust solutions are needed to ensure the coherence of the data. The power efficiency problem is essential among the important challenges, since in many cases the sensors are deployed in a location where regular maintenance would be difficult to perform (Hoye et al. 2022). Energy-efficient designs and solar-powered options are some of the innovations that extend their operational life. Sensors meant for harsh environments should be long-lasting and resistant to water immersion or physical wear from dense forests. Such sensors require rugged construction, water resistance, and adaptive calibration to continue their service with reliability. Finally, high-volume data generated by multimodal sensors creates other challenges—power consumption, storage, and data transfer—and thus a real-time management of such data with cloud-based systems is being enhanced with the addition of edge computing (Duc 2020).

Figure 3 highlights the rapid advancements in sensor technology over the past decade. Power efficiency has improved from 50% in 2015 to 90% in 2023, driven by innovations such as energy-efficient designs and renewable energy integration. Simultaneously, data storage capacity has grown exponentially, increasing from 100 to 1200 GB, enabling sensors to handle the demands of multimodal and high-resolution data collection. These advancements signify the growing capability of sensor technologies to support extended deployments and comprehensive ecological monitoring, especially in remote or challenging environments.

Figure 4 shows that the following types of sensors are increasingly being adopted in the conducted bioacoustic research: from 50% of adopting acoustic sensors in 2015 to 90% in 2023, and for cameras, it has gone up from 20% in 2015 to 80%, environmental sensors have also increased from 5% in 2015 to 70% in 2023. Whereas physiological sensors were hitherto very underutilized at only 5% in 2015, today they have reached an impressive 50%, underlining their role in linking vocalizations to internal animal states. This trend shows an increasingly wide scope of multimodal sensor integration driven by technological improvement, cost reduction, and the growing demand for holistic ecological insights.

3 Machine Learning Algorithms for Bioacoustic Data Fusion

Integration of bioacoustic data with other multimodal sources has changed the way ecological monitoring, species identification, and behavioral studies are performed. ML algo-

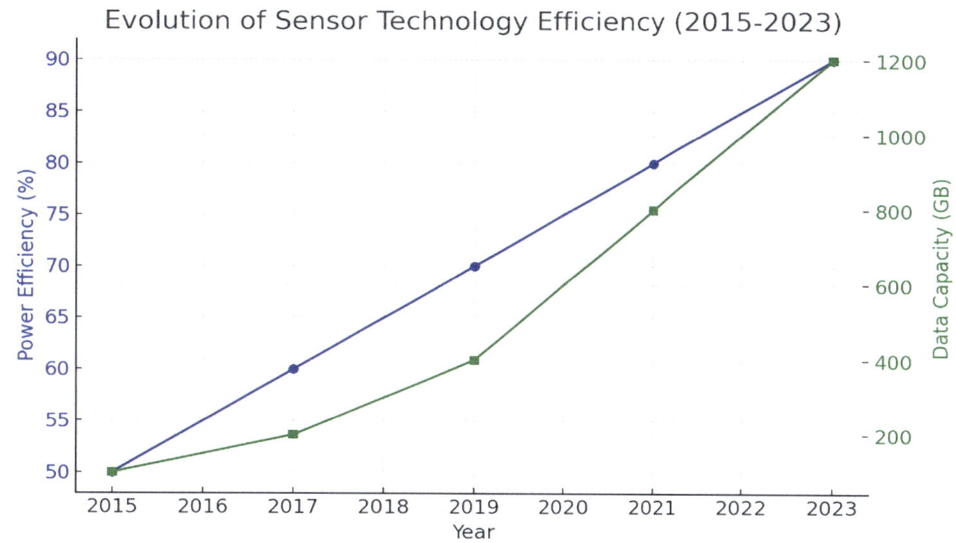

Fig. 3 Evolution of sensor technology efficiency

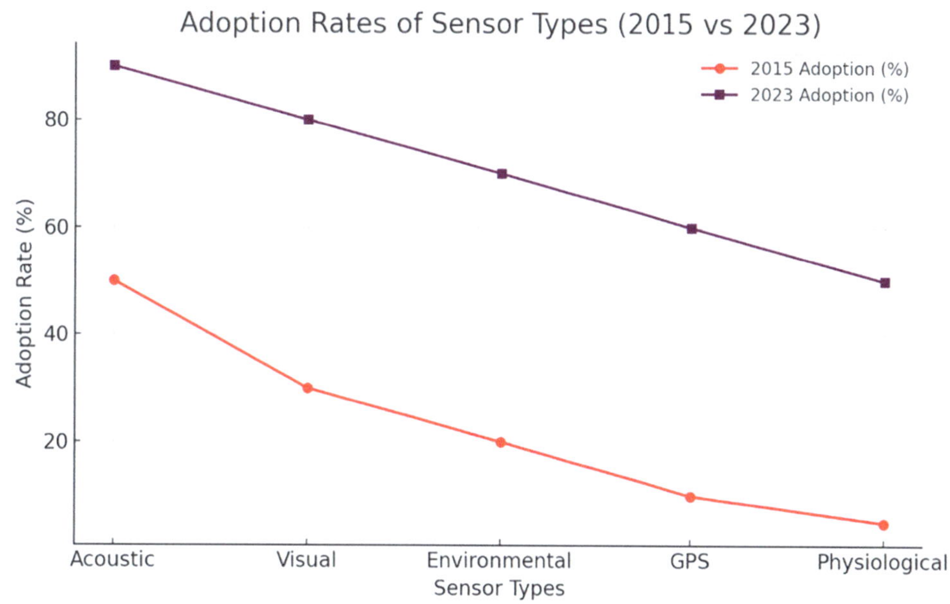

Fig. 4 Adoption rate of sensor types

rithms play a major role in this domain by helping researchers process and analyze diverse datasets for extracting meaningful patterns and insights. Application of various ML approaches has made the process of bioacoustic data fusion more accurate and scalable (Aly et al. 2021).

Most of the bioacoustics applications are based basically on the methods of supervised learning and, hence, require labeled data sets in order to train either a classifier or predictor models. Methods involving support vector machines have proven very powerful in species classification using their vocalizations because they can operate in high-dimensional feature spaces without rapidly deteriorating in performance. Random forests are robust against overfitting and generally combine different data types in an effective way, for example, acoustic, visual, and environmental data toward the goal of improving classification (Pancerasa et al. 2018). The CNN is specifically appropriate, as with most other deep neural networks, for handling tasks involving complex features hidden within spectrograms: tasks like species recognition or call detection. While on the other side, unsupervised learning is a process very helpful to have when one seeks to investigate unlabeled data and try to find their hidden pattern or group the entities in similar classes, as a new species' identification or similar behavior categorization—a work done with such machine learning techniques as k-means or hierarchical clustering. Principal Component Analysis (PCA) aids in reducing the dimensionality of large datasets, making it easier to visualize and analyze complex data structures (Tol et al. 2018).

Semi-supervised learning fills the gap between both supervised and unsupervised methods, which include a small amount of labeled data with a larger pool of unlabeled data. It is very useful in those ecosystems that are understudied or remote and have very little availability of annotated datasets. Algorithms such as self-training and graph-based methods find their wide applications in these scenarios for effective data fusion and analysis. Thereby, reinforcement learning is an active technique that optimizes decision-making processes by way of interactions with the environment. The algorithms instruct a drone or an underwater robot autonomously in bioacoustic research to collect multimodal data adaptively in real time and thus more efficiently and relevantly (Lappe et al. 2011).

Transfer learning is one of the powerful approaches where a model pretrained on other related data assists in quickening the training for the new bioacoustic datasets. This concept, in particular, is highly useful for applications involving rare species monitoring, since gathering large volumes of data is time-consuming. Therefore, this kind of knowledge transfer in similar tasks aids in increasing model performance using minimal data. In contrast, ensemble learning combines the predictions of independent models toward more robust and more accurate predictions in bioacoustic data analysis. CNNs combining the extracted acoustic features with a random forest that analyses the parameters of the environment would increase accuracy in the species detection at higher levels. An approach here is one by which some weaknesses of the model have the strengths of the other to perform better (Cross 2023).

Efficient methodologies of data fusion will be indispensable in integrating the multimodal bioacoustic data. Feature-level fusion combines features extracted from different modalities into a single feature set, such as merging acoustic data, which may include frequency and amplitude with environmental data including temperature and humidity. In decision-level fusion, the processing of each modality is done separately and then combined using, for instance, majority voting or weighted averaging. Hybrid fusion integrates both feature-level and decision-level strategies for multimodal data integration and hence ensures more reliable outcomes.

3.1 Challenges in Applying Machine Learning to Bioacoustic Data Fusion

Several factors are considered challenges for the application of machine learning on bioacoustic data fusion. First, data imbalance, from dominating species or behaviors into datasets, often results in biased models; this demands balancing techniques such as data augmentation or synthetic sampling. Computational complexity is one of the major challenges in that processing large, multimodal datasets requires considerable computation. Optimized algorithms coupled with advanced hardware are required in order to handle such requirements efficiently. The main challenges to deal with during preprocessing and model training are background noise, overlapping signals, and environmental variability. Advanced noise-reduction techniques and adaptive models may mitigate these effects, leading to more accurate and reliable analysis. Besides, in the field of bioacoustics, there is still a lack of labeled datasets. Semi-supervised learning and transfer learning are some of the approaches that can be used to overcome this limitation and make bioacoustic data fusion more accessible and effective for ecological research (Devi et al. 2025).

Figure 5 shows the trade-off of accuracy versus training time for different machine learning algorithms applied in bioacoustic data fusion. The CNNs have the highest accuracy of 92%, considered ideal for complex tasks such as species identification, while their training time is the longest among all, at 35 s, reflecting their computation-intensive nature. The main difference is that while SVM and random forest have a good balance of accuracy—85% and 88%, respectively—this is on considerably lesser training time: 20 s and 15 s, respectively. This makes these algorithms good in real-time and resource-constrained situations. However, k-means and PCA are fast on computation though less accurate. For such a reason, they position for complementary tasks like exploration or unsupervised data grouping.

Figure 6 describes some challenges which both the supervising and unsupervised learning encounter. In this supervised learning, the top contribution, which is an imbalance in data, at 70%, while for the supervised learning methods, it can be limited labeled data by 80%. For unsupervised learning, they are 60% for noise variability and 55% for computationally complex issues, reflecting their highest challenges in dealing with unsupervised and unstructured data. Both have their different challenges, and this therefore requires hybrid techniques that leverage the strengths of both supervised and unsupervised learning while mitigating their individual weaknesses.

4 Synchronization Challenges in Multimodal Bioacoustic Systems

Synchronization is a critical issue in multimodal bioacoustic systems, where data from acoustic recordings, environmental sensors, visual imaging, and behavioral observations must be temporally aligned for meaningful analysis. Precise synchronization is crucial to the integrity of multimodal data sets; misaligned data may lead to incorrect conclusions or missing important information. However, several technical and logistical issues arise in the synchronization process.

4.1 Temporal Resolution and Clock Drift

One of the key issues is that there is a different temporal resolution for different sensors. While audio sensors may sample

Fig. 5 Comparison of algorithm efficiency

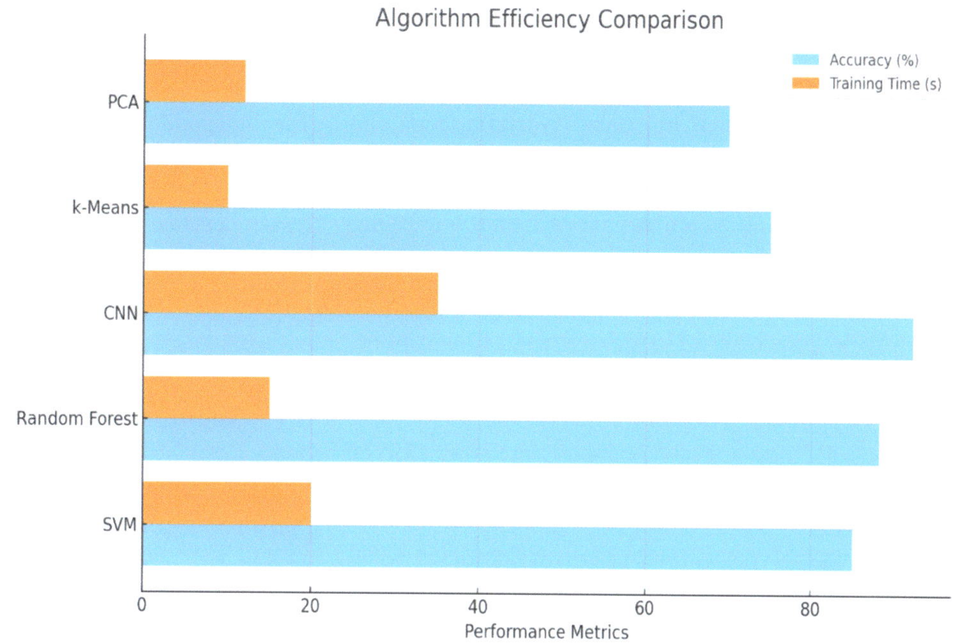

Fig. 6 Challenges in applying ML algorithms

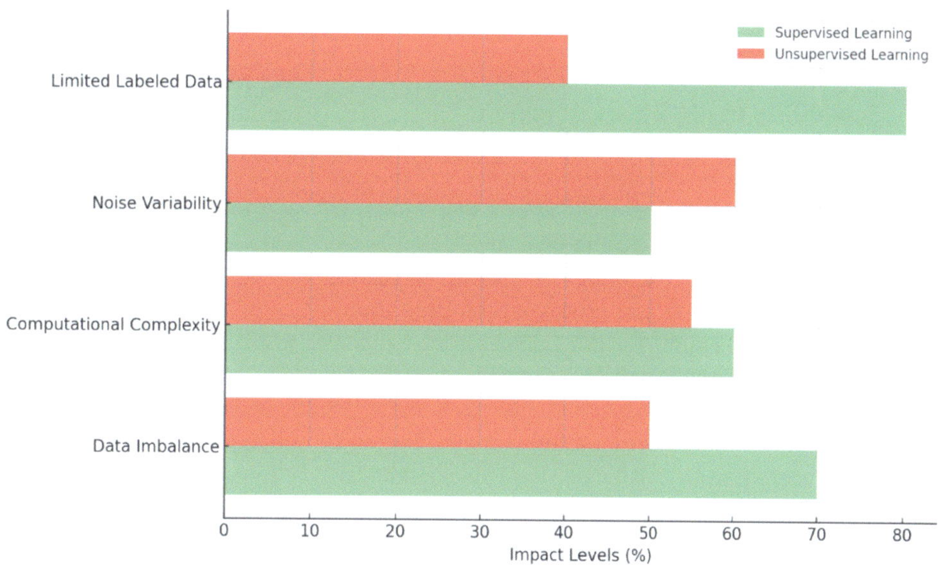

at high rates, in kilohertz or higher, environmental sensors or video cameras may be recording at much lower frequencies. In order to align these diverse data streams, down-sampling, interpolation, or resampling operations will be required—operations that should be done with great care in order to avoid information loss.

Besides that most device clocks tend to drift after some time. Even minor discrepancies in timekeeping can quickly add up, leading to serious misalignment in datasets that are gathered over long periods of time. In many cases, this clock drift needs to be mitigated by periodic synchronization with a trusted time reference, such as GPS-based systems, or post hoc corrections based on identifiable synchronization events in the data (Neha et al. 2024).

4.2 Variability in Conditions of Data Acquisition

Biomedical multimodal systems are normally very exposed and difficult field conditions, often influencing data acquisition, such as meteorological conditions, terrain features, and animal behavior. For instance, acoustic sensors continuously record sound, while cameras can only record during the daytime or based on motion trigger mechanisms. These discrepancies in data availability further increase complexity in the alignment of these data sets and require more efficient algorithms to handle gaps and inconsistencies (Tomaszewski and Kołakowski 2023).

4.3 Latency and Delays in Processing

Real-time processing involves latency introduced by hardware or software, which can break synchronization. For instance, the delays in the transmission or processing of video frames compared to audio streams can result in timing mismatches. Careful system design, including low-latency communication protocols and efficient processing pipelines, is required to overcome these issues.

4.4 Calibration and Reference Events

In practice, the most common temporal framework unifying modalities relies on the use of either calibration signals or reference events. Examples are a clapperboard in audiovisual recording and an unexpected, easily recognizable sound-a gunshot, for example-serving as a temporal anchor. It is often nontrivial to identify these reference events in ecological conditions when the data stretches over long periods or includes sparse recordings (Sinawang et al. 2021).

4.5 Scalability in Big Data Contexts

As the scale of bioacoustic studies continues to increase, this synchronization challenge scales from a few sensors to thousands and from hours of data to terabytes. Cloud-based platforms and distributed computing frameworks that enable automated synchronization and processing of large multimodal datasets are emerging. Ensuring these systems remain cost-effective and accessible remains an ongoing concern (Le et al. 2019).

Figure 7 gives an overview of the severity for each identified synchronization challenge. Data integration and alignment algorithms are again the most critical challenge, obtaining a score of 9, hence showing that there is substantial complexity to accurately align all the information within a set of multimodal data. Obtaining precise synchronicity given the features of temporal resolution and clock drift, or scalability when it comes to big data scenarios, also proves to have an important grade, with 8 points.

Figure 8 impacts of each challenge to the quality of the analyses of bioacoustic data. The highest impact score, 10, comes from algorithms of data integration and alignment, hence underlining critical issues in keeping data coherent throughout modalities. In summary, temporal resolution and also clock drift (both had a score of 9) and scalability within big data (with a score of 9) show a high rank; poor management of such issues will most likely badly spoil the analytical results.

5 Case Studies: Applications in Wildlife Monitoring and Conservation

The integration of bioacoustic data with other modalities is fast becoming an invaluable force in wildlife monitoring and conservation. In fact, these multimodal methods allow researchers to explore details on animal behavior, population dynamics, and ecosystem health. Some key examples show the applied use of these bioacoustic systems in real-world conservation:

5.1 Monitoring Endangered Species

Bioacoustic systems have been highly effective in the monitoring of endangered species through the capture and analysis of their unique vocalizations. Acoustic sensors, such as automated recording units, are deployed in remote habitats to record vocal activity over long periods. These recordings allow researchers to identify the presence of a species, estimate population sizes, and monitor changes in behavior without disturbing the animals. For example, research on the critically endangered Kakapo in New Zealand uses bioacoustic data to monitor individual birds during mating seasons. Acoustic data, along with environmental metrics such as temperature or vegetation cover, also helps in understanding habitat suitability and informs targeted conservation action (Winters and Michael 2013).

5.2 Tracking Migration and Habitat Use

Knowledge of the pattern and habitat preference is crucial to understanding species ecology and protecting key areas. Acoustic monitoring has been highly instrumental in detecting animal movements over long distances, like tracking the songs of migrating humpback whales across ocean basins. Such integration of acoustic data with GPS tracking and satellite imagery provides a complete understanding of migration corridors and environmental factors that influence habitat use. Such a multimodal approach would ensure that effective conservation strategies are pursued, such as the establishment of marine protected areas or the protection of migratory stopover points, in order to ensure the long-term survival of migratory species (Jannathul Firdous and Sabena 2024).

5.3 Understanding Ecosystem Health Through Soundscapes

Soundscape analysis adopts a holistic approach in the assessment of ecosystem health by evaluating the total acoustic output from an area. It generally provides that a rich, diverse

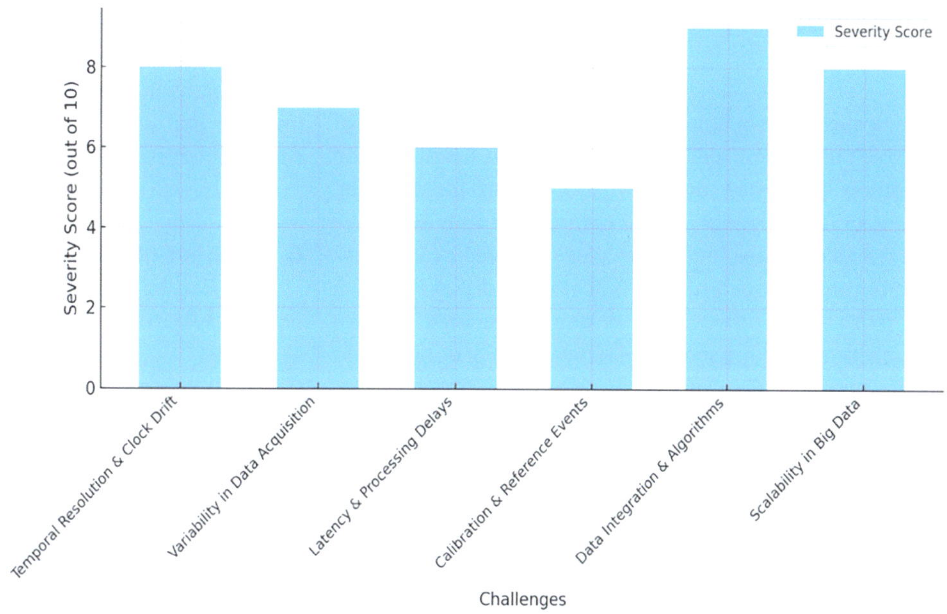

Fig. 7 Severity of synchronization challenges

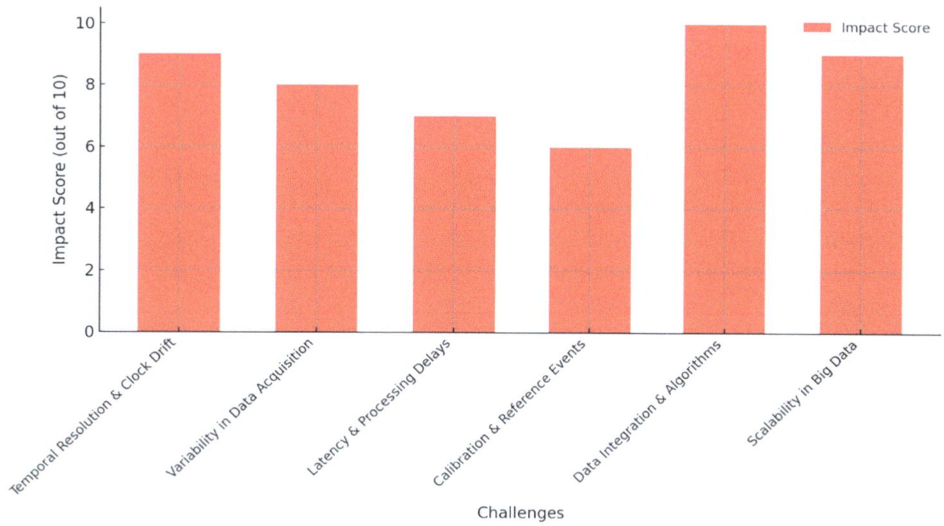

Fig. 8 Potential impact on data quality

soundscape may well equate to high biodiversity, whereas reduced acoustic diversity can indicate habitat degradation. In this respect, soundscapes in the Amazon rainforest have been applied to assess some impacts of deforestation and habitat fragmentation on biodiversity. Merging these with remotely sensed data allows establishing correlations between the dynamics of the soundscape and land use, thus providing action-ready interpretations useful for ecosystem restoration or planning of nature conservation (Manikandan and Jayashree 2024).

5.4 Mitigating Human–Wildlife Conflict

Systems that take control of bioacoustics and detect them play core tasks of reconciliation, where conflicts between local wildlife and people happen in major conditions. In the year of acoustic detection of an elephant, calls near various parts of Indian farmland created multiple onsets of farmers having a potential chance, lessened the damage brought into the crop, a direct clash of humans along these animals. The various technology addition of motion and heat sensing is integrated well; this, in sum with these systems, increases detecting success events; an efficient early warnings' setup has come into way. This proactive approach will also help in fostering coexistence, reducing risks to both humans and wildlife.

5.5 Assessing the Impact of Climate Change

Bioacoustics offers a unique perspective on the influence of climate change upon species and ecosystems. Such shifts in the timing, frequency, or intensity of calling-for instance, changes in frog choruses during the tropical rainy season—often reflect changing breeding cycles promoted by temperature and precipitation changes. Integration of acoustic data with those of climate and hydrological records permits researchers to make patterns and future predictions due to climate stressors. This information is priceless for the adaptation of conservation strategies to mitigate long-term climate change impacts on vulnerable species (Gonçalves and Rocha 2022).

5.6 Behavioral Studies in Captive and Wild Settings

Acoustic monitoring is, therefore, a powerful tool for studying animal behavior and understanding how animals communicate, interact socially, and respond to stressors. As an example, the study of vocalizations in orangutans has been used to elucidate the complex social behaviors and habitat preferences of these animals, enhancing reintroduction efforts for rehabilitated individuals. Bioacoustic studies in combination with complementary data from observations or physiological measurements—such as cortisol levels or heart rate—provide better insight into the behavioral contexts of vocalizations. These findings help design conservation programs with species-specific needs, therefore improving the success of in situ and ex situ conservation programs.

Figure 9 presents a comparison of the effectiveness and feasibility of various applications of bioacoustic systems for wildlife monitoring and conservation. Applications such as "Monitoring Endangered Species" and "Assessing the Impact of Climate Change" have the highest ranks in effectiveness, which means that they are very important for the conservation and assessment of ecosystems. These methods will provide very important data on long-term species survival and monitoring of environmental changes and thus are irreplaceable. On the other hand, feasibility scores for applications such as "Impact of Climate Change" are lower, considering technical and logistical hardships related to their deployment on extended surfaces for extensive periods. Therefore, "Behavioral Studies" and "Mitigation of Human-Wildlife Conflicts" are applications relatively balanced in consideration of the effectiveness-feasibility equation and thus able to stand for localized applications or where intervention should take place without too much delay. The large difference between these two metrics reflects a trade-off where very effective methods might be less practical due to resource or infrastructure limitations, and feasible solutions may have moderate impact. The development of technology, cost-effective solutions, and interdisciplinary collaborations could reduce this gap and improve the suitability of bioacoustic systems in various conservation scenarios.

6 Impact of Climate Change on Bioacoustic Research and Applications

Climate change is one of the leading reasons that will keep restructuring the face of ecosystems across the world. Bioacoustic research, within such a view, has much in depth to offer toward understanding and mitigating such impacts. This might involve studying changes in soundscapes or vocalization patterns that researchers employ to monitor changes in biodiversity, track the way species adapt, and determine the health of ecosystems regarding climate-induced changes. A number of key areas are indicative of the interaction between climate change and bioacoustics.

6.1 Changes in Species Distribution and Behavior

Climate change influences the way in which many species are distributed across regions and forces behavioral changes, particularly regarding migration or adaptation to their respective new environments. Such changes often materialize in their vocalization, making bioacoustics a very useful means by which such changes can be monitored. Birds and anurans might alter their period of breeding with temperature fluctuations that eventually alter calling behavior. Acoustic monitoring enables researchers to monitor these shifts over time and across regions for often critical data to inform conservation strategies. This could provide information on the protection of new habitats, restoration of ecological corridors, and other actions to support species survival in changing climates (Borra 2024).

6.2 Changes in Ecosystem Soundscapes

With an ever-changing climate, for instance, the acoustic activities of whole ecosystems, or ecosystem soundscapes, have increasingly been disrupted. Increased temperature, altered precipitation patterns, and increased severity of weather events can eventually result in reduced richness and diversity of acoustic signals. For example, drought conditions may lead to declines in insect and amphibian populations that result in quieter soundscapes, whereas flooding can temporarily increase acoustic activity in aquatic species.

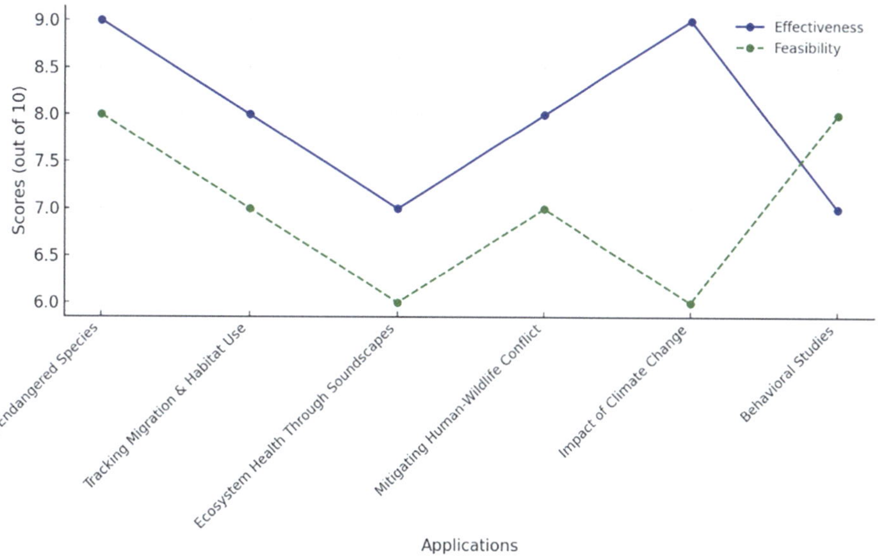

Fig. 9 Comparison of effectiveness and feasibility of conservation applications

These changes are extremely valuable indicators of ecosystem health, providing early warnings of ecological stress or degradation that may not otherwise be evident (Liebscher et al. 2024).

6.3 Monitoring of Species Vulnerability

Species vulnerability to the variations in climate varies regarding habitat preference and physiological limits. Bioacoustic tools are therefore of importance in determining such vulnerabilities by recording and analyzing species presence, activity, and vocalization. For instance, there are coral reef soundscapes driven by the acoustic activity of marine organisms, strongly associated with ocean warming and acidification. Acoustic monitoring will log declines in reef activity, be critically supportive of restoration programs, and emphasize the need to protect especially vulnerable species and ecosystems by formulating directed strategies against impacts due to specific climatic changes.

6.4 Acoustic Interference Induced by Climate

Acoustic disturbances caused by climatic change could interfere directly with animal communication. Increased rainfall, wind, or anthropogenic noise associated with extreme weather events could mask animal calls and hence inhibit communication that is critical for mating, territoriality, or predator warnings. Heavy rains in the tropics could mask frog mating calls and reduce their reproductive success. These bioacoustic studies of these disruptions have made it possible to generate habitat management plans that can secure the persistence of effective communication environments, which in turn allows species to survive despite all climate-induced adversities.

6.5 Long-Term Monitoring and Trend Analysis

The subtle and chronic nature of climate change impacts dictates the need for long-term data capture that identifies a pattern or trend. Bioacoustic research does this work in being able to continually monitor systems that provide dense data, giving insight into the way biodiversity is reshaping itself. Examples range from studies on birds' populations showing northward migration in the temperate region, for which bioacoustic data confirm such displacements. This is such long-term monitoring that complements other available research methods, thereby rendering a full understanding of climate impacts on both ecosystems and biodiversity.

6.6 Support for Mitigation and Adaptation Efforts

Bioacoustic research directly informs efforts at mitigating and/or adapting to the impacts related to climate change. Acoustic monitoring data inform policies and conservation strategies that go toward the protection of vulnerability in ecosystems and species. For example, applying bioacoustic data, restorations involving mangrove or coral reef systems can easily evaluate the effectiveness of interventions that make sure ecological functions are restored within an ecosystem. These applications highlight the proactive role bioacoustics can play in addressing climate challenges by underpinning evidence-based solutions to protect biodiversity and ecosystem health.

7 Conclusion

Bioacoustic research has emerged as a transformative field, offering innovative methods to study and conserve ecosystems through the integration of diverse data sources. This chapter discussed recent advances in the field of bioacoustic data fusion and multimodal analysis and underlined the fact that merging acoustic signals with other kinds of environmental and biological data is what really enriches insights into ecological processes. While bioacoustics could find potential applications with rapid progress in sensor technologies, artificial intelligence, and big data processing, they range from the conservation of endangered species to habitat monitoring and understanding the far-reaching impact of climate change on biodiversity. However, several practical challenges come up in managing heterogeneous datasets, ensuring synchronization across modalities, and scaling up computational resources for large datasets. However, breakthroughs in algorithms related to machine learning, real-time monitoring systems, and scalable cloud-based platforms are solving these challenges, therefore enabling the more effective and wider application of bioacoustic methods. Case studies presented in this chapter substantiate some tangible benefits that are being obtained through bioacoustic applications in wild-life monitoring, conservation, and ecological research. Moving forward, bioacoustics is in a good place to help in two of the most pressing crises: climate change and biodiversity loss. Further work will need to be built upon technological advances and across-discipline collaborations if its full potential is to be realized over and above the present limitations. Greater innovation and integration of bioacoustics into larger conservation strategies will enable researchers and practitioners to make even more significant contributions toward the conservation of natural ecosystems and the advancement of a sustainable future. This chapter emphasizes the value of bioacoustics as an essential tool in ecological discovery and stewardship for the environment.

References

Aly L, Silva H, Bernades G, Penha R (2021) Appropriating biosensors as embodied control structures in interactive music systems. Human Technol 17(1)

Bain M, Nagrani A, Schofield D, Berdugo S, Bessa J, Owen J, Hockings KJ et al (2021) Automated audiovisual behavior recognition in wild primates. Sci Adv 7(46):eabi4883

Borra VK (2024) Machine learning based identification of rainforest species using audio data. PhD diss, California State University, Northridge

Cross T (2023) Spectral, spatial and temporal complexity in electroacoustic composition: a statistical, theoretical and cognitive analysis of morphological form. PhD diss

Devi GD, Kumar N, Manikandan J, Rekha V (2025) Innovative brain tumor detection: stacked random support vector-based hybrid gazelle coati algorithm. Biomed Sig Process Control 101:107156

Di Stefano N, Vuust P, Brattico E (2022) Consonance and dissonance perception. A critical review of the historical sources, multidisciplinary findings, and main hypotheses. Phys Life Rev 43:273–304

Duc PNH (2020) Development of artificial intelligence methods for marine mammal detection and classification of underwater sounds in a weak supervision (but) big data-expert context. PhD diss., Sorbonne Université

Gonçalves FSSR (2022) Acoustic ecology in artistic creation: rediscovering underwater soundscapes. PhD diss, Universidade do Porto (Portugal)

Hoye T, Groom Q, Juslén A, Mandon C, Dinesen L, Rosenberg A, Hendriks RJ, Eggermont H, Vihervaara P (2022) Biodiversity monitoring knowledge gaps and research & innovation priorities. Biodiversa+. URL: https://www.biodiversa.eu/wpcontent/uploads/2022/12D2

Jannathul Firdous R, Sabena S (2024) A novel approach to coral species classification using deep learning and unsupervised feature extraction. J Spatial Sci: 1–28

Lappe C, Trainor LJ, Herholz SC, Pantev C (2011) Cortical plasticity induced by short-term multimodal musical rhythm training. PLoS ONE 6(6):e21493

Latif S, Cuayáhuitl H, Pervez F, Shamshad F, Shehbaz Ali H, Cambria E (2023) A survey on deep reinforcement learning for audio-based applications. Artif Intell Rev 56(3):2193–2240

Le M-L, Garvin CM, Barber JR, Francis CD (2019) Natural sounds alter California ground squirrel, *Otospermophilus beecheyi*, foraging, vigilance and movement behaviours. Anim Behav 157:51–60

Liebscher M, Dell'Orco A, Doll-Lee J, Buerger K, Dechent P, Ewers M, Fliessbach K et al (2024) Lifetime musical activity and resting-state functional connectivity in cognitive networks. Plos One 19(5):e0299939

Manikandan J, Jayashree K (2024) Enhancing lung nodule classification: a novel CViEBi-CBGWO approach with integrated image preprocessing. J Imag Inf Med 37(5):2108

Neethirajan S (2204) Decoding the language of chickens-an innovative NLP approach to enhance poultry welfare. *bioRxiv*: 2024-04

Neha F, Bhati D, Shukla DK, Dalvi SM, Mantzou N, Shubbar S (2024) U-net in medical image segmentation: a review of its applications across modalities. arXiv preprint arXiv:2412.02242

Nieri R, Anfora G, Mazzoni V, Stacconi R (2021) Semiochemicals, semiophysicals and their integration for the development of innovative multi-modal systems for agricultural pests' monitoring and control. J Mediterr Ecol 19, no. special issue:42

Pancerasa M, Ambrosini R, Winkler DW, Rakhimberdiev E, Rubolini D, Liechti F, Saino N, Casagrandi R (2018) Analysis of almost a hundred long distance migration paths of Italian and Swiss barn swallows (*Hirundo rustica*) reconstructed on the basis of light level geolocators. In: Proceedings of the '10th international conference on ecological informatics: translating ecological data into knowledge and decisions in a rapidly changing world. DEU, pp 268–268

Pandiyan V, Baganis A, Richter RA, Wróbel R, Leinenbach C (2024) Qualify-as-you-go: sensor fusion of optical and acoustic signatures with contrastive deep learning for multi-material composition monitoring in laser powder bed fusion process. Virtual Phys Prototyping 19(1):e2356080

Pearson M (2024) Pioneering urban biodiversity: using AI-sensors, eDNA and traditional methods to create a novel biodiversity monitoring toolkit and assessment framework. Master's thesis

Rappaport DI, Andrew Royle J, Morton DC (2020) Acoustic space occupancy: combining ecoacoustics and lidar to model biodiversity vari-

ation and detection bias across heterogeneous landscapes. Ecol Ind 113:106172
Sathi SKR (2024) Adaptive sensor placement inspired by bee foraging: towards efficient environment monitoring. arXiv preprint arXiv: 2411.15159
Sinawang PD, Soto F, Ozen MO, Akin D, Demirci U (2021) Progress and challenges in biomarker enrichment for cancer early detection. Progress Biomed Eng 3(4):043001
Tomaszewski L, Kołakowski R (2023) Mobile services for smart agriculture and forestry, biodiversity monitoring, and water management: Challenges for 5G/6G networks. Telecom 4(1):67–99. MDPI
Winter NR, Blanke J, Leenings R, Ernsting J, Fisch L, Sarink K, Barkhau C et al (2023) A systematic evaluation of machine learning-based biomarkers for major depressive disorder across modalities. medRxiv 2023-02
Winters IV RM (2013) Exploring music through sound: sonification of emotion, gesture, and corpora. McGill University (Canada)
Wolanin A, Guanter L, Camps-Valls G, Gómez-Chova L, Garcia GM, van der Tol C, Zhang Y (2018) Estimating gross primary productivity in crops with satellite data, radiative transfer modeling and machine learning. In: 10th international conference on ecological informatics

Enhancing Human–Computer Interaction with Bioacoustics: Techniques and Applications

D. Anitha, N. Kanimozhi, A. Sheryl Oliver, R. Usharani, and S. Thirumal

Abstract

Bioacoustic-driven HCI uses sound-based signals emanating from the human body, such as voice, breathing, and physiological noises, not only to enable intuitive and adaptive systems but also to create a noninvasive and natural modality for communicating. This new paradigm nicely complements the state-of-the-art input based on touch and vision. Allowing the integration of knowledge in sensing technologies, signal analysis, and artificial intelligence, bioacoustics enables the development of systems that can dynamically respond to the physical and emotional states of their users. This chapter delivers an in-depth investigation into bioacoustic-driven HCI, starting with a general overview of bioacoustics and its adoption within interaction design. We discuss the potentialities that stem from acoustic sensors, wearable devices, and ambient microphones to record human sound patterns. The discussion goes back in time to cover processing approaches that include feature extraction, noise mitigation, and machine learning models capable of interpreting bioacoustic signals. It includes discussions on applications involving voice-controlled systems, health monitoring through acoustic biomarkers, and emotional recognition. The chapter also points out some of the challenges: real-time responsiveness, data privacy management, and ethical concerns related to user-sensitive data. Emerging trends, such as the integration of multimodal inputs and personalization, are underlined to show future perspectives in the field. By providing an overview of technologies, methodologies, and applications, this chapter offers inspiration for innovative ideas and practical adoptions in the field of bioacoustic-driven systems toward new and more intuitive paradigms of human-centered interactions.

Keywords

Bioacoustic · Human–computer interaction · Multimodal inputs · Exploration · HCI

1 Introduction to Bioacoustics in HCI

1.1 Definition and Scope

Bioacoustics is a generalist branch standing at the crossroads of biology and acoustics. It investigates production, transmission, and reception of sound in living organisms. In human–computer interaction, bioacoustics can be defined as an emerging research area where physiological sounds produced by the human body are used as sound carriers to communicate and control computing systems. The key notion is to take advantage of the inherent human acoustic expressions to create natural, noninvasive, and adaptive interfaces. Accordingly, bioacoustics cover very broad grounds in HCI: from simple voice recognition to health monitoring based on acoustic biomarkers and affective computing. Now, it may also enable modern systems to sense users' physical and emotional states and thus offer them newer paradigms

D. Anitha · N. Kanimozhi (✉) · A. Sheryl Oliver · R. Usharani
Department of Computational Intelligence, School of Computing, SRM Institute of Science and Technology, Kattankulathur, India
e-mail: kanimozn@srmist.edu.in

D. Anitha
e-mail: anithad@srmist.edu.in

A. Sheryl Oliver
e-mail: sheryloa@srmist.edu.in

R. Usharani
e-mail: usharanr2@srmist.edu.in

S. Thirumal
Department of CSE, School of Engineering, Vels Institute of Science, Technology and Advanced Studies, Chennai, India
e-mail: thirumal.se@velsuniv.ac.in

K. Subbaraj et al. (eds.), *Computational Bioacoustic Artificial Intelligence*, Sustainable Artificial Intelligence-Powered Applications, https://doi.org/10.1007/978-3-032-06088-4_13

toward personalization and contextually sensitive interaction experiences.

1.2 Historical Context and Evolution of Bioacoustics

The roots of bioacoustics can be understood to trace back to the early studies on animal communication and echolocation in the mid-twentieth century. Researchers first focused on how animals make and use sound for locomotion, communication, and mating. This foundational work laid the groundwork for bioacoustics as a scientific discipline.

At this time parallel, in the 1970s and 1980s, speech recognition, along with advancements in signal processing, investigated human-produced sounds as forms of input into computational systems. Application domains included ultra-primitive voice-operated switches but also sound signals for heart and lungs diagnostics. Long-term advances in microphone technology, computational power, and machine learning algorithms have transformed bioacoustics from a niche scientific study into a cornerstone of modern HCI research. Nowadays, bioacoustics not only serves as an informant in developing assistive technologies but also inspires innovation in areas such as games, virtual reality, and emotion-aware systems (Uesaka et al. 2023).

1.3 Relevance to Current HCI Paradigms

State-of-the-art HCI paradigms develop systems that have to be natural, accessible, and adaptive to user needs. Bioacoustics contributes to these directions by providing a modality that is per se intuitive and noninvasive. The traditional interface, which used to rely on tactile or visual input, should be complemented with the human-generated sound in all bioacoustic-driven applications. Applications such as virtual assistants using voice, like Siri and Alexa, provide a very practical utility in daily life arising from bioacoustics. Besides voice, breathing or heart rate could become an acoustic signal enabling continuous health monitoring or the recognition of stress for systems adapting to physical and emotional conditions of users. Bioacoustic-driven interfaces may allow persons with various impairments in mobility or visual capability to overcome certain difficulties while interacting. Further, bioacoustics represents a significant enabling aspect in rapidly developing areas such as affective computing, where systems perceive emotional displays to respond personally. In the gaming and virtual reality worlds, bioacoustics allows immersion by way of voice or physiological signals, dynamically in control. In other terms, bioacoustics allows one of the most promising, potent, and polysemic frames for a renewed, innovative HCI which tries to answer the demands toward a much more human-centered interface—a system capable of capturing nuance within the user's self-expressive possibilities (Kobayashi and Matsushima 2014).

Figure 1 shows the progressive development of bioacoustics and its adoption into HCI applications over the decades. Starting in the 1970s, the progress in bioacoustics has been gradually building up through advances in acoustic sensing, signal processing, and computational techniques. The HCI applications, although starting a little later, also picked up quite significantly because of the integration of bioacoustic technologies into user interfaces. By the 2020s, both are almost parallel, meaning there is a very strong interrelation between technological development and application diffusion in interactive systems.

Figure 2 shows the comparison of relevance of bioacoustics with traditional methods for a few important HCI domains. Indeed, in some of these HCI domains, namely voice recognition, health monitoring, and affective computing, bioacoustics seems to outperform traditional methods as far as the subtleties in acoustic features could be interpreted by the former. Regarding accessibility and gaming, while there is still a lead ceded by traditional methods, strong competition has been growing from approaches using bioacoustics for greater adaptability and personalization. These results further support the importance of bioacoustics as a transformative force in modern HCI.

2 Technologies and Methods

2.1 Bioacoustic Signal Sensing Devices

Bioacoustic signal sensing devices form the backbone of systems that rely on human-generated sound for interaction. These are designed to capture acoustic signals with high precision while minimizing the interference caused by environmental noise. The microphone, conventionally used in voice applications, has gone through a metamorphosis of development, with some featuring advanced noise-canceling features that ensure clarity in dynamic environments. Contact sensors, which generally detect vibrations right from the body, are quite useful in picking low-frequency signals, such as heartbeats or breathing sounds. These are generally integrated with wearable technologies that continuously provide physiological activity, such as chest straps or smart textiles. Moreover, wearables containing acoustic sensors—like smartwatches—play a significant role in health monitoring based on the detection and analysis of bioacoustic patterns. Passive while gathering sound, and ambient acoustic sensors are deployed commonly in smart environments, enabling context-aware interactions without the user's need to wear any device. These provide a diversified toolkit for collecting rich acoustic data in diverse settings, thus

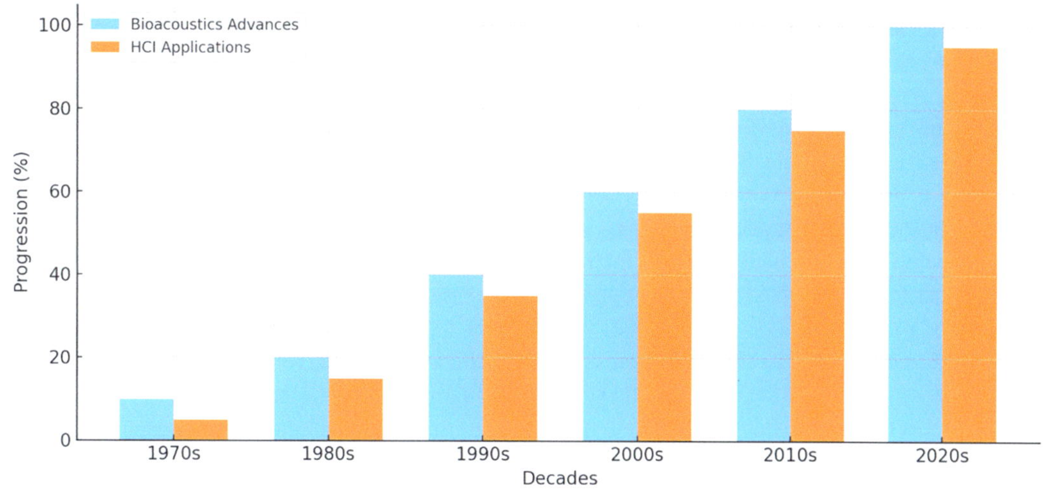

Fig. 1 Evolution of bioacoustics and HCI applications over time

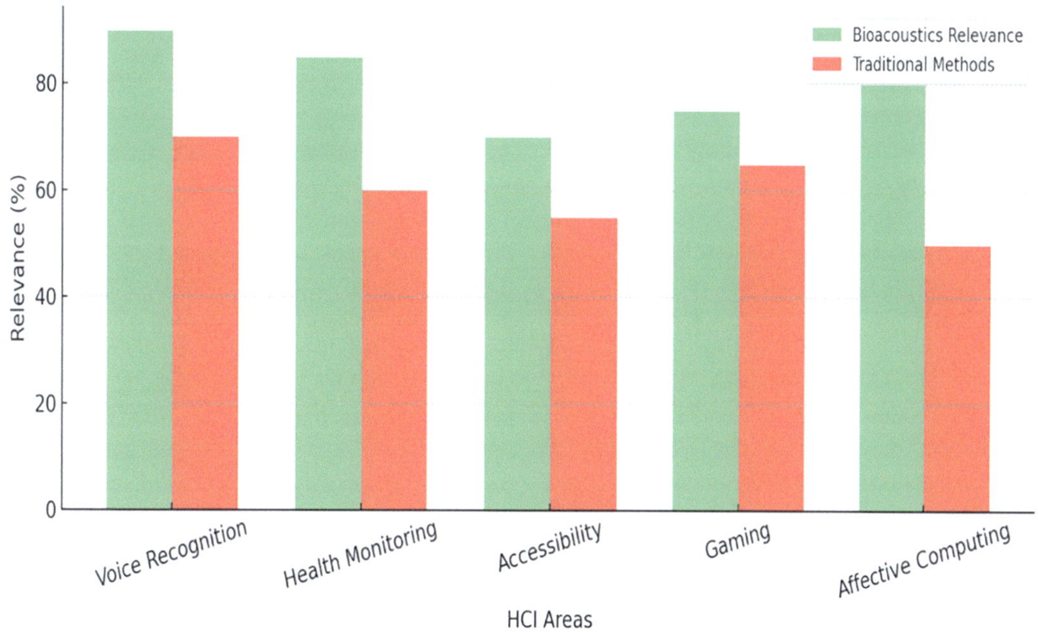

Fig. 2 Comparison of bioacoustics versus traditional methods in HCI areas

enhancing human–computer interfaces (Erceg and Palamas 2023).

2.2 Feature Extraction and Signal Processing Techniques

Feature extraction from the raw signal is a major processing task in bioacoustic data to make it analyzable. Among the major challenges in this area, noise reduction and ensuring signal clarity are normally carried out using techniques such as spectral subtraction and adaptive filtering. These methods remove background interferences and leave a clean representation of the acoustic signal. While time-domain analysis is generally performed on such features as amplitude and signal duration, it yields a basic insight into the nature of the sound (Herman et al. 2015). On the other hand, frequency-domain analysis—through methods like Fourier transform or wavelet transform—provides a much closer look into spectral properties like pitch and harmonics that are responsible for voice interpretation and complex signals. Also, pattern recognition algorithms pick out specific acoustic signatures that allow the system to detect abnormalities, such as problems in breathing or changes in pitch in speech. In conclusion, combining these technologies into one transform raw bioacoustic data into meaningful and structured insights, becoming the basis of interaction and decision-making in systems supported by HCI (Arce-Lopera et al. 2021).

2.3 Machine Learning for Bioacoustic Interpretation

There could presently be a game-changing tool in efforts being invested in the interpretation of all bioacoustic data: bioacoustic data analysis, deriving insights from possibly complex acoustic patterns. Larger datasets can be used for training models that track minute variations in sound inconspicuous to an individual. Advanced neural networking, such as transformers with simple RNNs, speaks to the ability of new advanced development speech recognition systems literally to process and understand spoken word with high accuracy. Emotion detection will consider pitch modulation, rhythm, and tone as acoustic features to make inferences on the emotional state of the speaker, hence, to make the systems contextually aware (Gul and Khan 2023). In health monitoring applications, for instance, conditions from respiratory distress to cardiac anomalies can be classified using acoustic biomarkers with the use of supervised learning models, while unsupervised learning techniques uncover hidden patterns that could give new diagnostic insights. These machine learning algorithms are improved continuously with more diverse datasets to which they are exposed, hence their robust performance across a wide range of applications. Machine learning bridges the gap between raw bioacoustic signals and actionable insights, hence making bioacoustics a reliable and versatile component of human–computer interaction systems (Gupfinger 2020).

Figure 3 depicts the progressive increase in accuracy of microphones and wearable devices for bioacoustic signal sensing throughout the period of the year 2000–2025. They have consistently been ahead of wearable devices because microphones have had better noise cancelation and signal processing. There was rapid improvement by wearables and a closing of the gap to a decent level by 2025, reflecting increased dependability with respect to these wearables in health monitoring applications requiring continuous and noninvasive sensing.

Figure 4 displays the performance comparison between both traditional methods and machine learning methods across different applications in the bioacoustic domain. There is a certain degree through which machine learning outperforms tradition methods in every aspect of speech recognition and emotion detection. Health monitoring and pattern recognition are their weakest points because they have failed to process such complex or high-dimensional acoustic data. The results highlight the transformation that machine learning has affected in extracting nuanced insights from bioacoustic signals; this makes it the go-to approach for modern HCI systems.

3 Ethical and Privacy Issues in Bioacoustic HCI

3.1 Data Ownership and Consent

The integration of bioacoustics in HCI raises critical questions regarding data ownership and consent from users. These data are indeed very personal and sensitive, often carrying much private information in voice patterns, heartbeats, and respiratory sounds. Unlike general user data, bioacoustics signals can provide health conditions, emotional states, and behavior patterns that could lead to harm if misused. Therefore, clear policies about data ownership are needed to let people keep control over their biometric information. Fully informed consent will be one of the cornerstones in the ethics of data collection. Users should be informed about what bioacoustics data is collected, how it will be used, and who will have access to it. Furthermore, consent mechanisms should be transparent and revisitable, allowing users at any given time to revoke permissions (Mace et al. 2013). In addition to explicit consent, there is a growing need for models of implied consent

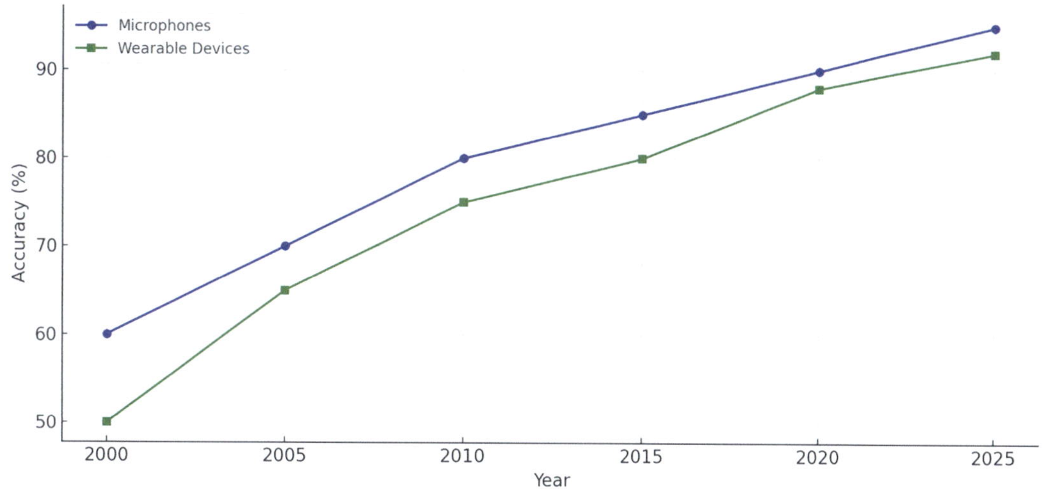

Fig. 3 Accuracy of bioacoustic sensing devices over time

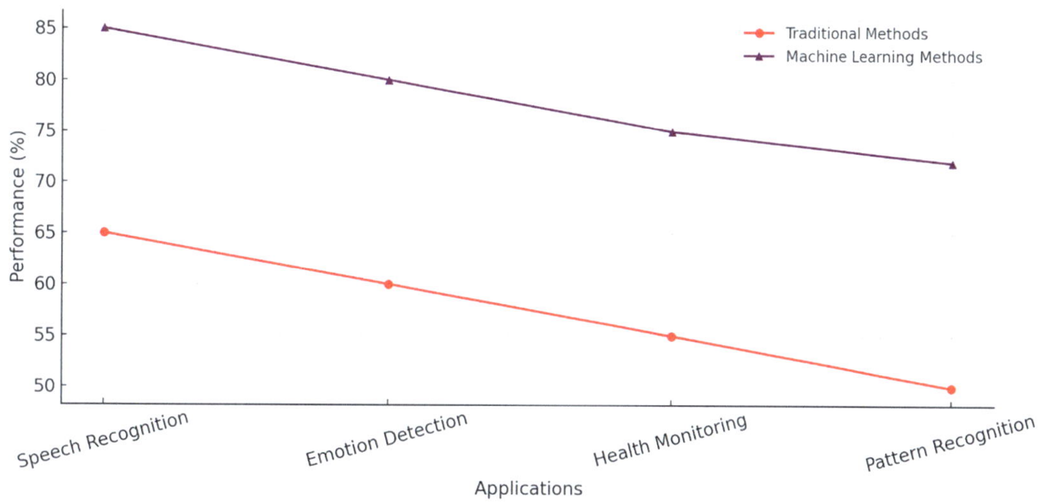

Fig. 4 Performance comparison of traditional versus machine learning methods in bioacoustic applications

that respect user autonomy while responding to the many situations where explicit permissions are not possible. For example, in smart environments using ambient bioacoustics sensors, default privacy settings should ensure minimal data collection and anonymization. Such organizations should also provide data governance frameworks to ensure the utilization of bioacoustics data for purposes consented to by the provider and restricted from unauthorized access. While the General Data Protection Regulation, among others, provides guidelines toward ensuring compliance, there is a pressing need for continuous innovation within the practices that handle data to meet the challenges presented by bioacoustics data in unique ways (Zilli 2015).

3.2 Privacy-Preserving Techniques in Acoustic Data Handling

In this regard, privacy-preserving techniques have evolved to be an integral requirement when handling bioacoustic data. Among the most followed approaches is data anonymization, entailing the removal of identifiable information from the raw data so that it cannot be traced back to a certain individual. A number of techniques, such as masking voice, obfuscation of frequency, or conversion of acoustics into nonreversible features, can effectively protect user identity while still retaining the functional aspects. It has been found that edge computing is an efficient technique of preserving privacy (Longdon 2023). Instead of sending raw bioacoustic data to the main servers, processing will take place on a local device itself. The sensitive data does not go outside the environment of users, reducing breaches and other forms of unauthorized surveillance with considerable importance. With FL, each node in the federation can jointly train machine learning models without actually moving their private data, leveraging advantages in large-scale sets but with guarantees of data privacy. In practice, encryption techniques form one of the most important means for bioacoustic data protection during its transmission and storage. For instance, homomorphic encryption is an advanced class of cryptographic methods that enables operations on encrypted data without exposing its content and hence ensures privacy even in cloud-based processing scenarios. The security of the data is further enhanced by the use of access control mechanisms, preventing it from being made available to unauthorized users or systems (Glodek et al. 2015).

Another alternative can be guaranteed through promising differential privacy, blurring individual contribution by adding noise into the data. Such an approach opens much promise for computing aggregate outputs by using bioacoustic data and guaranteeing data protection at an individual level. However, how to achieve an appropriate trade-off between the privacy and usefulness of data contributions is still a subject of active research. It takes time for effective implementation and adaptation. By implementing these strategies, the bioacoustic HCI system may address ethical concerns and foster trust among users toward wider adoption, but it has to be balanced against safeguarding privacy.

Figure 5 shows the perceived importance of a number of ethical considerations regarding the treatment of bioacoustic data. Ranked first position is the "Informed Consent," meaning the utter seriousness taken in making sure the end is well-informed and retaining user control. "Data Ownership" can also be considered another strong need to be satisfied, so that the subject remains authoritative in respect to their bioacoustic information. "Data Governance" and "Unauthorized Access Prevention" stay near the top, both of which speak to the same need: strong frameworks and security measures should be in place to ensure that sensitive data is well cared for. These results underline, from an ethical standpoint, the foundational role of trust in ensuring that bioacoustic HCI technologies will be adopted.

Figure 6 also puts to a test the effectiveness of various techniques for preserving privacy. "Encryption" holds the highest rating, pointing to its vital role in data security both during transmission and storage. "Edge Computing" also stands at a high rating because it can keep data local and reduce exposure to breaches. "Anonymization" is effective but slightly less so, since it might limit advanced data analysis. The privacy technique of "Differential Privacy" is rated a bit lower due to the fact that striking a balance between privacy and utility is problematic; it generally contributes to solving all the privacy problems in bioacoustic HCI systems.

4 Multimodal Integration in HCI

4.1 Combining Bioacoustic Signals with Visual and Tactile Inputs

Multimodal integration in human–computer interaction involves the use of multiple senses concurrently, such as bioacoustic, visual, and tactile signals, to build more robust, adaptive, and natural systems. Bioacoustic signals, for instance, speech or physiological sounds, are usually enriched when combined with visual and tactile inputs. Such fusion enhances the accuracy, usability, and overall effectiveness of interactive systems, especially in complex environments. For instance, facial expressions detected by visual sensors in speech-based applications, integrated with bioacoustic signals such as tone or pitch, naturally enhance system understanding of user intent and emotion. These combined inputs are crucial for applications involving virtual assistants; both the verbal cues and facial gestures bring about contextual improvement. In the case of accessibility technologies also, input methods like haptic feedback supplement the bioacoustic data with the intention of helping people who cannot hear or have speech impairments (Pantidi et al. 2014). For example, a wearable device can utilize bioacoustic sensing in the detection of vibration of vocal cords and then combine the same with tactile feedback to provide real-time alerts to the user regarding the pitch or volume of one's voice. This could mean that when the bioacoustic inputs are degraded because of noisy environments, the supplementary channels could be visual data from lip reading or tactile feedback to maintain continuous communication. These multimodal systems increase error tolerance and make HCI more flexible toward real-world scenarios, where variability is inevitable (Johansen et al. 2022).

4.2 Synergies Among Various Sensing Modalities

This synergy will go beyond the simple aggregation of bioacoustic, visual, and tactile modalities by leveraging strengths of each sensory input to overcome the limitations of others, creating a holistic and cohesive interaction model. Bioacoustic signals, for example, are very good at capturing dynamic real-time information, such as spoken commands or physiological changes. On the other hand, the visual modality provides spatial and contextual awareness, such as recognizing gestures or facial expressions. Haptic signals are great for providing direct and immediate feedback to the user in a way that can make interactions more intuitive and engaging. These all together are going to be especially influential in areas like AR/VR. In these, the bioacoustic inputs enable the interaction by voice commands, while tracking of the body movements is due to visual sensors and the tactile systems, providing realistic

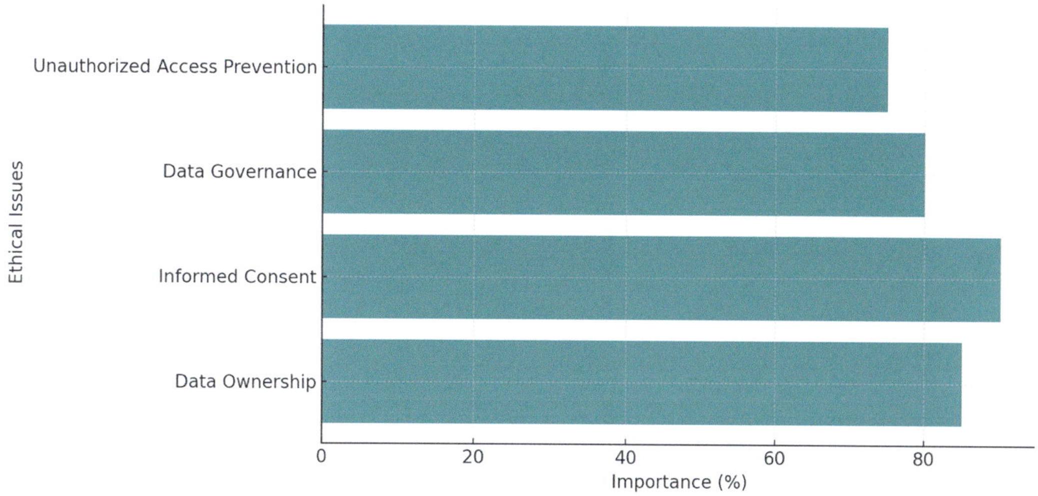

Fig. 5 Importance of ethical considerations in bioacoustic data

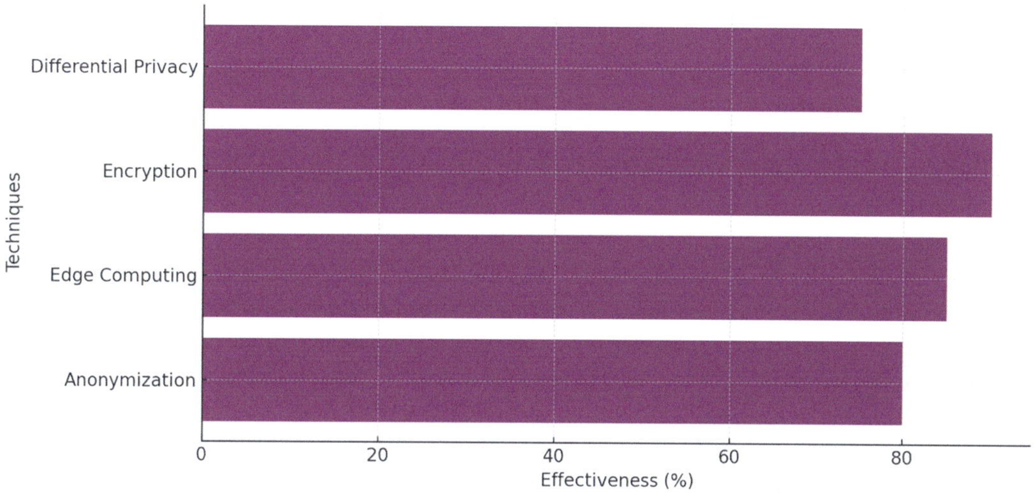

Fig. 6 Effectiveness of privacy-preserving techniques in bioacoustic HCI

feedback through haptic devices (Johansen et al. 2022). Such integrations open a continuous and more natural user experience to numerous applications in gaming, virtual collaboration, and training by using simulations. Medicine is one field that greatly benefits from this integrative approach in human–computer interaction. At any case, patient monitoring systems integrating direct biosensor heart rate with contextual interpretation of visual facial cues enhanced by wearable device tactile stimulation can deliver current health metrics. The result is then a much deeper layer necessary for better diagnostics and efficiency. The layered approach allows higher-level approaches concerning diagnostic accuracy and responsiveness associated with critical events (Praveen and Pabitha 2023).

Figure 7 shows that the most contribution regarding accuracy comes from the visual inputs, 40%, due to their rich spatial and contextual information such as gestures and facial expressions. The bioacoustic signals contribute to 35% because they excel in dynamic and real-time information such as speech and physiological sounds. The tactile input allows for only 25% because it principally avails mechanisms of feedback. On the other hand, tactile inputs are just as important in terms of engagement, 35%, since they offer immediate and direct feedback, hence making interactions more immersive. The share of visual inputs is also quite big—35%—to make users engaged by intuitive visual cues. Bioacoustic signals contribute a bit less, 30%, but it is important for natural and conversational interaction.

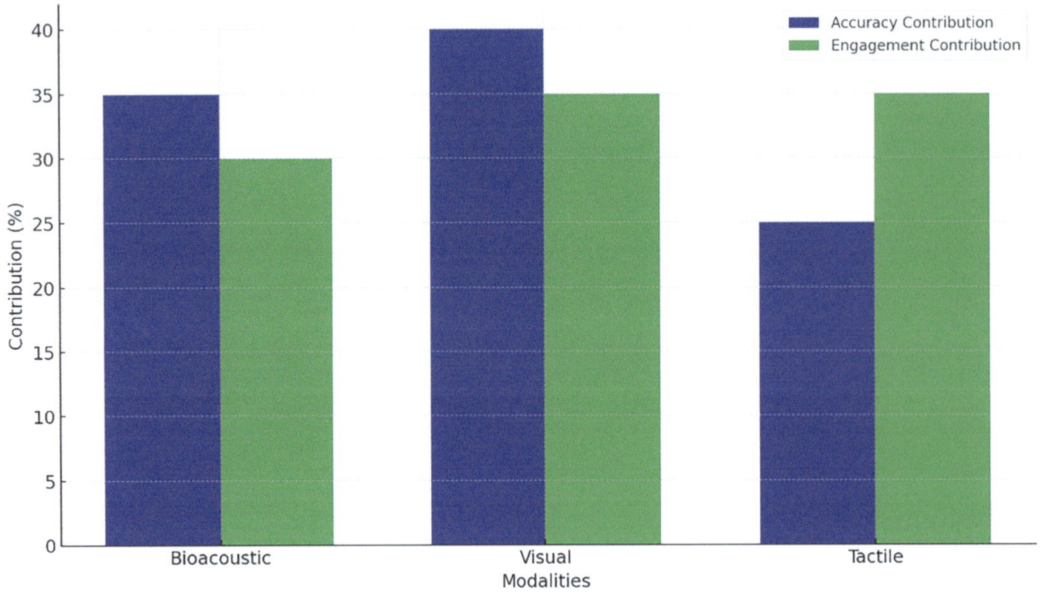

Fig. 7 Comparison of modalities in multimodal HCI

5 Cultural and Accessibility Implications

5.1 Designing Systems for Diverse Linguistic and Cultural Contexts

Human–computer interaction systems that use bioacoustics have to consider linguistic and cultural varieties in diverse usage situations. There are substantial phonetic, syntactic, and prosodic differences among languages, which sometimes influences the bioacoustic systems, especially those that depend on speech recognition and emotion detection. For instance, tonal languages, such as Mandarin, carry meaning in pitch variations, and systems would have to adjust their algorithms in signal processing (Bergström and Hornbæk 2019). Similarly, regional accents and dialects introduce variability that would need to be dealt with if inclusivity and effectiveness were to be attained. Cultural factors also shape the way different people interact with technology. Such attributes are the intonation of the voice, rhythm of speech, and pause usage, which vary among different cultures and may affect how the user's intent is perceived. Designing culturally sensitive systems requires large, varied datasets and robust training methods in order to capture this variability. Systems must also be respectful of cultural norms for data privacy and usage. Moreover, the reasons for vocal data sensitivity include certain cultural aspects that make high demands on privacy for their trust to be built, resulting in its acceptance. Other very important reasons include localization, which goes well beyond the simple translation of text-based outputs and involves changes in the behavior of the system in relation to what cultural expectations predict. For example, the tone and formality in the system's response should correspond to the cultural norms so as not to create any impression of unnaturalness and discomfort while interacting with the system (Stanney and Salvendy 1998).

5.2 Increasing Accessibility for People with Disabilities

The bioacoustic HCI systems hold huge potential to make lives more accessible for people with disabilities by providing methods of interaction other than conventional ones, which include sound and voice. In cases where there is motor impairment, voice-controlled systems offer the possibility of hands-free control of devices, enhancing autonomy and convenience. In turn, visually impaired people will find voice-based interfaces, replacing visual input with audio, a very effective means of interaction. Such technologies, including speech synthesis and speech recognition, are of great value for people with speech disorders (Samuel et al. 2024). Advanced systems catch on even to very weak and incomprehensible voice patterns by the user, so the system may understand their will and help them communicate their intention clearly. When speech might be impossible, systems based on other types of signals, especially bioacoustic signals, such as throat vibration or even humming, may allow for alternative modes of communication. Accessibility features for speech also cover the needs of persons with hearing impairments. Such systems would bridge the gaps in communication by integrating bioacoustic inputs with visual or haptic feedback. For example, speech-to-text transcription in real-time or avatars that use sign language provide an accessible medium for interaction. Moreover, the information conveyed through haptic devices using vibration can be integrated into bioacoustic data as a means to alert the user to sounds occurring within one's environment. To be inclusive, designers have to be user-centered and involve people with disabilities right from the development stage. Such collaboration makes systems not only functional but intuitive and empowering. By placing cultural sensitivity at the forefront and making the systems accessible, bioacoustic HCI systems can serve a wider population and hence promote equity and inclusion in technology (Babu et al. 2023).

The comparison between general HCI systems and bioacoustic HCI systems shows that the latter performs better within both cultural and accessibility contexts. While addressing cultural diversity and linguistic adaptability, bioacoustic HCI systems are far more effective because they process various accents, tonal variations, and cultural nuances with greater precision. Also, concerning the accessibility point of view, bioacoustic systems offer especially improved performance by voice control and auditory feedback for users with motor and visual impairments. In addition, their adaptability is further underlined with their improved support for hearing impairments achieved by integrated visual and tactile feedback mechanisms. These results are really indicative of the potential that bioacoustic HCI systems have in consideration of inclusivity and sensitivity from a cultural perspective. Basically, these systems come with an advantage over other generic HCI technologies, considering aspects regarding personalization and effectiveness (Fig. 8).

6 Real-Time Processing and Computational Challenges

6.1 Optimizing Algorithms for Latency and Efficiency

Any good bioacoustic-driven HCI will necessarily hinge on real-time processing, whereby systems need to process inputs instantaneously and act on them. This makes it one of the major challenges for achieving low latency with high accuracy. Bioacoustic signal processing algorithms are needed for optimization in their performance on big datasets, filtering of noise, and extraction of relevant features. The key methods of optimization involve the use of light algorithms that are modified for speed without compromising functionality. FFT algorithms, for example, are typical in the spectral analysis of bioacoustics, since this is computationally efficient. The same goes for machine learning models designed for edge devices: quantized neural networks allow for real-time processing by reducing the computational load (Nagayama and Takada 2023).

Another key optimization strategy involves parallel processing. By performing tasks in parallel across more than one core or processor, systems can perform features extraction, classification, and real-time feedback all at once. Besides that, adaptive algorithms may dynamically adapt resource allocation to the conditions of a workload, minimizing delays and further enhancing responsiveness.

6.2 Hardware Constraints and Creative Solutions

Among the most important issues for real-time bioacoustic processing systems, hardware limitations rank high. In wearable devices and edge computing systems, widely used in bioacoustic HCI, there are considerable limitations regarding power, memory, and computational budget. To overcome these limiting constraints, some innovative solutions have emerged. Energy-efficient processors are replacing the processors in many bioacoustic devices that support low-power execution. Traditional processor architecture, such as RISC, is used here, but much enhanced and optimized in performance for higher processing speed with minimum energy consumption. Accelerators like GPUs and TPUs are one of the pivotal hardware that accelerates machine learning applications or higher-order signal processing applications (Lin and Fernando 2023).

The coming of edge computing has changed the way in which bioacoustic data is being carried out. Rather than sending all raw data to the server, it involves the local preprocessing by edge devices, thus reducing latency and bandwidth. This, put together with the miniaturization of sensors, allows carrying out real-time and high-fidelity bioacoustic sensing and processing. Also, deeper integration with custom hardware, including application-specific integrated circuits and FPGAs, allows degrees of customization for specific tasks unprecedented until recently. This will enable the implementation of tailored, compact systems to deliver very particular demands imposed on certain bioacoustic human–computer interaction applications. Algorithms are continuously in refinement, but the introduction of new, emerging hardware solutions will further bring down the challenge toward real-time processing in bioacoustics-driven HCI. These developments open the way to responsive, efficient, and scalable systems for a wide variety of applications and user needs (Mathew et al. 2024).

7 Applications

7.1 Voice-Driven Interfaces

Among all applications, voice-driven interfaces represent one of the most popular usages of bioacoustics-driven HCI. These systems use bioacoustics signals, such as speech and vocal

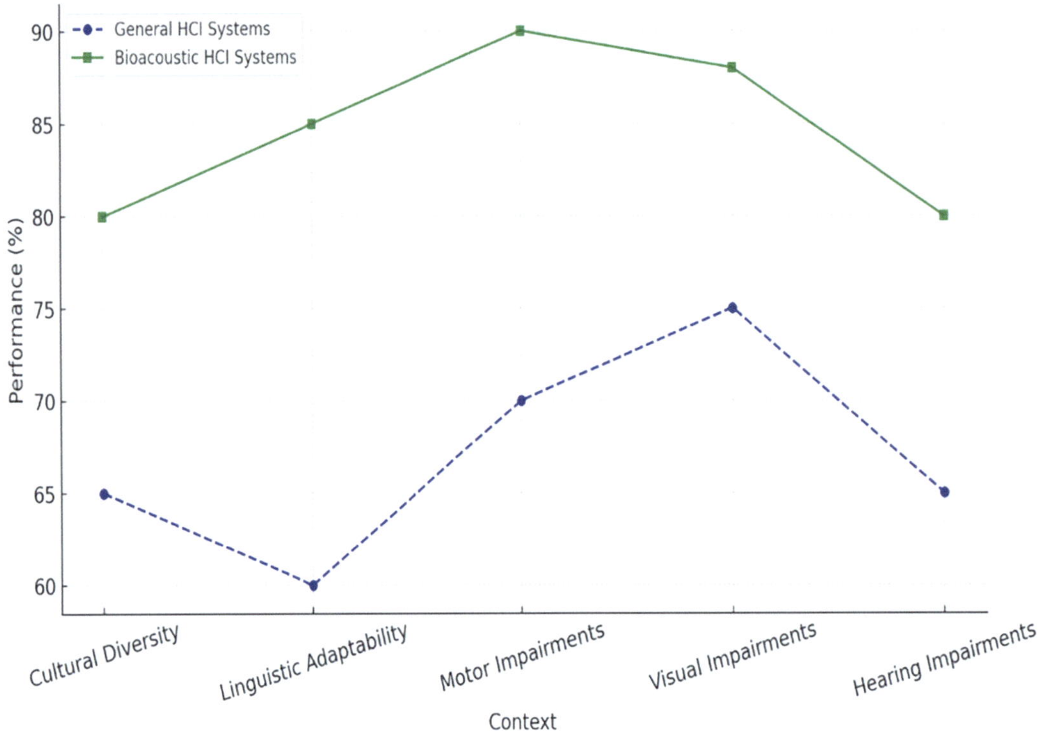

Fig. 8 System performance across cultural and accessibility scenarios

commands, to offer seamless interaction between users and machines. The applications include virtual assistants like Alexa, Siri, and Google Assistant, which can set reminders, search the Web, or control smart home devices by the user with natural language. Advanced voice-driven systems understand nuances, accents, and even intent-very important in hands-free interaction. Beyond personal devices, this interface finds wide application in enterprise environments: voice-controlled customer service platforms, accessibility tools for people with serious motor disabilities, and more (Wang 2025).

7.2 Emotion and Stress Detection Based on Acoustic Biomarkers

Bioacoustic signals are of great value in the detection of emotional states and stress levels, serving as a window to the user's psychological condition. Variations in tone, pitch, speech rate, and vocal intensity provide acoustic biomarkers that systems analyze for emotions such as happiness, sadness, anger, or stress. These capabilities are particularly useful in applications like mental health monitoring, where the early detection of signs of stress or depression may lead to timely intervention. Emotion detection is also put into practice in customer service, where automated systems can adapt their responses according to the emotional state of the user, thus improving customer satisfaction (Manikandan and Jayashree 2024).

7.3 Healthcare and Rehabilitation Applications

The transformative potential of bioacoustics HCI systems could exist in the domain of healthcare and rehabilitation. Generally, the bioacoustic system monitors the various physiological signals related either with the breathing sounds or to those vibrating in the process, thereby facilitating diagnosis and managing mechanisms. For example, detection analysis of acoustic cough sounds will result in determining diseases such as pneumonia, asthma, and other respiratory problems associated with a patient's human organs (Napier et al. 2024). Similarly, voice analysis is being investigated as an early detector of neurodegenerative diseases like Parkinson's, which may manifest first in altered vocal patterns. In rehabilitation, bioacoustics systems can provide personalized therapy. For patients recovering from speech impairments, such as post-stroke, real-time feedback regarding their vocal exercises is possible. Wearable devices also help persons with physical disabilities use their non-verbal utterances or throat vibrations to communicate with assistive technologies fitted with bioacoustics sensors.

7.4 Gaming and Immersive Technologies

Bioacoustic-driven HCI enhances the realism and engagement of players in video gaming and other virtual environments through voice commands. Gamers can give orders, move within the virtual place, talk with other members, or issue orders during a game using their voice. It also comprises emotion detection systems that adjust game parameters based on the detected emotional state of the user to increase the feeling of immersion. For instance, the difficulty level of a game can be changed upon recognition of stress or frustration in a player's voice to further facilitate personal and enjoyable experiences. Speaking of virtual and augmented reality, bioacoustic HCI adds one more level of interaction that feels natural and life-like. Coupled with emotion detection, voice-driven commands make these systems capable of providing users with interaction with virtual environments that is very similar to real life. Thus, such systems are very suitable for training simulations, education, and entertainment (De Boeck and Vaes 2024).

Figure 9 shows the adoption rates of bioacoustic-driven HCI applications, representing their ever-increasing importance in a number of domains. All those refer to voice-driven interfaces, leading the position as shown in Fig. 9, with 85% of the share because of the enormous usage in virtual assistants and hands-free interaction technologies. Healthcare and rehabilitation come next with an adoption rate of 80%, indicating how important this bioacoustic system is in medical diagnostics and therapy on a personalized basis. Emotion and stress detection technologies have a very high current adoption rate, at about 75%, hence drawing in more reasons of relevance regarding their roles in both mental health monitoring and the enhancement of customer service. At 70% is where gaming and immersive technologies are, revealing how bioacoustic systems may take on major roles within personalized and engaging user experience creation. The graph, generally, underlines the potentiality and flexibility of Bioacoustic HCI systems with regard to catering to highly varied user needs.

8 Challenges and Limitations

8.1 Signal Variability and Noise Handling

In general, one of the main challenges associated with bioacoustics systems is signal variability and real-world noise. Acoustic signals may refer to speech, heartbeats, breathing sounds, and so on, which differ a lot among individuals due to their age, health condition, and conditions of surroundings. This high variability of signals complicates the development of algorithms that generalize well to diverse users. Another big concern is the noise, mainly in dynamic or outdoor settings, that easily masks and interferes with the process of detection and interpretation of bioacoustics signals. Background chatters, machinery noise, or possibly overlapping speech degrades system accuracy. Advanced noise reduction techniques like adaptive filtering and deep-learning-based denoising algorithms are required for enhanced signal clarity. These, however, mostly require huge computational resources, adding up to system design complexity (Devi et al. 2025).

8.2 Ethics and Privacy Issues in Bioacoustic Data Utilization

The collection and analysis of bioacoustic data raise severe ethical and privacy concerns. Bioacoustic signals, such as speech patterns or physiological sounds, can contain very private information about the emotional condition, health condition, or behavioral pattern of a person. Making sure users provide informed consent for collecting data is thus a key ethical requirement. Misuse of bioacoustic data in unauthorized surveillance and further data breaches increase the related privacy concerns. In this vein, systems shall be designed to provide means for limiting these risks, including data anonymization, encryption, and performing all the processing on a device to limit the raw data disclosures. It should have transparent policy for usage of data along with following the regulatory standards set by GDPR or HIPAA standard for user trust.

8.3 Technical Limitations in Real-Time Processing

The most significant challenge facing bioacoustic systems is real-time processing, expected in general by those applications that require immediate feedback or instant decisions. Feature extraction and classification steps within the computational pipeline for processing bioacoustic signals introduce latencies that can be quite significant, especially when such processing has to be accomplished on resource-constrained devices. Limitations due to hardware include lower processing power, memory, and energy, making things more difficult. While recent breakthroughs in edge computing and dedicated hardware-like GPUs and TPUs alleviate a few of these, getting low latency without sacrificing much performance remains an endeavor. Moreover, increased interest in integration with multiple modalities—visual or even tactile input—provides further challenges for the developers themselves in designing algorithms, but also pushes the limits of what software and hardware can do together effectively. Conveyance of these challenges is of paramount importance: a multidisciplinary

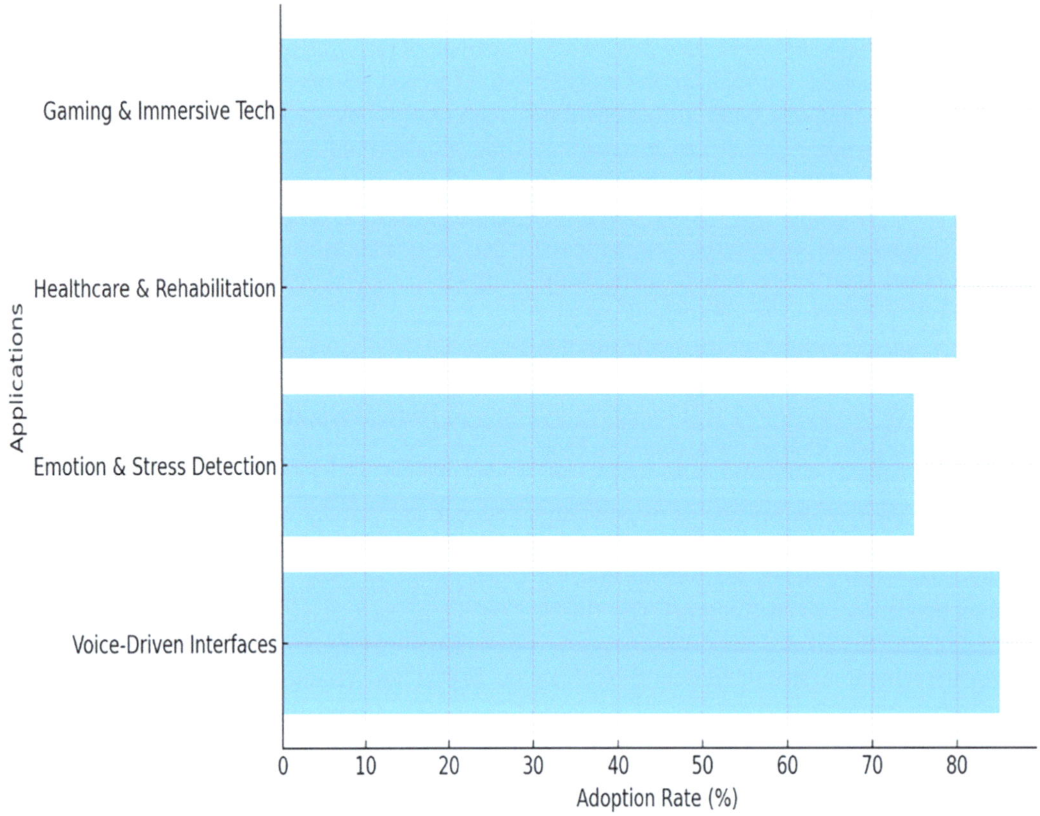

Fig. 9 Adoption rates of bioacoustic-driven HCI applications

process of innovation in signal processing and machine learning, hardware design, and ethical practices with respect to using bioacoustically-driven human–computer interaction systems should make them effective and responsible (Nagayama and Takada 2023; Story et al. 2024; Chulev 2024).

Figure 10 highlights the relative impact level generated by key challenges of bioacoustic-controlled HCI systems: the impact level of variability in signals and noise management reaches a mark of 85%. This would be supported by the significant hurdle posed by attempting to deal with rich acoustic variations while reducing noises, hence restricting its feasibility in different natural settings. Limitation by real-time processing is almost inevitable, at about 80%, depicting significant computational overhead together with particular hardware constraints that limit performance at low latency. While ethical and privacy concerns are somewhat lower, at 75%, these too are critical in view of the sensitivity of bioacoustic data, which demands adequate measures with regard to data protection. All the above challenges cumulatively indicate a development requirement in signal processing, ethical governance, and computational efficiency to enhance effectiveness and promote the adoption rate of bioacoustic HCI systems.

9 Conclusion

Bioacoustic-driven HCI is a transformative frontier in technology that bridges the natural signals of the human body with computational systems. This chapter reviewed the basic concepts, technologies, and various applications of bioacoustic HCI, together with the ethical, cultural, and technical challenges forming the basis for its development. Integrating bioacoustic signals with multimodal inputs has immense potential for enhancing interactivity, accessibility, and personalization across domains. Bioacoustic HCI has kept rewriting the ways of human–machine interaction, from voice-driven interfaces and emotion detection to healthcare diagnostics and immersive technologies. Real-time processing algorithms, novel hardware, and techniques for preserving privacy have been critical in tamping down the computational and ethical challenges; however, overcoming limitations with respect to signal variability and system adaptability requires ongoing

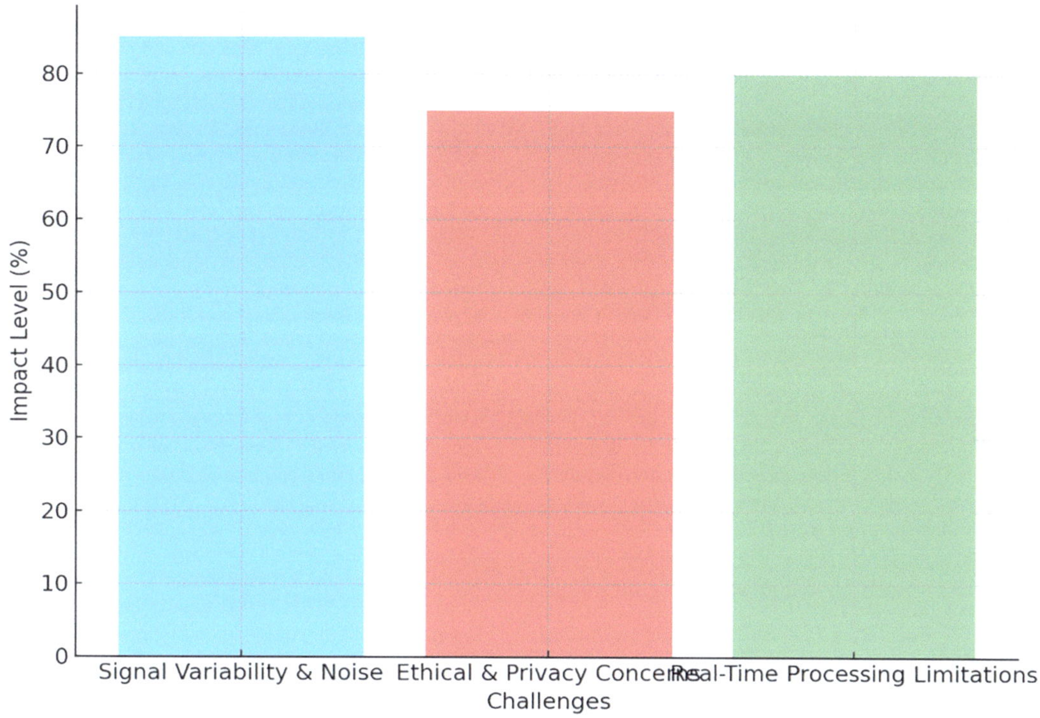

Fig. 10 Challenges and limitations in bioacoustic HCI

effort. Moreover, the cultural and accessibility implications of bioacoustic HCI are strong incentives for the inclusion of diverse linguistic, cultural, and physical needs. By using bioacoustic signals, systems can cater to underrepresented populations and increase their usability across the world. As research and development progress, the future of bioacoustic HCI lies in seamless integration with other sensing modalities, considering ethical considerations well in advance, and utilizing artificial intelligence for further refinement in terms of accuracy and responsiveness. This is an interdisciplinary field that could bring changes to HCI paradigms by making human-centered technologies more intuitive, efficient, and inclusive than ever. By understanding the challenges and opportunities presented in this chapter, researchers, developers, and stakeholders will contribute toward responsible and innovative growth in bioacoustic-driven systems for human–computer interaction, confirming relevance and impact beyond the years.

References

Arce-Lopera C, García-Muñoz P, Restrepo-Quiceno S, Gómez-Marín D, Londoño GA (2021) Designing data visualization assistance for a bioacoustics labeling software. In: HCI international 2021-posters: 23rd HCI international conference, HCII 2021, Virtual Event, 24–29 July 2021, Proceedings, Part III 23. Springer International Publishing, pp 475–482

Babu R, Kannappan J, Viswanathan Krishna B, Vijay K (2023) An efficient spam detector model for accurate categorization of spam tweets using quantum chaotic optimization-based stacked recurrent network. Nonlinear Dyn:1–18

Bergström J, Hornbæk K (2019) Human–Computer interaction on the skin. ACM Comput Surv (CSUR) 52(4):1–14

Chen W (2025) On-skin interaction system and smart wearable research based on innovative gesture input. In: International conference on human-computer interaction. Springer, Cham, pp 387–406

Chulev J (2024) Improving musical instrument classification with advanced machine learning techniques. arXiv preprint arXiv:2411.00275

Ciaburro G (2025) Automated home security system based on sound event detection using deep learning methods. In: Modern advancements in surveillance systems and technologies. IGI Global Scientific Publishing, pp 273–302

De Boeck M, Vaes K (2024) Human augmentation and its new design perspectives. Int J Design Creativity Innov 12(1):61–80

Devi GD, Kumar N, Manikandan J, Rekha V (2025) Innovative brain tumor detection: Stacked random support vector-based hybrid gazelle coati algorithm. Biomed Sig Process Control 101:107156

Erceg M, Palamas G (2023) Towards harmonious coexistence: a bioacoustic-driven animal-computer interaction system for preventing ship collisions with North Atlantic Right Whales. In: Proceedings of the tenth international conference on animal-computer interaction, pp 1–10

Glodek M, Honold F, Geier T, Krell G, Nothdurft F, Reuter S, Schüssel F et al (2015) Fusion paradigms in cognitive technical systems for human–computer interaction. Neurocomputing 161:17–37

Gul S, Salman Khan M (2023) A survey of audio enhancement algorithms for music, speech, bioacoustics, biomedical, industrial and environmental sounds by image U-Net. IEEE Access

Gupfinger R (2020) The design of musical instruments for grey parrots: an artistic contribution towards animal-human-computer interaction. University of Art, PhD diss

Herman I, Impett L, Wollner PKA, Blackwell AF (2015) Augmenting bioacoustic cognition with tangible user interfaces. In: Foundations of augmented cognition: 9th international conference, AC 2015, held as part of HCI international 2015, Los Angeles, CA, USA, 2–7 August 2015, Proceedings 9. Springer International Publishing, pp 437–448

Johansen SS, Van Berkel N, Fritsch J (2022) Characterising soundscape research in human-computer interaction. In: Proceedings of the 2022 ACM designing interactive systems conference, pp 1394–1417

Kobayashi HH, Matsushima J (2014) Basic research in human–computer–biosphere interaction. Buildings 4(4):635–660

Lin YH, Noel O, Fernando N (2023) Animal hunt: AI-based animal sound recognition application. In: International conference on human-computer interaction. Springer Nature Switzerland, Cham, pp 465–474

Longdon J (2023) Visualising forest sound: justice-led ecoacoustic data interaction. In: Extended abstracts of the 2023 CHI conference on human factors in computing systems, pp 1–5

Mace M, Abdullah-al-Mamun K, Naeem AA, Gupta L, Wang S, Vaidyanathan R (2013) A heterogeneous framework for real-time decoding of bioacoustic signals: applications to assistive interfaces and prosthesis control. Expert Syst Appl 40(13):5049–5060

Manikandan J, Jayashree K (2024) Enhancing lung nodule classification: a novel CViEBi-CBGWO approach with integrated image preprocessing. J Imag Inf Med 37(5):2108

Mathew R, Tigwell GW, Peiris RL (2024) Deaf and hard of hearing people's perspectives on augmented reality interfaces for improving the accessibility of smart speakers. In: International conference on human-computer interaction. Springer Nature Switzerland, Cham, pp 334–357

Nagayama K, Takada R (2023) Unvoiced vowel recognition using active bio-acoustic sensing for silent speech interaction. In: International conference on human-computer interaction. Springer Nature Switzerland, Cham, pp 150–161

Napier T, Ahn E, Allen-Ankins S, Schwarzkopf L, Lee I (2024) Advancements in preprocessing, detection and classification techniques for ecoacoustic data: a comprehensive review for large-scale passive acoustic monitoring. Expert Syst Appl:124220

Pantidi N, Moran S, Bachour K, Rodden T, Zilli S, Merrett G, Rogers A (2014) Field testing a rare species bioacoustic smartphone application: challenges and future considerations. In: 2014 IEEE international conference on pervasive computing and communication workshops (PERCOM WORKSHOPS). IEEE, pp 376–381

Praveen R, Pabitha P (2023) A secure lightweight fuzzy embedder based user authentication scheme for internet of medical things applications. J Intell Fuzzy Syst 44(5):7523–7542

Samuel KV, Krishna P, BV et al (2024) Exploration of sentiment analysis in twitter propaganda: a deep dive. Multimed Tools Appl 83:44729–44751. https://doi.org/10.1007/s11042-023-17383-6

Stanney K, Salvendy G (1998) Aftereffects and sense of presence in virtual environments: formulation of a research and development agenda. Int J Hum-Comput Interaction 10(2):135–187

Story B, Gillespie P, Derryberry G, Derryberry E, Fefferman N, Maroulas V (2024) DialectDecoder: human/machine teaming for bird song classification and anomaly detection. Ecol Inf 102657

Uesaka L, Khatiwada AP, Shimotoku D, Parajuli LK, Pandey MR, Kobayashi HH (2023) Applications of bioacoustics human interface system for wildlife conservation in Nepal. In: International conference on human-computer interaction. Springer Nature Switzerland, Cham, pp 172–181

Zilli D (2015) Smartphone-powered citizen science for bioacoustic monitoring. University of Southampton, PhD diss

Ethical Considerations and Broader Impact

Ethical Considerations and Societal Impact

R. Saranya, G. Vijayakumari, S. Srividhya, V. Karthi, and Mahmoud Ahmad Al-Khasawneh

Abstract

The computational bioacoustic Artificial Intelligence (AI) combines machine learning with bioacoustic data to assess animal vocalizations for conservation actions and ecological monitoring. This technological development raised the importance of societal and ethical issues. Privacy is a key concern since sensors put in natural environments may inadvertently record human noises. This is because there is a risk of data exploitation and spying. In addition, problems with animal privacy and the ethical treatment of species arise, especially when invasive techniques such as tagging are employed. Ownership of the data is an additional significant factor, particularly when performing research in areas where indigenous communities reside. The use of bioacoustic data comes from the particular location should maintain the control over it to ensure the importance of the data. The spices or ecosystem which has received less attention will face the risk of biasing conversation effects, compared to the species models which are trained largely for a particular face. In order to improve the conservation efforts at the societal level through the use of bioacoustic AI, by providing a real time data to support environmental regulation. The developed countries and universities may lead this field due to difference in admission of technology which could downgrade the researchers from developing nations. Once the researchers performed the AI automates the events and offers a new opening for AI experts in conservation and increase the concerns about job displacement. Intensive use of AI operations has huge data processing with a carbon impact on environment that needs to be balance with sustainability goals. To ensure a control AI development and usage, ethical outlines under the governance point of view needs to be established. Hence, the transparency in decision-making process, inclusive community engagement, and accountability for AI-driven results are important. The public awareness of conservation shall be done through educational platforms but ensure that inaccurate information does not spread. Apart from benefits of studying biodiversity, the development and implementation of computational bioacoustic AI ultimately frame a strong ethical foundation for the benefit of both the scientific community and the general public profit. This chapter explores the positive approach of bioacoustic AI for both scientific discovery and environmental sustainability by determining the issues related to data ownership and access, privacy, algorithm bias, and community involvement. Addressing these issues can ensure that the social impacts are positive and bioacoustic AI will serve both ecological and social welfare.

Keywords

Computational bioacoustic AI · Environmental sustainability · Data exploitation · Conservation ·

R. Saranya (✉) · V. Karthi
Department of EEE, Kangeyam Institute of Technology, Nathakadaiyur, Tirupur, India
e-mail: saranyaprabhu@gmail.com

V. Karthi
e-mail: karthi.eee@kitech.edu.in

G. Vijayakumari
Department of ECE, Kangeyam Institute of Technology, Nathakadaiyur, Tirupur, India
e-mail: viji.ece05@gmail.com

S. Srividhya
Department of Civil, Kangeyam Institute of Technology, Nathakadaiyur, Tirupur, India
e-mail: srividhya132@gmail.com

M. A. Al-Khasawneh
School of Computing, Skyline University College, University City Sharjah, Sharjah, UAE
e-mail: mahmoud.alkhasawneh@skylineuniversity.ac.ae

K. Subbaraj et al. (eds.), *Computational Bioacoustic Artificial Intelligence*, Sustainable Artificial Intelligence-Powered Applications, https://doi.org/10.1007/978-3-032-06088-4_14

Biodiversity · Job displacement · Community engagement · Ethical consideration

1 Introduction

Computational bioacoustic AI is an innovative multidisciplinary area that analyzes, infers, and classifies natural soundscapes and vocalizations by fusing bioacoustic and artificial intelligence. The integration of computational methodologies with bioacoustic has employed here to understand the acoustic communication of living organisms. With the advancement of machine learning, deep learning, and signal processing, this field is changing the way in which the scientists and practitioners investigate ecosystems, animal behavior, and environmental changes. Bioacoustic encompasses the recording and analysis of biological sounds produced by animals, plants, and ecosystems. These acoustic signals transmit vital information about species behavior, ecological health, and biodiversity.

1.1 Overview of Computational Bioacoustic AI

Computational bioacoustic AI is a cross-cutting area, which incorporates AI, ML, and bioacoustic monitoring techniques in assessing and analyzing the vocalizations of the animal for the purposes of conservation. The data are collected by the application of various instruments that record sounds in the wild environments where such animals call. Such calls include avian and terrestrial mammals, amphibians, and aquatic organisms. They carry great ecological value. These recordings can be used in identification processes based on species through AI algorithms and based on machine learning methodologies, such as deep learning. They also enable utilization in monitoring biodiversity and ecosystem health assessment.

Bioacoustic AI has various benefits over traditional methods of ecological monitoring. For instance, biodiversity can be seen continuously and in real time in remote and often inaccessible locations with the use of AI systems. Such systems can focus on huge amounts of acoustical data much more intensively than human researchers might, thus enabling thorough analyses and providing information that would otherwise be difficult to obtain, for example, (Deichmann et al. 2018) artificial algorithms may determine the species through differences in the call by using subtle cues and monitor the species movements or detect even behavior changes triggered by environmental stressors or seasonal variation.

Technology's advent opens up wide opportunities for conservation and environmental sustainability projects. But that does not come without the need to critically evaluate its ethical and social implications alongside. Computational bioacoustic AI is not in any way different from some of the recent technological innovations, raising questions on issues to do with privacy, ownership of data, bias in algorithms, and broader societal impacts resulting from its use.

1.2 Purpose and Scope of the Chapter

This chapter reviews the ethics and societal issues linked with using computational bioacoustic AI in conservation science. On the one hand, it is impressive but also calls for caution so that these positive benefits do not translate to negative consequences. The scope of this chapter, therefore, includes further explorations of ethical issues in the area of bioacoustic AI.

(a) **Privacy Issues**: The most significant ethical issue associated with bioacoustic AI is the unintentional recording of human auditory phenomena. As those devices are meant to work in natural environments to monitor animal vocalizations, inadvertently they may record human dialog or actions, which are highly problematic issues of privacy breaches and surveillance.
(b) **Ownership and Governance of Bioacoustic Data**: An additional issue pertains to the ownership of the data obtained in situ. A significant portion of this data has been collected from indigenous territories and relates to ecosystems that are distinctive to those regions. This raises the fundamental question of who possesses the authority to manage, utilize, and disseminate the data.
(c) **Algorithmic Bias**: Intrinsic to all the frameworks that carry the artificial intelligence tag, biases in bioacoustic models arise from the nature of datasets on which their training is conducted. Thus, some species are represented more than others, causing problems for conservation and likely leading to outcomes that are often not acknowledged enough, or worse, under-researched.
(d) **Social and Environmental Impact**: The rapid increase of artificial intelligence being used is associated with job losses, not only in the conserved states but also in developed countries in which indigenous researchers sideline the use of automation that continues to perform its traditional functions. In environmental terms of carbon footprints, tied to computation powers related to AI, great assessment has to be done in huge quantities for sustainability reasons.

This chapter further goes on in discussion into how such ethical issues would be resolved by proper governance, interactive community participation, and increased public awareness. Structures instituted for AI-driven conservation must be transparent, accountable, and inclusive in the aim to provide for an equitable spread of benefits derived from bioacoustic AI in society.

We plunge deeply into these issues and their relationship in the remaining parts of this text, detailing especially how balance could be affected here between innovation and ethical interests. We propose recommendations, with the purpose of filling some of the gaps facing the challenges presented by such computational bioacoustic AI. In so doing, we strive toward a rounded outline of an ethical framework conducive to the creation and roll-out of the technology but for the benefit of either ecological systems or human social systems.

2 Technological Context and Advances

The approaches that investigators analyze and employ on acoustic data have progressed as a result of the integration of bioacoustic with artificial intelligence approaches. Conventional bioacoustic analysis relies on rule-based techniques and manual annotation, which are frequently difficult and likely to human error. The advancement of artificial intelligence techniques such as machine learning, especially with the development of supervised, unsupervised, and deep learning algorithms, has transferred the approaches of the researchers toward automate anomaly detection, sound classification, and feature extraction.

2.1 Integration of Machine Learning with Bioacoustic

The integration of bioacoustic with machine learning dramatically changes how researchers approach studies of ecological monitoring and conservation measures. Previously, bioacoustic data collection was solely a process of passive recordings of natural environments through an audio source, followed by the subsequent manual analysis. It proved to be usually laborious and time-consuming, yet effective. This development brought artificial intelligence and machine learning to bioacoustic at a new level of processing large quantities of acoustic data very efficiently.

This technique has proven to be useful in pattern detection in bioacoustic datasets, especially in deep learning methodologies. (Xie et al. 2021) Deep learning learns species-specific vocalization classification using extensive datasets; therefore, it can find slight variations in auditory sounds that an analyst cannot possibly pick out. For example, the CNN model, originally for image recognition, has been used to classify audio data such that species can be identified using their unique vocal profiles.

Its application goes beyond simple classification of sounds in bioacoustic applied in the field of ML. It is much more complex, including temporal and spatial patterns, for example, tracking shifts in vocalization frequency, strange behaviors, and even the threat posed by ecosystems. For example, through the continuous learning patterns from data, ML models will automatically be able to report biodiversity as well as the status of health in ecosystems immediately. All these have important implications for managing natural resources in a more adaptive and forward-looking administration and directly affect the conservation effort.

2.2 Applications in Conservation and Ecological Monitoring

This has greatly enhanced the possibility of observing biodiversity, changing environments, and endangered species in realms of conservation and ecological observation through the integration of bioacoustic monitoring with machine learning.

1. **Biodiversity Monitoring**
Bioacoustic artificial intelligence: An excellent tool to monitor inaccessible or distant areas. Biodiversity bioacoustic Artificial Intelligence Scientists record vocals from many species otherwise hard or impossible to detect in the field by placing sensors throughout ecosystems. Artificial intelligence algorithms operate on this data, providing information on species distribution, viability of populations, and habitat usage, without disturbing those ecosystems being watched.

2. **Endangered Species Monitoring**
Bioacoustic AI will track population and trace the behavioral studies of endangered species. For example, the bioacoustic system has tracked Sumatran orangutan, whose status is critically endangered, as well as the Kakapo parrot, with similarly critically endangered status. Data from tracing the vocal patterns and population dynamics may then give information concerning rates of reproduction and survival threats to offer a chance for intervention.

3. **Ecosystem Health Assessment**
Sound tracking enables grading ecosystems based on changes in sound produced by animals. Such changes which may involve reduced calls or even loss in patterns can act as one indicator of stress, and the factors involved are usually change in habits and pollution. The researchers using such machine algorithms for the changes analysis can, therefore, enable

early indicators of stress so that some reversal action will be timely.

4. **Identify illegal activity**
The systems will help track illegal activities like poaching, logging, and fishing using these bioacoustic with AI. They will be detecting human sounds and alerting the authorities of such dangers to these places. Sounds such as a chainsaw or gunfire within a forest reserve will set off alarms.

5. **Real-time environmental monitoring**
With bioacoustic AI, dynamic conservation management becomes possible with the real-time monitoring of the environmental condition. Continuous analysis of information provided through bioacoustic monitoring enables giving feedback concerning the health of the ecosystem, thus making it possible for well-informed conservators to make prompt actions to respond and combat threats that may happen.

The various machine learning techniques available to detect the speech and their classification (Lastname et al. 2021a; Mutanu et al. 2022; Napier et al. 2024) are discussed in the Table 1

3 Ethical Considerations in Computational Bioacoustic AI

The rapid development and deployment of computational bioacoustic AI have raised a number of ethical issues that need to be addressed so that the technology can be used for the purpose it was meant to be: conservation and ecological monitoring without causing harm. This section explores some of the critical ethical concerns associated with bioacoustic AI, including privacy issues, the ethical treatment of animals, and data ownership in sensitive environments.

3.1 Privacy Issues

Human Noise and Data Exploitation Risk

One of the most prominent ethical concerns associated with the use of bioacoustic AI is that unintentional recordings of human voices and other activities of people in their natural habitats are involved. Sensors and microphones left in the field to record animal sounds may at times record the sounds of humans, such as a conversation, private discussion, or even private activities. This raises issues of privacy as sensitive or personal data is recorded without the consent of people. Although bioacoustic data is mainly intended to monitor wildlife, the interference with human sounds poses a serious concern in maintaining ethical boundaries in the matter of privacy.

And besides there lies another danger that is the threat of human-recorded data exploited. For instance, there are human data which applied sometimes execute surveillance or the other application that violates rights accorded to a human's personal privacy. For instance, in the acoustic sensor's application found in the urban or even semi-urban environment, there lies the danger at most times, the information garnered can misapply and the issues arising may be relating to consent, privacy as well as profiling.

Recording in the Natural Environment Implications

Recording is primarily used for monitoring biodiversity and indeed involves issues of potential research objectives and personal privacy when done in natural settings. In the vast majority of situations, bioacoustic sensors are deployed in basically untouched areas of wilderness or natural preserves with very minor human presence; yet recorded data may nonetheless capture the sounds of communities within a local area, other researchers visiting those areas, or recreational travelers.

The key ethical question is: In what ways can one collect bioacoustic data while minimizing the risks for people's privacy, especially of the individuals and groups probably not having given consent to these activities being recorded? That dilemma becomes particularly relevant where there are indigenous or otherwise marginalized communities who might not fully be aware of the breadth to which their activities are captured and analyzed.

3.2 Animal Privacy and Ethical Treatment

Invasive Techniques such as Tagging

Long-term monitoring is common in conservation biology through the use of invasive research techniques such as animal tagging (Stowell et al. 2019). These may include the attachment of physical or electronic tags on animals for tracking their movements and behavior's, raising ethical questions regarding animal privacy and welfare. Although intended to serve as information that may be critical to conversations, tagging generally results in the catching and handling of animals to cause physical impairment or stress.

This has decreased the demand for tagging since AI systems can passively monitor animals through their vocalizations. However, how much should researchers invade these areas to collect data? As one could realize, even though it offers better behavioral data, the technique would interfere with the animal's natural behavior and might lead to physical or psychological harm to the animal.

Table 1 Techniques of machine learning used in bioacoustic

Machine learning technique	Description	Application in bioacoustic
Convolutional neural networks (CNNs)	A deep learning architecture primarily used for spatial data processing (images), which can be adapted for spectral patterns in acoustic signals	Species classification: CNNs can classify animal vocalizations based on unique sound patterns, helping to identify species from bioacoustic recordings
		Sound pattern recognition: detecting specific calls or environmental sounds like footsteps, chainsaws, or distress calls
Recurrent neural networks (RNNs)	Neural networks designed to handle sequential data, making them ideal for time-series analysis. RNNs, particularly Long Short-Term Memory (LSTM) networks, can capture temporal dependencies	Temporal analysis of vocalizations: RNNs can track the changes in vocalization frequency over time, identifying patterns in animal behavior such as mating calls or territorial sounds
Random forests (RF)	An ensemble learning method that combines multiple decision trees to improve classification accuracy. Random forests work well with high-dimensional data	Classification of acoustic events: RF can classify a wide range of animal sounds, separating them from background noise and identifying specific species vocalizations
		Anomaly detection: RF can be used to detect unusual sounds in an ecosystem, such as poaching activities or the presence of invasive species
Support vector machines (SVMs)	A supervised learning model that finds the optimal hyperplane to separate different classes. SVMs are highly effective in classification problems with limited data	Species identification: SVMs can classify animal calls based on frequency and amplitude features
		Soundscape classification: identifying and categorizing complex soundscapes (e.g., differentiating between bird, mammal, or amphibian calls)
K-nearest neighbors (KNN)	A simple, instance-based learning algorithm that classifies data points based on the majority class of their nearest neighbors	Sound classification for species: KNN can be used for identifying species by comparing new vocalization samples to a library of known species calls
		Environmental noise detection: KNN can classify human-made sounds or other non-natural noises to monitor human impact on ecosystems
Autoencoders	Neural networks used for unsupervised learning, where the model is trained to compress and then reconstruct input data. Autoencoders are useful for anomaly detection	Noise reduction: autoencoders can be used to clean acoustic data, removing background noise and enhancing the clarity of animal vocalizations
		Anomaly detection: identifying rare or anomalous sounds, such as distress calls, poaching, or environmental hazards

A factor critical for using AI in bioacoustic monitoring is the animals' welfare. The researcher will need to ensure he does not disturb the normal habitat and behavior of such a species. (Stahl and Stahl 2021; Ayling and Chapman 2022) Then, an important factor for guidelines on this ethical side will be considering the potential impacts of research methods on the welfare of animals, especially when a method applied deals with endangered species in which the impact caused on one may create effects to its population sustainability.

Respect for Animal Welfare

Computational bioacoustic ensures the technology used is non-harmful and does not cause suffering to the species under observation. Reduced physical damage through the non-invasive procedure of using microphones or underwater hydrophones to track the animals is thus very essential. Even a non-intrusive observation can have impacts such as causing a disturbance in the animals' natural behaviors as the animals might perceive the gear as unusual and are therefore distracted by it.

The main ethical consideration will be that the good this bioacoustic AI contributes toward conservation outweighs the evils (Hayes et al. 2024; Hickok 2021). The deployment must be designed such that the disruption in the ecosystem, particularly in sensitive or critical habitats, is minimized.

3.3 Data Ownership and Rights of Native Communities

Ownership Issues in Indigenous and Sensitive Areas

In bioacoustic research, the question of ownership over data presents immense complexity, particularly when applied to indigenous or local areas. There are indigenous peoples who have profound cultural and spiritual ties with their land and its ecosystems across many parts of the world. This means that once bioacoustic data is collected in such locales, questions regarding the ownership of such data arise as well as its use.

In areas of traditional indigenous land rights, bioacoustic data collected needs to be managed in such a way that the peoples will have a say in decisions related to the use and dissemination of such data. The people may feel a concern over biodiversity data commercialized without suitable incentives or recognition. In addition, the bioacoustic data obtained in those places should respect indigenous knowledge systems and should not be used to misrepresent or distort their relationship to the land.

The main ethical issue is to ensure that appropriate rights of definitive ownership along with governance for bioacoustic data can be designed for indigenous as well as ecologically sensitive areas. Researchers should, therefore, seek approvals from local communities before engaging in data gathering, transparency about the intentions of exploiting the data gathered, as well as any benefits derived would be shared justly to the community involved (Wood et al. 2021).

Importance of Localized Data Governance and Regulation

As a response to the issues brought about by ownership discussed above, localization of control and regulation over data must be created as a way of guaranteeing proper data management of bioacoustic information. This entails giving local communities that are indigenous in nature the power to control and manage their own data obtained in their territory. It ensures that individuals living in and interacting with the ecosystems being studied are consulted over the use of the information and in setting conservation policy.

In cases where there is indigenous knowledge of the species that exist within their territories, bioacoustic data should be a tool that encourages cooperation (Montesano et al. 2023; Ziemba et al. 2024). The scientists and conservationists must show respect to such communities and be involved in designing the monitoring projects, recognize their knowledge, and ensure that there is equitable benefit sharing between the community as benefits from the conservation activities.

This way, artificial intelligence in bioacoustic must be guided by ethical standards respecting the rights of indigenous people. All the procedures on data collection, processing, and distribution must take place under strict ethical guidelines that address local concerns, thus upholding sustainable conservation practices.

There are highly complex ethical issues in computational bioacoustic for using artificial intelligence in regard to the issues of privacy, the welfare of animals, and owning data. This needs to be looked into as new bioacoustic AI technologies develop so that their applicability can be responsible enough (Aderibigbe et al. 2023). The conservation capability would improve with respect to the individuals, animals, and rights of the community being upheld based on the ethical consideration of benefits.

4 Bias and Representation in AI Models

When ML models consistently favor or disadvantage particular groups, viewpoints, or results, bias and representation problems in AI emerge. These problems typically arise because of biases in the training data, (Sanchez et al. 2021; Dhiman et al. 2024) the architecture of the model, or the interpretation and application of the outputs.

4.1 Risk of Bias in Conservation Efforts

When conservation efforts are biased, priorities, techniques, or resources are systematically skewed in a way that might not lead to fair or optimal conservation results. These biases, which might arise from a lack of facts, societal ideals, or outside forces, impede fair resource management and successful biodiversity protection. The sources of errors and bias are represented in Fig. 1.

1. **Geographic Bias**

It is common practice to concentrate conservation efforts on particular places, such as biodiversity hotspots, while ignoring less well-known regions that may nonetheless be ecologically significant.

Risk:

- Ecosystems in locations that are not hotspots are not adequately protected. Loss of biodiversity in regions that have received less attention or are less interesting.

Example

We tend to pay more attention to tropical rainforests like the Amazon than to less obvious ecosystems like grasslands and wetlands.

Fig. 1 Sources of errors and bias

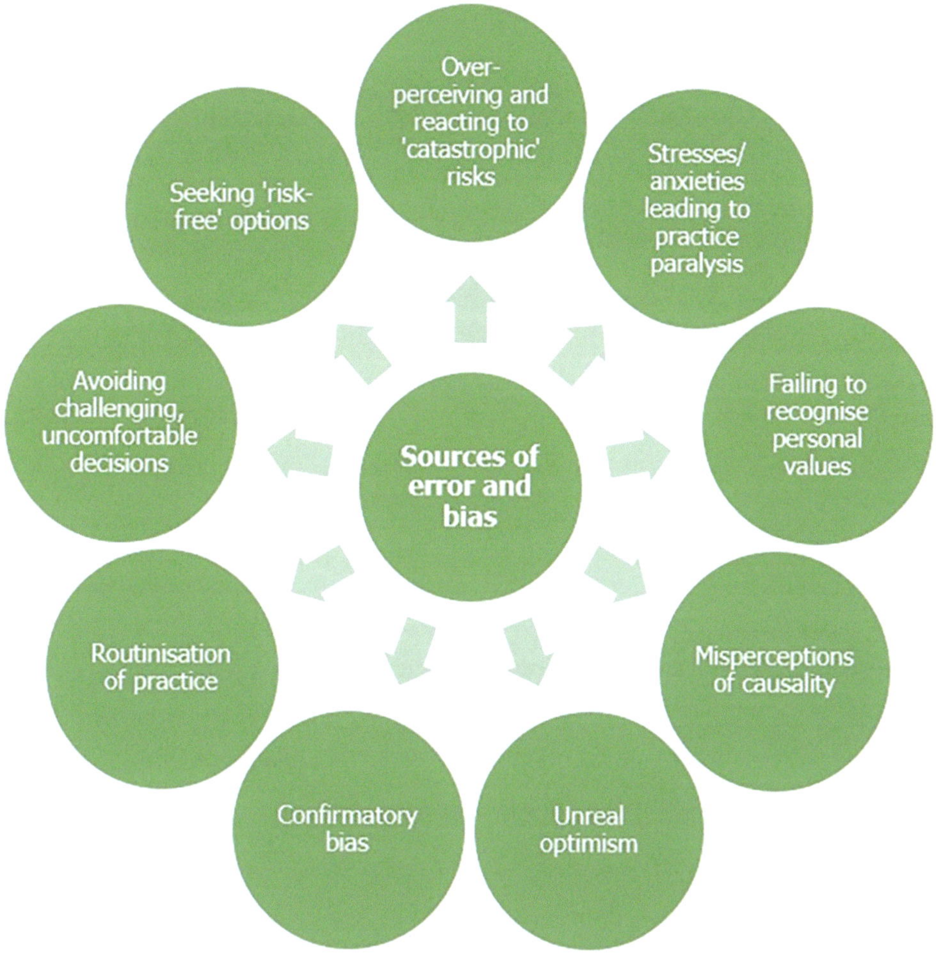

2. **Taxonomic Bias**:

While disregarding less "appealing" species like insects or fungus, conservation efforts disproportionately center on some species, especially charismatic megafauna (e.g., tigers, elephants).

Risk:

- Unbalance in resource allocation results in neglect of species vital for the operation of ecosystems.
- Loss of fundamental or Keystone species that might not be as flashy or obvious.

Example

Pollinators like bees frequently receive less attention than large mammals, despite their vital role in food security and environmental health.

3. **Cultural and Socioeconomic Bias**

It's possible that conservation goals overlook the requirements and expertise of local or indigenous communities, favoring instead the values and points of view of groups who are more politically influential or more financially secure.

Risk:

- The isolating of local populations, which might result in resistance or a lack of cooperation.
- There were possibilities that were lost to utilize indigenous knowledge and environmentally responsible techniques.

Example

Establishing protected areas without including local inhabitants would displace the people and trigger conflict as happened in some of the African wildlife game reverses.

4. **Methodological Bias**
Conservation efforts can be skewed by biases in research methodology or data collecting, such as favoring more accessible or financially well-off regions.

Risk:

- Exaggeration of patterns in the distribution of species or ecological processes.
- Using conservation funds in the wrong way.

Example

Biodiversity in less-visited or distant places may go unnoticed because studies are overly concentrated in and around urban research facilities.

5. **Policy and Funding Bias**
Environmental concerns take a back seat to donor goals or political ambitions in conservation initiatives.

Risk:

- Prioritize the visibility of donors over the long-term ecological benefits in short-term programs.
- Giving more weight to programs that reflect global trends than to those that address specific regional need.

Example

Other pressing concerns for biodiversity, such habitat fragmentation, may be neglected if funds are primarily focused on climate change.

4.2 Addressing Underrepresented Species or Ecosystems

The success of any conservation effort depends on AI models accurately representing all relevant ecosystems and species, including those that are underrepresented. It is crucial to diversify datasets by adding data from neglected species, habitats, and geographic regions in order to address underrepresented ecosystems or species in AI models (Chemnad and Othman 2024). To accomplish this, we can use methods for modeling ecosystems at the level of the individual ecosystem, as well as remote sensing tools like satellites and LiDAR. Collaboration with ecologists and Indigenous communities can help fill a few of the existing gaps in the knowledge, and transfer learning can use the data about highly studied species to be able to infer about those not as well-studied. Equitable representation of species and ecosystems, frequently overlooked despite critical contributions to biodiversity, can be achieved by rebalancing datasets and executing bias audits on AI models (Hallamaa and Kalliokoski 2022). These measures assist in the development of equitable and science-based conservation policies that focus on the unexplored ecosystems as well as species for ecological stability.

Case Study Examples

1. ***Coral Reefs and Remote Sensing***
Integration of satellite imagery into AI models revolutionizes the cartography of coral reefs as it facilitates easy identification and monitoring of inaccessible areas or those which have been poorly surveyed. The high-resolution spatial data ascertains the position of the reefs, their health status, and temporal sequence in monitoring changes with the aid of these models. These improve shortcomings in global reef mapping, focus conservation efforts, and facilitate decisions over protection measures of threatened ecosystems as shown in Fig. 2.

Insect Monitoring

These artificial sensors that use AI to collect and analyze the sound created by insects can help track their populations. They then make decisions supporting activity and diversity monitoring to obtain information on the network pollination and the overall ecosystem health. The system captures insect wingbeat signals using acoustic sensors to identify species based on their spectral frequency profiles as shown in Fig. 3a Bioacoustic—Based Insect Detection Setup. A compact IoT-based sensing module deployed in the field for real-time insect detection and wireless data transmission is represented by Fig. 3b IoT—Enabled Field Monitoring Sensor Unit.

4.3 Impacts of Biased Data and Algorithmic Limitations

Critical aspects such as conservation, health care, recruitment, or in serious matters like criminal justice where skewed data and the boundaries posed by algorithms in an artificially intelligent framework can deeply play a role in forming judgment and perpetuating system anomalies.

1. ***Reinforcement of Existing Biases***
Because of the patterns that tend to repeat in the training data, artificial intelligence models trained on biased datasets can learn and reproduce as well as amplify already existing societal biases that can bring forth biased and discriminatory

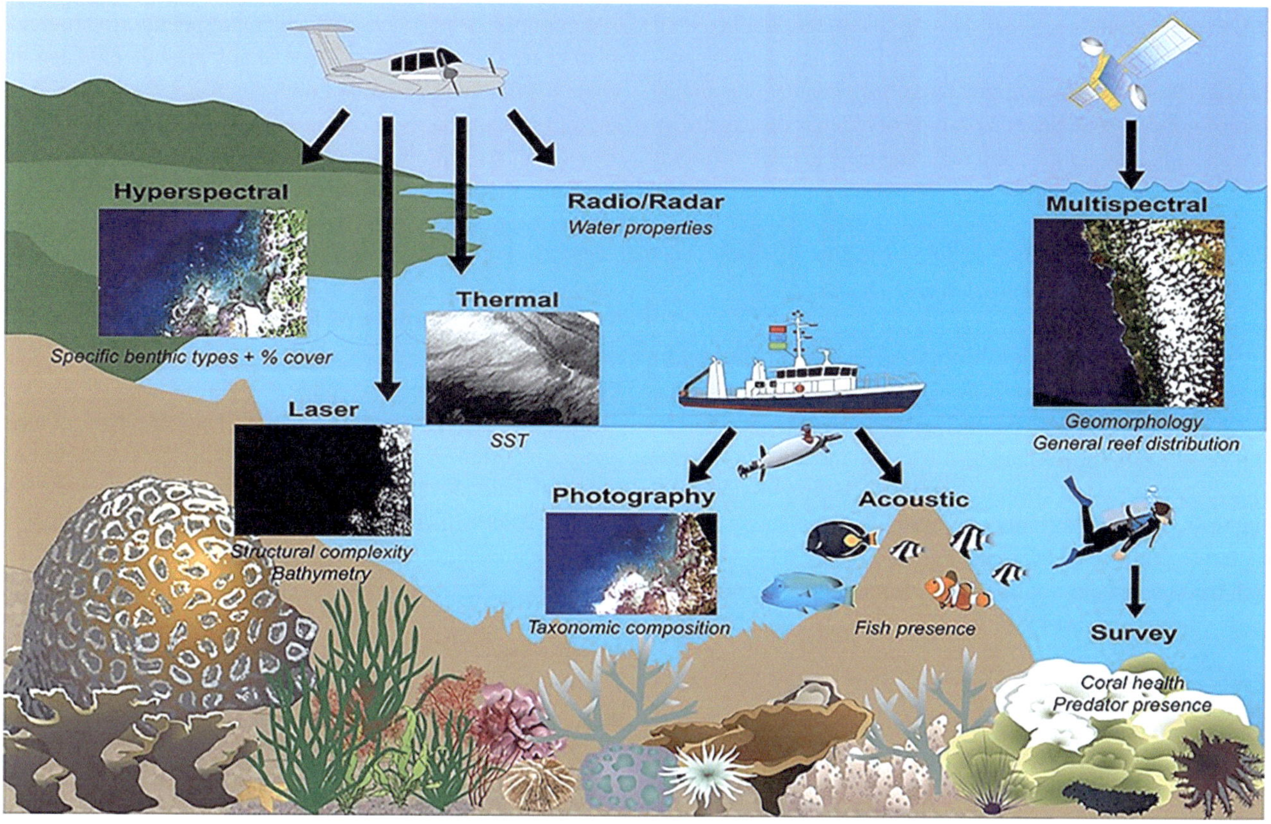

Fig. 2 Case study of coral reefs and remote sensing

Fig. 3 Representation of insect monitoring
(**a**) Bioacoustic—Based Insect Detection Setup.
(**b**) IoT—Enabled Field Monitoring Sensor Unit

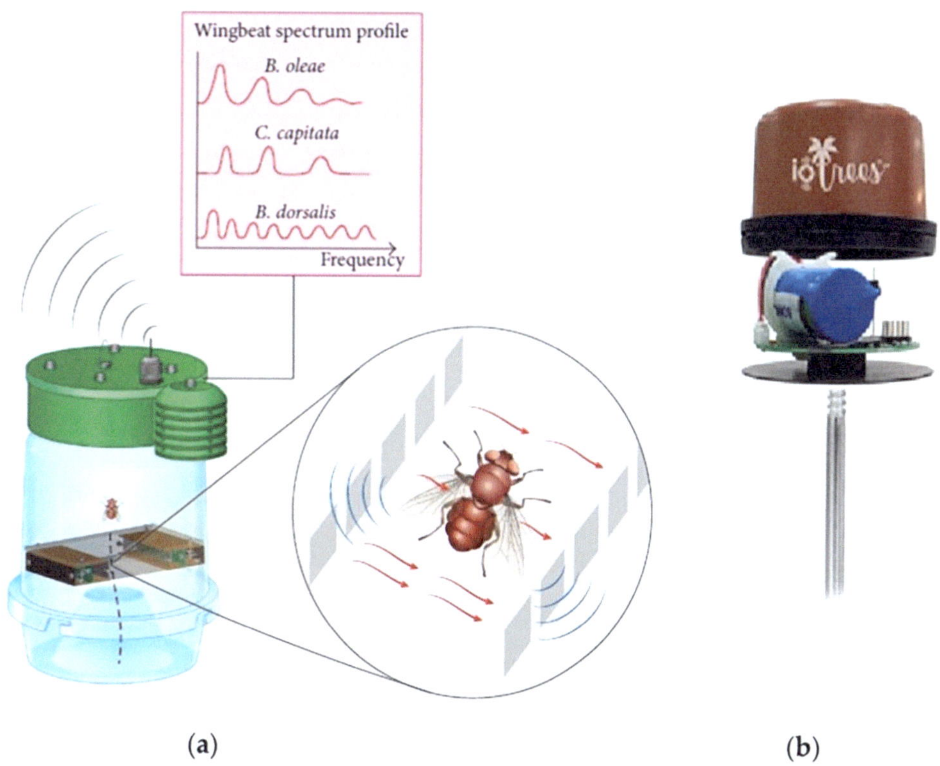

outputs. For example, if considering historical data favoring male candidates, a recruitment algorithm based on historical data would reject qualified female applicants, thus maintaining gender inequalities. This type of bias has the capability of escalating inequalities and reducing trust in these artificial intelligence technologies.

2. ***Misrepresentation of Real-World Dynamics***
Poorly developed algorithms or biased data could lead to the exclusion of vital ecological factors and thus might result in an imperfect or, at times, even wrong perception of ecosystems, species, or populations. For instance, charismatic organisms that are so well known may be favored in models of artificial intelligence for conservation, thereby excluding less obvious but biologically important organisms like pollinators and soil bacteria. This might lead to inefficient utilization of resources and maladjusted distribution of conservation that, in turn may cause further damage to the biodiversity.

3. ***Reduced Generalization and Reliability***
Artificial intelligence models that have been trained on limited or unbalanced datasets tend to fail in generalizing well, hence poorly performing in a variety of real-world contexts. For instance, a healthcare algorithm trained on data from a single demography will yield erroneous results for populations underrepresented, so lowering the dependability and equality of its applications.

4. ***Ethical and Legal Risks***
Such artificial intelligence systems have the chance to create biased results, hence raising ethical and legal issues, which forms the origin of discriminatory action. System and algorithms which are developed using artificial intelligence can therefore influence the human decision-making process; problems also persist in aspects of automation, decision, and transparency. Unless there are proper rules governing their systems, decisions may be unfair or discriminatory, and may encourage stereotypes, infringe the user's privacy, or even create concerns about human rights. All this is partly because artificial intelligence learns in a process dubbed machine learning. Any data fed into the system by the user will then train a machine learning algorithm. They can either be used to make predictions or classify different categories. Based on the data that it has already processed, the system could train itself and increase its functionality throughout the course of time. On the other hand, these systems frequently function in a manner much like a black box, which makes it difficult to comprehend the process by which judgments are produced or to guarantee that they are free of any prejudice. When the system is trained on a pool of data that is not broad or diverse enough to suit the system's needs, machine learning might provide an ethical challenge (Bleher and Braun 2023). This is despite the fact that machine learning is an awesome type of technology.

5 Societal Impacts of Bioacoustic AI

Bioacoustics AI does have huge social implications because it is actually a technology that would make everyone change their perception of the environment, will support conservation and help develop human welfare. Applying machine learning and artificial algorithms on animal sound analysis (Oikarinen et al. 2019), it is, in fact, opening a channel for scientists to study nature in ways that couldn't previously be done before, through bioacoustic AI observation; one is on its impact on the conservation of biodiversity. This will allow for real-time monitoring of wildlife populations and ecosystems through AI-enhanced bioacoustic, thereby making it possible to monitor at-risk species, changes in habitats, and early indicators of ecological distress. This ability promotes more anticipatory approaches to conservation, thereby reducing the likelihood of biodiversity decline and sustainable environmental methodologies.

Bioacoustic AI is making tremendous contributions toward monitoring the state of ecological well-being. It can track changes in the soundscape from pollution and climate change, and that gives a vital understanding of how the very same changes are influencing ecosystems and human health. Such functionality can benefit by formulating better policies in policy and interventions to be developed that may contribute toward reduction of environmental degradation. This can also lead to more sustainable approaches at the broader level of society. In agriculture, forestry, and even urban development, bioacoustic artificial intelligence will aid in preventing human–wildlife conflict. It will be achieved through monitoring and controlling interactions between wildlife and agricultural areas so that crops are not damaged. Further, the technology applies to improving human health, mainly concerning identifying zoonotic diseases typically spread from animals to human beings through behaviors or sounds changed by animals. This democratizes access to advanced scientific tools for communities, researchers, and local governments in how they make their decisions to be data-driven decisions protecting the health and the environment. Thus, this has a potential positive impact: making conservation as well as resource management in a future, more equal and diverse context. But ethical considerations based on concerns for data privacy and propensity of abuse of technology mean scrutiny is warranted to determine that there are no negatives lurking around without killing all the positives.

5.1 Contributions to Conservation Efforts at the Societal Level

Bioacoustic, endowed with artificial intelligence, greatly contributes to the creation of mass awareness on environmental conservation and biodiversity. AI interpretation of animal vocalizations permits the monitoring of ecosystems and wildlife in real time, transforming cluttered ecological information into actionable insights that can be understood by a wider audience. This technology, therefore, links scientific pursuit with the public at large, thus making a presentation of the importance of biodiversity conservation more interesting and understandable (Regona et al. 2024; Oyedele et al. 2021). One of the major contributions bioacoustic AI makes is toward identifying and monitoring the endangered species' vocalizations. Indeed, this places it as an effective awareness-raising tool about such animals' condition. For instance, animal calls associated with rare species or ecosystem conditions can be recognized through AI technology, and thus one will be able to depict why urgent conservation steps are necessary. This might mean the findings are shared in real time with the public through interactive platforms, social media, or documentaries, which might give increased urgency and a personal feeling toward conservation initiatives.

Furthermore, bioacoustic artificial intelligence can facilitate the design of exciting sonic experiences by which people can enjoy sounds in cities or elsewhere, a little more remote in nature. By working with digital media, educational programs, and public sound presentations, one can connect with the calls by species they may never have heard in their entire life. This encourages sympathy and emotional attachment with nature, which in turn leads them to take up activities such as joining wildlife preservation schemes, reducing their ecological footprint, or community-based environment projects.

Besides, the AI technology of bioacoustic is also empowering educational programs about the observation and documentation of ecological changes that are resulting from human activities. For example, it can detect the change in animal sounds resulting from the disruption by noise pollution, habitat destruction, or climate change. Bioacoustic AI, by reducing the sophisticated complexities of conservation to relatable data on these issues, makes the issues more relatable and accessible to the general population, thus bringing about more support for projects aimed at reducing human interference with the ecosystem.

Bioacoustic AI brings scientific information into the lives of ordinary people, intensifies the emotional attachment toward wildlife, and forms educational programs wherein conservation is highly required. Bioacoustic AI opens up a new dimension to listening and understanding nature and lets the people and the community engage more proactively with the cause for environmental protection that aids in developing a more ecologically aware society.

5.2 Accessibility and Challenges for Researchers in Developing Countries

For researchers in developing states, at least, the easy access to bioacoustic AI and all its associated challenges is multifaceted, from techno-financial infrastructural restrictions to opportunities for improvement and even "progress." (Kvsn et al. 2020) Bioacoustic AI offers much for biodiversity conservation and surveillance. Implementation in developing states faces many obstacles.

Problems: This area lacks adequate availability of advanced technologies and infrastructure. Basic requirements of the most adequate availability of advanced technologies and infrastructure are not in place in developing countries for implementing bioacoustic artificial intelligence. High-performance computing, cloud services, and advanced AI models consume many resources scarce in this area. It happens in many developing countries; many scholars face obstacles such as inadequate hardware, poor quality of internet connectivity, or limitations in accessing applications and artificial intelligence platforms through which substantial data analysis or training models for machine learning can be done.

Financial Constraints: This cost can be too high for getting microphones, recording equipment, and sensors. Other than the initial cost of buying it, there are other expenditures like maintaining and using it because the repair and upgrades plus storing data require money as well. Furthermore, the acquisition cost of AI software or the cost of paid cloud services for data analysis can be too expensive for a research institution with limited financial resources. This means that researchers must use outdated tools or work with limited sets of data, which consequently affects the quality of their work.

Lack of Trained Professionals: Bioacoustic AI requires knowledge in ecology and machine learning, which is not always available in developing countries. The lack of trained professionals may be a problem in developing countries, who can design, use, and interpret AI-based bioacoustic research. The few options for training in AI and bioacoustic exacerbate this problem by creating a gap in the skills that researchers need.

Accessibility and Data Sharing: Collecting bioacoustic data is mostly fieldwork in many cases. This can prove a very challenging activity, particularly in distant or inaccessible locations. After the data collection, sharing such large audio files and data over weak internet connections will also become challenging. High-quality, pre-labeled datasets required to train AI models may also become an issue, mainly because they originated from institutions that are developed, from developed countries. Without easy access to data, it is nearly impossible for research in developing countries to produce valid models and conclusions.

Ethical and Cultural Obstacles: Examples may include ethical issues, for instance, local community privacy and rights in research, and so on. Cultural oppositions exist to the application of certain research or types of technology in certain areas wherein the local people are neglected by the conservation goals and also by the benefits from bioacoustic AI.

5.3 Areas for Improvement

However, there are many ways in which bioacoustic AI can be made accessible to academics from developing countries:

1. **International Co-operations**: International co-operations among the research institutions of developed and developing countries may bridge technological and financial gaps. Tapping shared resources, human capital, and infrastructure may help overcome regional constraints to bioacoustic research. The leverage in this regard would be immense, particularly from NGOs, institutions of learning, and the government toward providing funds, training, and new technologies.
2. **Open Source**: The best of these opportunities is an open-source AI software as well as the repository for bioacoustic data. It has opened access to the highly sophisticated tools for developing country scholars, without incurring costs on proprietary software. Opportunities of open source will thus help scholars in developing regions to interact with the international community, hence raising chances for a successful research project.
3. **Low-Cost Technologies and Inventing Solutions**: Progressions in technological development enables the effective and affordable bioacoustic monitoring equipment. Scientist can make cheaper sensors, microphones, and storage devices, while improving functions in the resource-poorer environment. This deters undue advantage the scientists from developing regions have as they try to collect quality data using costly devices.
4. **Crowdsourced Data Collection**: Bioacoustic research is highly crowdsourced. Applications and platforms in which people record and upload the sounds of wildlife are some fundamental tools that collect huge data to be analyzed using AI. The approach gets these people who live nearby to take interest in the cause of conservation, thereby becoming interested in the process itself.

The hurdles around the inclusion of bioacoustic AI are significant but not insurmountable. International collaboration, engagement of open-source technologies, efforts to build capacity, and advances in affordable equipment would be some avenues for tapping this great potential among researchers in those countries to implement impactful conservation works. By reducing these barriers, bioacoustic technologies will be made more accessible while ensuring that the power of this tool is widely disseminated to contribute to the biodiversity conservation efforts in some of the world's most vulnerable areas.

5.4 Job Displacement Concerns and the Role of AI Experts

The fears about AI displacing jobs, particularly about the types of jobs and areas involved in bioacoustic and other lines of study, are that technology might automate people out of work or render their expertise obsolete. The more frequently AI machines are shown to perform what was previously only human endeavor, the more frightening is the specter that employment opportunities will be replaced. However, a part of what makes all that anxiety so allayed are the roles for AI specialists (Sachan et al. 2024). In addition to developing and upholding these systems of artificial intelligence, a professional is also burdened with heavy reskilling the workforce and creating jobs. Such professionals contribute to the success in ensuring that technology may actually enhance human labor and not replace it with machines, hence fostering an atmosphere in which automation may exist with human aptitude. Further, with the increasing artificial intelligence, new domains of occupation are expected to emerge demanding human supervision, moral debates, and inventive solution finding abilities, which ensure job loss is going to be compensated with new, more complex roles.

6 Sustainability Challenges

6.1 Carbon Footprint and Environmental Impact of Data Processing

The carbon footprint and environmental impact of data processing in bioacoustic AI are very high because the technology is highly adopted for the monitoring, conservation, and environmental research of wildlife. Bioacoustic AI refers to the collection and processing of large volumes of audio data from ecosystems with machine learning models. While holding a high potential in conservation of biodiversity, the environmental costs are simultaneously implied due to the energy in which data is stored and processed and models' training.

- Energy Consumption in Data Processing:

Training and running artificial intelligence models, especially those relying on large data sets such as bioacoustic recordings, consume many resources. It often requires high-performance computing and cloud-based infrastructures to support complex algorithms and large amounts of data. Data centers contain the computational resources needed to process AI and are one of the major consumers of electricity. Most of the energy in these data centers is not renewable, thus resulting in carbon emission. As the energy consumption of the bioacoustic AI system is anticipated to rise due to increased capacity for large datasets and complex models, the environmental implication is also expected to worsen.

- Carbon Emissions:

Emissions of carbon associated with the utilization of carbon directly related to bioacoustic AI involve carbon consumption relating to underpinning basic infrastructure within the data center. Deep learning architectures typically form an artificial neural network and are thus computationally intense to allow training or even fine-tuning. New research suggests that a single large AI model will consume as much carbon as the total emissions from many cars over a lifetime. In such a scenario where AI is increasingly being integrated with environmental monitoring and conservation initiatives, addressing the carbon footprint of AI becomes important for sustaining sustainability in this technology arena as well.

- Environmental Impact of Data Storage:

Apart from the energy consumed in processing activities, bioacoustic AI consumes a huge amount of data storage space for information, which is bound to increase its impact on the environment. Bioacoustic data is continuously being collected and archived across ecosystems, which therefore creates large amounts of data that must be properly stored and handled. Such requirements necessitate robust storage infrastructures, often housed in data centers with large energy needs. The length of time the data is stored and the rate at which it is retrieved for analytical use increase environmental impact, especially where infrastructure used for storage is not designed to optimize energy use.

Although bioacoustic AI brings great opportunities to boost conservation efforts and further understanding of ecosystems, there is a need for careful assessment of carbon emissions and broader environmental impacts. That would be achieved by more efficient algorithms, renewable energy resources, and new ways to employ edge computing.

6.2 Balancing AI Operations with Sustainability Goals

The power of AI technologies, despite promising the world the solution to such problems, however, relies on tremendous amounts of computation, thereby contributing significantly to energy consumption and environmental issues. To harmonize the operations of artificial intelligence with objectives of sustainability, there should be strategies that reduce its ecological impact while enhancing the benefits of the AI technology. A potential approach is that of improving AI models so that they become more energy efficient. The researchers could reduce the energy usage connected with both training and inference processes of AI applications by making algorithms that require less computing power. Techniques such as model pruning, quantization, and distillation enhance the efficiency of AI systems while keeping performance levels up, thus integrating sustainability into the AI design framework.

Another critical approach is to shift toward renewable sources of energy that would, in turn, ensure the infrastructure for AI working, which includes data centers and cloud computing systems. Several companies already invest in green energy projects where they power their servers through solar, wind, and hydropower that reduce carbon emissions associated with AI computation by several folds. Further paradigms for distributed computing, including edge computing, can be envisaged to reduce environmental impacts since data processing is now conducted locally, thus requiring lesser quantities of data transfer and storage within central infrastructures. Artificial intelligence can drive the sustainability goals directly by enhancing better resource management. For instance, AI technologies can be used in boosting energy efficiency in the various industries for example in manufacturing and cut wastes with predictive analytics capabilities and enhancement of climate models for application in effective policies concerning the environment. In so doing, artificial intelligence will drive innovation across various industries into issues of sustainability.

Table 2 Challenges and mitigation strategies for educational platforms

Challenge	Mitigation strategy
Resource constraints	Partnering with local and international organizations to pool resources and share expertise
Cultural barriers	Developing culturally relevant and multilingual content
Engagement fatigue	Employing interactive and innovative content delivery methods
Access inequality	Providing offline or low-tech educational solutions and infrastructure investments

7 Challenges in Utilizing Educational Platforms for Conservation Awareness

Despite their potential, educational platforms for conservation face numerous challenges (Lamba et al. 2019):

- **Resource Constraints**: Limited funding and resources can hinder the development and maintenance of educational initiatives.
- **Cultural and Language Barriers**: Educational content may not be culturally or linguistically accessible to all audiences.
- **Engagement Fatigue**: Maintaining sustained interest in conservation topics can be difficult, especially in digital formats.
- **Access Inequality**: Many communities may lack access to internet-based platforms or modern educational facilities (Table 2).

7.1 Measuring the Impact of Educational Platforms

The effectiveness of educational platforms can be evaluated through:

- **Knowledge Gains**: Assessing increases in public understanding of conservation topics.
- **Behavioral Changes**: Tracking changes in environmental practices, such as recycling rates or energy use.
- **Community Engagement Levels**: Evaluating the extent of community involvement in conservation activities.
- **Policy Influence**: Gauging the influence of educational initiatives on conservation-related policy changes.

7.2 Case Studies: Combating Misinformation

Case Study 1: Addressing Vaccine Misinformation

During global vaccination campaigns, misinformation about vaccine safety and efficacy spread rapidly. In response, governments and organizations collaborated with trusted healthcare professionals, conducted transparent public briefings, and utilized fact-checking mechanisms to provide accurate information and combat myths.

Case Study 2: Misinformation in Environmental Conservation

False claims about climate change have historically delayed environmental action. Conservation groups have countered this through educational campaigns, transparent data, and collaborations with reputable scientists to provide the public with accurate environmental data.

7.3 Challenges in Ensuring Accuracy

Despite efforts, there are significant challenges in maintaining accuracy and combating misinformation:

- **Information Overload**: The vast amount of data available makes it difficult to discern credible information.
- **Confirmation Bias**: People are more likely to believe information that confirms their existing beliefs, even if it is false.
- **Rapid Spread of Misinformation**: Social media platforms allow misinformation to spread faster than corrections.
- **Lack of Trust in Institutions**: Skepticism toward institutions can undermine even well-intentioned public awareness efforts.

8 Conclusion and Future Directions

Computational bioacoustic brings a new aspect of artificial intelligence in the field of ecological monitoring and wildlife conservation. Bioacoustic AI offers the potential for revolutionizing biodiversity and ecosystem monitoring because machines can understand animal sounds. Besides the promises, some important issues concerning ethical and societal considerations arise to ensure that this technology be developed responsibly. There are privacy concerns because the deployments of sensors in natural environments unknowingly record human sounds. The problem is data exploitation and surveillance. This issue becomes highly sensitive when data are collected on indigenous people's land because of problems with getting their consent and owning property over the data. However, even though bioacoustic monitoring might be less invasive than many other more traditional approaches, such as tagging, there still are ethical concerns related to the treatment of animals and ensuring that activities do not interrupt natural behaviors or otherwise adversely affect welfare.

Among other things, the second critical challenge is related to how biases in AI models lead to misrepresentation of species or ecosystems thus skewing conservation outcomes. In addition to the impacts mentioned above, there are other effects on traditional employment monitoring and environmental concerns as far as the energy requirements of AI technologies are concerned. Lastly, governance of applications of bioacoustic AI presents very urgent issues as far as following application with transparency but also sensitive to issues touching the rights of local people and indigenous communities. Sustainability should be a focus for the development and deployment of artificial intelligence systems; these involve challenges in association with sustainability—the environmental impacts and long-term ecological effects. Bioacoustic AI, for example, will have a significant impact on biodiversity preservation, but application needs to be such as it serves the principles of sustainability, so as not to aggravate the problems which it aims at solving.

Future Directions and Recommendations for Ethical AI Use in Bioacoustic

Issues relating to privacy, ownership of data, animal welfare, and sustainability will need to be embraced by internationally recognized ethical guidelines guiding the use of bioacoustic AI moving forward. Such guidelines would mandate a sense of stake for all the stakeholders—from local communities, conservationists, and indigenous groups—in bioacoustic research design and implementation. AI models must be developed, understanding bias and containing many species and ecosystems, so that this should ensure conservation works for all. Often, AI systems need to be checked to find out and correct biases to avoid a hot spot specified by the species and places.

This should also include sustainable AI, for instance, developing energy-efficient algorithms and hardware with a minimal carbon footprint for AI technologies. The development of such AI should, therefore, balance the benefits that AI offers with a high environmental impact: the technology should be aligned with broader sustainability goals. That knowledge needs to be made use of by a very large-scale collaboration between researchers, conservationists, and the local community. Open initiatives for data sharing with interdisciplinary research would help in making the bioacoustic data more democratized toward equitable and coordinated global conservation efforts. The monitoring and accountability framework for long-term evaluation of the bioacoustic AI ethical, environmental, and social impacts will similarly be important for sustained improvement in harmony with the principles of ethics.

References

Aderibigbe AO, Ohenhen PE, Nwaobia NK, Gidiagba JO, Ani EC (2023) Artificial intelligence in developing countries: bridging the gap between potential and implementation. Comput Sci IT Res J 4(3):185–199

Ayling J, Chapman A (2022) Putting AI ethics to work: are the tools fit for purpose? AI Ethics 2(3):405–429

Bleher H, Braun M (2023) Reflections on putting AI ethics into practice: how three AI ethics approaches conceptualize theory and practice. Sci Eng Ethics 29(3):21

Chemnad K, Othman A (2024) Digital accessibility in the era of artificial intelligence—bibliometric analysis and systematic review. Front Artif Intell 7:1349668

Deichmann JL, Acevedo-Charry O, Barclay L, Burivalova Z, Campos-Cerqueira M, d'Horta F et al (2018) It's time to listen: there is much to be learned from the sounds of tropical ecosystems. Biotropica 50(5):713–718

Dhiman R, Miteff S, Wang Y, Ma SC, Amirikas R, Fabian B (2024) Artificial intelligence and sustainability—a review. Analytics 3(1):140–164

Hallamaa J, Kalliokoski T (2022) AI ethics as applied ethics. Front Comput Sci 4:776837

Hayes P, Fitzpatrick N, Ferrández JM (2024) From applied ethics and ethical principles to virtue and narrative in AI practices. AI Ethics, 1–23

Hickok M (2021) Lessons learned from AI ethics principles for future actions. AI Ethics 1(1):41–47

Kvsn RR, Montgomery J, Garg S, Charleston M (2020) Bioacoustic data analysis–A taxonomy, survey and open challenges. IEEE Access 8:57684–57708

Lamba A, Cassey P, Segaran RR, Koh LP (2019) Deep learning for environmental conservation. Curr Biol 29(19):R977–R982

Montesano FS, Biermann F, Kalfagianni A, Vijge MJ (2023) Balancing or prioritising for sustainable development? Perceptions of sustainability integration among professionals. Sustain Dev 31(3):1921–1936

Mutanu L, Gohil J, Gupta K, Wagio P, Kotonya G (2022) A review of automated bioacoustic and general acoustics classification research. Sensors 22(21):8361

Napier T, Ahn E, Allen-Ankins S, Schwarzkopf L, Lee I (2024) Advancements in preprocessing, detection and classification techniques for ecoacoustic data: a comprehensive review for large-scale passive acoustic monitoring. Exp Syst Appl 124220

Oikarinen T, Srinivasan K, Meisner O, Hyman JB, Parmar S, Fanucci-Kiss A et al (2019) Deep convolutional network for animal sound classification and source attribution using dual audio recordings. J Acoust Soc Am 145(2):654–662

Oyedele A, Ajayi A, Oyedele LO, Delgado JMD, Akanbi L, Akinade O et al (2021) Deep learning and boosted trees for injuries prediction in power infrastructure projects. Appl Soft Comput 110:107587

Regona M, Yigitcanlar T, Hon C, Teo M (2024) Artificial intelligence and sustainable development goals: systematic literature review of the construction industry. Sustain Cities Soc 105499

Sachan VS, Katiyar A, Somashekher C, Chauhan AS, Bhima CK (2024) The role of artificial intelligence. In: HRM: opportunities, challenges, and ethical considerations. Educational administration: theory and practice 30(4):7427–7435

Sanchez FJB, Hossain MR, English NB, Moore ST (2021) Bioacoustic classification of avian calls from raw sound waveforms with an open-source deep learning architecture. Sci Rep 11

Stahl BC, Stahl BC (2021) Concepts of ethics and their application to AI. Artificial intelligence for a better future: an ecosystem perspective on the ethics of AI and emerging digital technologies, 19–33

Stowell et al (2019) Automatic acoustic identification of individuals in multiple species: improving identification across recording conditions. J Royal Soc Interface 16(153):20180940. https://doi.org/10.1098/rsif.2018.0940

Wolters et al (2021) Proposal-based few-shot sound event detection for speech and environmental sounds with perceivers

Wood et al (2021) Survey coverage, recording duration and community composition affect observed species richness in passive acoustic surveys. Methods Ecol Evol 12(5):885–896. https://doi.org/10.1111/2041-210X.13571

Xie J, Colonna JG, Zhang J (2021) Bioacoustic signal denoising: a review. Artif Intell Rev 54:3575–3597

Ziemba EW, Duong CD, Ejdys J, Gonzalez-Perez MA, Kazlauskaitė R, Korzynski P et al (2024) Leveraging artificial intelligence to meet the sustainable development goals. J Econ Manag 46(1):1–76

Artificial Intelligence in Agriculture: Technology-Powered Strategies for a Sustainable Farming Future

Shallu Duggal, Shivani Sood, Shilpa, Kamal Nain Sharma, Monika Sethi, and Chander Prabha

Abstract

Precision agriculture involves utilizing data-driven strategies and cutting-edge technologies to improve crop management. By leveraging machine learning and artificial intelligence (AI), targeted pest control, enhanced soil management, real-time crop monitoring, and efficient irrigation can be achieved. These advancements aim to minimize waste, boost crop yields, and ensure optimal resource utilization. Additionally, AI technologies optimize agricultural supply chain operations, improving logistics, cutting costs, and promoting sustainability. This data-focused method is revolutionizing agriculture by creating smarter and eco-friendly farming techniques. Precision agriculture, through enhanced production, assists farmers in meeting the rising food demand with a limited environmental footprint.

Keywords

Precision agriculture · Sustainable farming · Crop management · Artificial intelligence

1 Introduction

The improvement of artificial intelligence has revolutionized some of industries, which includes agriculture. A number of the most successful AI technologies that have the capacity to revolutionize the agriculture enterprise have been positively recognized and studied in this article. With the aid of the give up of this decade, they'll in all likelihood be 9 billion humans on the earth, and there are plenty of troubles that affect food manufacturing. The agriculture quarter isn't always able to satisfy the demands of the converting surroundings. Erosion, but there are issues with water, illnesses and pests, and a lack of workers. AI is idea to be a first-rate tool for tackling these issues that are critical to preserving food security and advancing sustainable development. AI has several uses in agriculture, along with as deliver chain optimization, animal management, crop monitoring, and precision farming that allows you to count on yields, become aware of plant illnesses, and make powerful use of water, fertilizer, and pesticides, superior analytics, and device gaining knowledge of algorithms are increasingly being applied to the evaluation of ancient and present records. For instance, there are AI-based totally technologies that use satellite imagery and faraway sensing methods to evaluate crop fitness, soil situations, and even forecast weather, allowing farmers to arrange their assets and save you waste. Because they help conserve sources and improve the capacity to cope with detrimental conditions, they're particularly beneficial in controlling the outcomes of climate change. Synthetic intelligence (AI)-pushed selection assists systems and also assist farmers by using guidelines on while to sow, a way to water, and a way to harvest. With the use of these technologies, farmers may additionally make correct and properly informed choices that reduce

S. Duggal · S. Sood (✉) · Shilpa · K. N. Sharma
Lovely Professional University, Phagwara, Punjab, India
e-mail: shivani.31165@lpu.co.in

S. Duggal
e-mail: shallu.30949@lpu.co.in

M. Sethi · C. Prabha
Chitkara Institute of Engineering and Technology, Rajpura, Punjab, India
e-mail: monika.sethi@chitkara.edu.in

C. Prabha
e-mail: chander.prabha@chitkara.edu.in

K. Subbaraj et al. (eds.), *Computational Bioacoustic Artificial Intelligence*, Sustainable Artificial Intelligence-Powered Applications, https://doi.org/10.1007/978-3-032-06088-4_15

environmental damage, decrease enter fees, and boom crop yields.

AI-based systems, for example, analyze vast databases to identify patterns and make predictions. Optimal farming techniques consider crop type, soil quality, and climate conditions. Farmers who have access to meaningful and up-to-date information, these technologies empower them. To make informed choices that enhance agricultural productivity and yield, precision farming, which is widely recognized for its effectiveness, has recently been the subject of extensive research. It highlights the increasing importance of artificial intelligence in various industries. Intelligence in the field of smart agriculture. Precision agriculture utilizing by leveraging data analytics and machine learning, farmers can efficiently keep track of fertilizer application, crop health, and other significant aspects. Water is essential for life and vital for the survival of all living beings.

According to a study conducted by Liakos et al. (2018), the use of machine learning algorithms can significantly improve the accuracy of medical diagnoses. Learning approaches can be utilized to Liu et al. (2020) achieve precision farming, resulting in the implementation of the new technology resulted in substantial improvements in crop yield and substantial reductions in production costs. Additionally, there were reductions. Real-time monitoring and analysis of farm data is made possible. With the assistance of Ai tools, farmers can able to react quickly to unexpected situations and environmental concerns like pest infestations, floods, and droughts (Liu et al. 2020).

The essential flexibility of artificial intelligence makes that an important tool for reducing risks caused by uncertain farming conditions. AI is very important not only for looking after crops but also for good pest control and complete soil health checks. Drones combined with AI and IoT sensors help gather, in real-time, information about soil characteristics such as moisture content, pH balance, and nutrient levels. This data allows farmers to apply precise, information-based actions to improve crop yield and boost soil richness. Also, AI solutions for pest control can spot outbreaks early and use deep learning algorithms to predict their spread.

As evidenced by research, AI outshines traditional methods of pest control in delivering better results by accurately identifying dangerous pests and diseases and therefore enabling prompt action which later minimizes the use of chemical pesticides.

AI helps to yield healthier consumer products: less pesticide usage entails less environmental pollution and lower costs of farming. AI is transforming agricultural supply chain management—an integration wider than just the field. Improved logistics, decreased food wastage, and timely delivery of agricultural produce to markets are among the benefits accrued due to AI algorithms and predictive analytics. As in source (Jha et al. 2019), AI technologies can predict consumer demand, and also spur distribution systems which cause fewer disruptions in the supply chain, thereby enhancing food security. An appraisal from 2021 (Demirel and Kumral 2021) sheds light on one of the abilities of pest management, carrying out early detection of pest infestations via pc vision and deep gaining knowledge of; it is able to trigger well-timed suspicious interest earlier than the overexposure damages the crops.

A research paper conducted in 2022 showcased the efficiency of artificial intelligence (AI) in managing irrigation systems where advanced technology not only ensured maximum crop yields but also dramatically reduce water consumption, especially in areas experiencing water scarcity, to mitigate the effects of drought (Ramdinthara et al. 2022). Furthermore, AI plays a role in agricultural economics by predicting future trends, market trends and commodity prices. This information helps to minimize financial risks and guarantees economic sustainability, providing farmers with the necessary tools to plan their agricultural activities effectively allocation initiatives, thereby highlighting the significance of AI in predictive analytics. Modeling, a study conducted in 2024, demonstrates how farmers can obtain valuable insights into their operations. Market demands and price fluctuations enable them to make well-informed choices that increase profitability. In a highly competitive market, these tools empower farmers to work effectively, guaranteeing a consistent income and ensuring a steady supply of food for consumers. Despite the vast potential of artificial intelligence in agriculture, there are still challenges to overcome. There are still a few unresolved matters that need to be addressed. For example, small-scale farmers, who are essential to the agricultural industry in many impoverished regions, play a crucial role in ensuring food security. Nations frequently encounter technological, financial, and infrastructure obstacles that hinder their progress and limit their access to advanced artificial intelligence tools. Furthermore, the execution of AI raises important ethical concerns, including those related to data privacy. Ensuring equal access to resources and considering the environmental impact are both important factors to consider.

It is of utmost importance. Policy-makers, technology companies, and agricultural stakeholders should collaborate to address the challenges faced by the industry. Work together to overcome these challenges, ensuring that artificial intelligence solutions are implemented effectively. Values, diverse, and responsible. Overall, artificial intelligence has the potential to revolutionize various industries and improve efficiency in many aspects of our lives. We promise to revolutionize agriculture by providing innovative solutions that enhance productivity and sustainability. Environment, effectiveness, and output. With advanced tools at their disposal, the team was able to complete the project ahead of schedule. Farmers can tackle modern agricultural issues, such as crop diseases

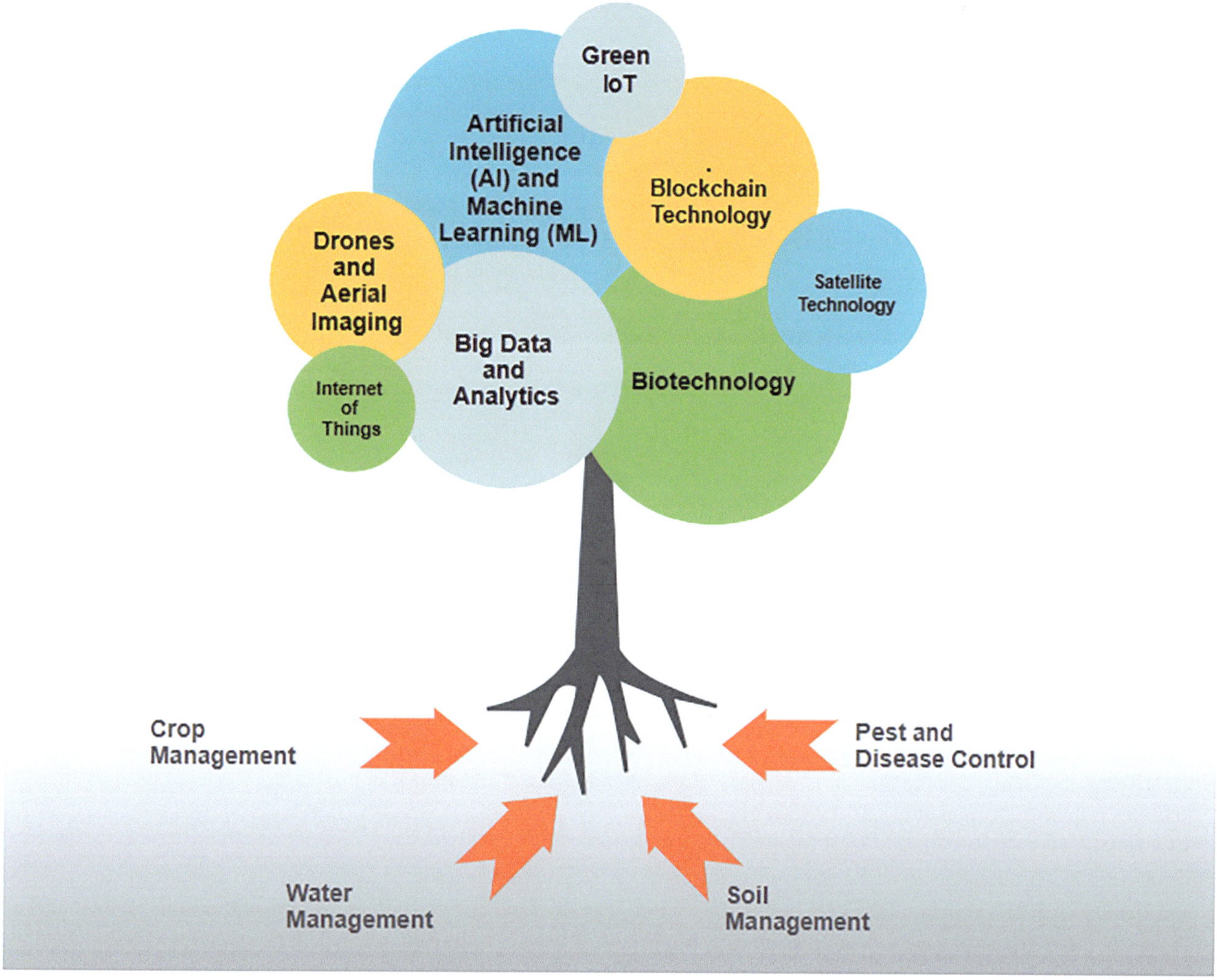

Fig. 1 Various technologies and their use in agriculture

and pest infestations, through proper disposal methods. From streamlining supply chains and implementing effective pest management strategies to embracing precision agriculture techniques, the company is committed to driving innovation and efficiency in the agricultural sector. Agricultural neglect. Artificial intelligence provides farmers with the ability to reduce expenses, improve decision-making, and minimize risks. Uncertainties related to the economy and the environment pose potential risks. Nevertheless, prior to AI can be broadly implemented, but further research and innovation are crucial for its successful implementation. Addressing ethical concerns enhance accessibility for smallholder farmers and lessen long-term environmental effects. This research examines a variety of topics. The field of agriculture has seen advancements in machine learning and deep learning techniques, with the goal of improving farming practices. A comprehensive overview of different technologies used in agriculture is presented in Sect. 2 and its application in different agricultural regions are discussed in detail. Section 3. The findings and opinions are discussed in Sect. 4, while the conclusion is presented in Sect. 5 that enumerates the revolutionary potential of artificial intelligence in agriculture.

2 Technologies for Agriculture

Technologies are playing their part in the agriculture and their goal is to make agriculture as a field more technical-oriented. Technology has assisted in overcoming numerous challenges like the change in climate, shortage of labor, and other issues connected with the agriculture.

Various advanced technologies have been integrated with agriculture as shown in Fig. 1. The leveraging of these technologies makes advancement in agriculture as more sustainable practices. These technologies play vital role. The following technologies are significantly influencing the improvements in agriculture:

- Internet of Things: An example of what the Internet of Things is taking in agriculture includes smart farming, which offers enhanced data collection and decision-making. Smart farming adopts sensors for detecting moisture, temperature, humidity, and soil and assists in real perceptions of irrigation, fertilizer, and crops. Additionally, IoT helps to reduce the manpower as well as monitor and optimize the resources (Kumar et al. 2024).
- Drones and Aerial Imaging: Drones are equipped with cameras and sensors to monitor crops and assist in operations management. They can capture HD-level images and identify crop health status, stress, and pests. Drones are also incorporated with spray systems that are used in pesticides and fertilizers on agricultural fields.
- Artificial Intelligence (AI) and Machine Learning (ML): AI and ML are used in agriculture data analysis for decision-making while increasing efficiency. AI algorithms utilize and incorporate the prediction algorithms on the history of agriculture to come up with crop diseases and even climate for agriculture. They are also used in identifying the image for the detection of early signs of diseases in plants (Kumar et al. 2024; Abiri et al. 2023).
- Blockchain Technology: Blockchain improves the food supply chain, allowing to identify the true provenience and quality of food products. Blockchain has the feature of supply chain transparency and smart contracts. Supply chain transparency of the blockchain system creates a record of products from the farms and down to the consumer hence allowing immutable records of the product's origin, production practices, and quality. Blockchain can be used to handle transactions in the agricultural sector, for example, payment between the farmer and the supplier since it would handle transactions securely and accountability (Kumar et al. 2020).
- Big Data and Analytics: In the agriculture fields, a huge amount of data is being generated, and an analysis of the data helps in decision-making. Data analysis helps to predict the input optimization required for fields such as water, and fertilizer, Based on the analysis, farmers plan their activities based on weather predictions and climate changes (Kumar et al. 2020).
- Biotechnology: Biotechnology uses genetic engineering to produce new crop varieties and enhance the practice of agriculture. Biotechnology produces GMOs that are effective for pest control, disease control, and management of the physical environment. Also, biotechnology uses CRISPR for disease control.
- Green IoT: Green IoT improves the sustainable irrigation systems to optimize the resources to be used such as water and soil and enhances the energy-efficiency farming process. In green IoT, sensors are deployed for real-time monitoring of remote areas and ensure that sufficient water is received by fields.
- Satellite Technology: Satellite technology can assist considerably in imagining areas and situations, for example in extensive farming. Remote sensing is used in tracking water usage as well as land degradation, while climate monitors help in identifying the effects of drought, weather and produce impacts on agriculture (Kumar et al. 2024).

2.1 Machine Learning Algorithms for Agriculture

Machine learning plays a crucial role in agriculture for smart farming, automating irrigation processes, increasing efficiency, and improving resource utilization. It helps in crop management, water management, soil management, pest and disease control, and supply chain management In this (Meshram et al. 2021; Araújo et al. 2023; Attri et al. 2023) section, various algorithms used in the agriculture field are discussed and shown in Fig. 2.

- Support Vector Machine: In the agriculture field, SVM is used for classification and regression. SVM helps to handle high-dimensional data and deal with noisy data. Table 1 highlights the various application areas where is being used.
- Naïve Bayes: In the agriculture field, Naive Bayes is used for classification based on different features. It helps in disease prediction and classification of crops and soil. Table 1 highlights the various application areas where Naïve Bayes is being used.
- Linear Regression: Linear regression defines the relationship between dependent and independent variables. This is used for analysis and regression purposes. Table 1 discusses the application of linear regression.
- Decision Tree: A decision tree is used for classification and regression. It helps in crop management, pest control, etc. Table 1 pens down the application areas of agriculture
  - AdaBoost: AdaBoost is an ensemble learning model that is used to integrate the weak classifiers that help in agriculture application as discussed in below Table 1.
  - Stochastic Gradient Descent (SGD): SGD is an optimization algorithm that helps us minimize the expected loss when we choose the set of parameters.

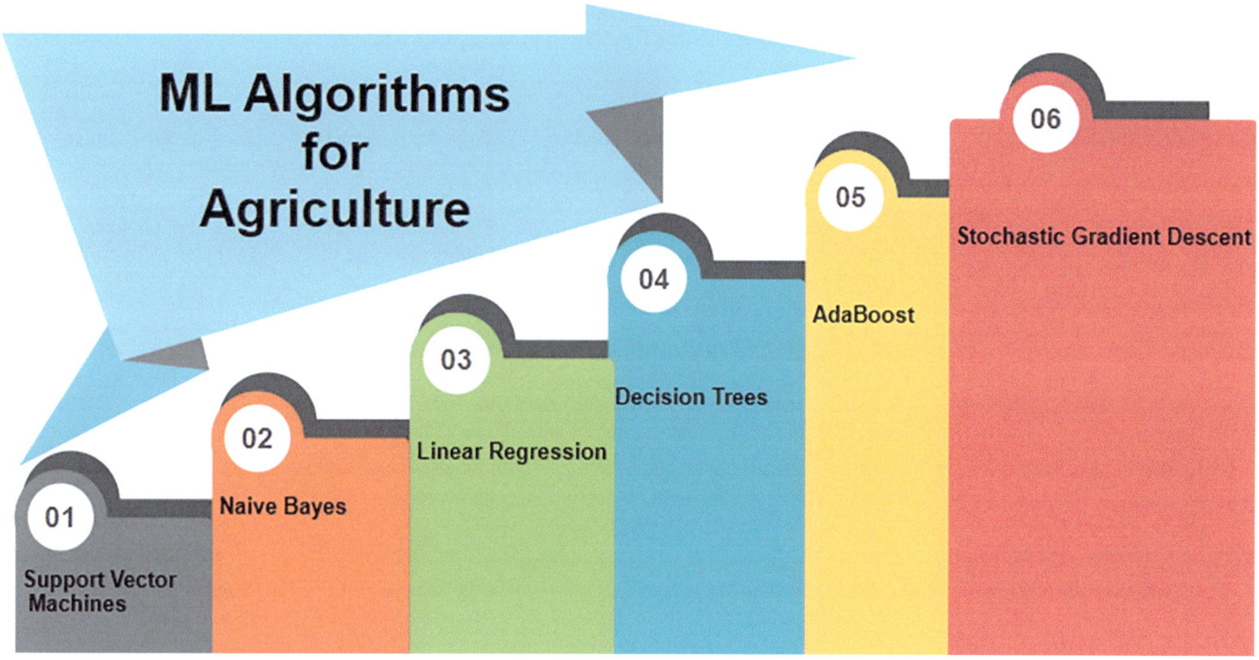

Fig. 2 Machine learning algorithms

This is used in disease diagnosis, while in crop management, for example, in soil management it can be classified as wet dry, or clay base, the amount of water wastage to be used to grow the crop among other important things that lead to good crop management. It makes it easier to monitor real-time organization liability.

2.2 Deep Learning for Agriculture

Deep learning has emerged as the greatest disruption to the agricultural industry since the Industrial Revolution allowing for data-driven farming. Through exploiting sensor data, satellite data, drone data and deep learning models can help farmers and agricultural experts analyze complex features and patterns and give recommendations (Attri et al. 2023). Some of the famous deep learning algorithms and their use in agriculture are as shown in Fig. 2:

- Artificial Neural Network: ANN is used in the agriculture field to improve efficiency, yield increase, and enhancement in decision-making. This algorithm helps in crop management, disease prediction, soil classification, and water management applications such as crop disease detection and yield prediction and improves the scalability level. ANNs are applied to such applications as crop yield prediction, soil quality assay, and disease diagnosis. These models analyze other relations between inputs for instance climate conditions, texture of soil, and plant characteristics. It helps in forecasting crop yields about input data that include rainfall, temperature, and soil type. It is effective for both, volume of data and identifying concealed complex patterns in the data.
- Long Short-Term Memory Networks: A type of RNNs which is built to address the issue of long-term dependencies. They also predicted crop yield, weather conditions, and methods of using them for irrigation using data collected from previous studies. It predicts weather or irrigation requirement for crops, future crop yields, etc. It can manage to operate on longer sequences on the data, which makes it suitable, especially in seasonal or long-term forecasts.
- CNN: CNNs are deep learning models mostly utilized for classification problems, and particularly for image classification problems. In agriculture, CNNs are used for the classification and diagnosis of diseases, pests, and nutrient deficiencies from drone and satellite imagery. It identify crop diseases, pest, and diseases through image recognition and analysis from a drone or satellite. It is very effective in new pattern recognition and new anomalies on images are easily detected since they are beyond human vision.
- Open CV: Open Source Computer Vision Library is an open source library for real-time applications. OpenCV is very useful in the contemporary agriculture; this change has brought a way through which farmers get real-time solution for checking crop health, resources control in addition to other issues like harvesting and fruit counting. This makes it very useful in precision agriculture since it can perform so many tasks that assist farmers increase

Table 1 ML/DL-based application areas of agriculture

S.no.	Application areas	Definition	Benefits	Example
SVM-based application areas of agriculture				
1	Crop classification	SVM is used to classify the crops based on satellite images and drone images	Helps the farmers in monitoring crops as well as managing resources	Classification of rice, wheat, maize
2	Pest and disease detection	SVM is used to detect diseases in plants based on different features such as color, texture, and shape	Reduce crop losses and minimize the use of pesticide in crop	Late blight in tomato
3	Soil quality management	SVM helps classify the soil to identify its suitability for the crop	Soil fertility based on chemicals use for crop.	Sandy, loamy, and clayey soils
4	Yield prediction	SVM is used for yield prediction based on different features such as weather conditions, soil health, and historical data	Allow for effective harvesting and efficient sale	Rice yield prediction
5	Water management	This algorithm helps predict the water requirement for crops based on the soil's moisture	Optimization of water usage, waste management	Moisture level classification of soil and supply of water
Naïve Bayes-based application areas of agriculture				
6	Crop recommendation	An algorithm is used to predict the crop by analyzing the probability level of crop growth based on features such as soil properties, climate and historical data	Crop selection and improvement in productivity and effective sustainability	Suitable crop recommendation such as wheat and maize
7	Pest and disease detection	An algorithm is used to detect the disease based on different features such as leaf color, texture, and shape of a leaf	Early detection of diseases in crops	Tomato blight or wheat rust
8	Soil quality management	An algorithm is used to classify the soil such as clay, candy, and loamy based on pH level and moisture level	Accurate soil classification for effective crop	Soil classification
9	Yield prediction	Predict the different levels of yield using weather data and irrigation patterns	Planning in harvesting and improvement in farm management	Rice prediction using rainfall and fertilizer
10	Water management	An algorithm predicts the soil status such as dry, wet, or optimal	Efficient water management	Classify the soil moisture level and provide the water
Linear regression-based application areas of agriculture				
11	Crop yield prediction	LR defines the relationship between crops such as yield and factors such as rainfall, fertilizer, temperature, and soil	Control the factors for yield improvement	Historical rainfall, soil nitrogen levels, and temperature
12	Soil properties	An algorithm is used to define the relationship between soil and crop quality	Improvement in crop productivity	Soil properties such as pH, organic matter, texture, and crop performance
13	Pest and disease prediction	An algorithm is used to define the relationship between environmental factors and disease incidence	Reduce crop loss	Wheat fields based on humidity and temperature data
14	Fertilizer optimization	An algorithm predicts the fertilizer effect on yield	Reduce the environmental effect	Increase in nitrogen level
15	Water management	An algorithm is used to identify the usage of water and define the relationship between irrigation and crop growth	Optimize the water resources	Rice production
Decision tree-based application areas of agriculture				
16	Crop yield prediction	The algorithm splits the crop data based on different parameters such as temperature and decides how it leads to different yield	Optimize the crop production	Fertilizer optimization and rainfall for the prediction of crop and yield
17	Soil properties	An algorithm is used to classify the soil to help the farmers determine the crop growth	Optimize the crop productivity	Classify and recommend the soil
18	Pest and disease prediction	Help to classify the leaf diseases based on color, pattern, and temperature	Early detection of diseases	Grapes in mildew

(continued)

Table 1 (continued)

S.no.	Application areas	Definition	Benefits	Example
19	Fertilizer optimization	It helps to identify the fertilizer for the enhancement of crop growth, and environment conditions identification	Reduce the wastage of fertilizer, and minimize the damage to the environment	The right blend of nitrogen, phosphorus, and potassium fertilizers
20	Water management	It identities the water requirement for the crop based on different factors such as temperature and soil moisture	Efficient irrigation, reduce the cost and water	Rice paddies
AdaBoost-based application areas of agriculture				
21	Crop yield prediction	An algorithm proposes a model to predict the rainfall, temperature, and yield of fertilizer	Optimize the usage of resources	Wheat and rice yield
22	Soil properties	Combines the weak classifiers and identifies the soil based on pH level and for predicting the soil	Improve the crop health by healthy and suitable soil	Classify the soil as wet, dry, sandy, loamy and clay
23	Pest and disease prediction	Help to classify the leaf diseases based on color, pattern, and temperature trained the labeled images to classify based on color, pattern, and texture and to identify whether a disease exists or not	Early detection of disease and prevention	Potatoes late blight
24	Crop quality grading	The algorithm classifies the crops based on quality, color, size, and text pattern	Increase the grading efficiency and reduce the labor cost	Apple or tomato grading based on appearance
25	Water management	An algorithm based on historical data identities the water usage for crops and predicts the weather conditions such as rainfall	Reduce water wastage, help in decision-making	Yield of maize or wheat prediction

their yields while being sustainable. OpenCV is actively involved in reinventing farming practice as technological advancement continues to prevail in the agricultural sector.

- LeNet: CNNs including LeNet and other models are widely used in activities like image-based plant disease identification, crop segmentation, and fruit ripeness determination. Because of its ability to learn and recognize patterns in image data, LeNet can be employed in analyzing images of agriculture to improve farming. LeNet supports producers and specialists in agricultural practices by automating and enhancing multiple operations. Thus, considering that LeNet produces quite a high accuracy in image recognition tasks, this model is still relevant to modern agriculture.
- VGG16: VGG16 is a deep learning model that is highly capable of solving problems in agriculture such as plant disease identification, fruit categorization, and crop analysis. Its property to extract complex features from images makes YOLO very helpful in automating jobs and increasing the productivity of farming businesses. Even though the VGG16 requires a lot of computational resources it is very accurate and can be used in highly varied applications for precision agriculture.
- MobileNetV2: MobileNetV2 which is a highly efficient deep learning model is meant for mobile and edge smart devices. So, MobileNetV2 in agriculture can be applied to different tasks, including disease diagnosis, crop identification, and even using in precision farming. It is designed to achieve high, accurate results using a minimum of computing resources, making it significant for real-time use in agriculture.

3 AI in agriculture

Without AI and robotics, traditional farming is labor- and time-intensive, requiring a lot of work for seeding, irrigation, and land preparation; it also depends more on human resources; it is less accurate in terms of weather, soil conditions, and fertilizer use; it takes longer to monitor crop health (Wakchaure et al. 2023). Precision agriculture benefits greatly from artificial intelligence (AI), which makes it possible to use a variety of predictive models that assist farmers in making defensible judgments (Linaza et al. 2021). Two concepts that are frequently discussed in agriculture are precision farming (PF) and smart farming. PF farming is a method that relies on gathering and analyzing certain data to help farmers make better decisions. It employs various technologies to manage crops, irrigation, soil, pest control (Sood and Singh 2021; Sood et al. 2022), livestock, and supply chain management. Smart farming, on the other hand, is centered on leveraging technology to allow equipment to function alongside farmers while also making some decisions on their own.

Farmers are constantly able to intervene to modify the judgments made by these devices, which acquire and process information in real time to make decisions. The main key areas where AI can be leveraged in agriculture are for crop management, water management, soil management, pest and disease control, and supply chain management. They came to the conclusion that ML technologies will undoubtedly improve agriculture, enabling Farm Management Systems to develop into real-time decision support systems (DSS) through the use of AI algorithms (Liakos et al. 2018). Crop management covers a wide range of applications, such as weed and pest identification, plant stress detection, the installation of smart farms, and yield forecast and estimation that are directly related to the nation's economic development (Attri et al. 2024).

3.1 Crop Monitoring

This paper (Dasgupta et al. 2020) proposed a system that aimed at improving agricultural practices by recommending suitable crops for specific soil types and detecting weeds in crop fields. The paper proposes a wireless sensor network (WSN) architecture to collect soil data, which is analyzed using machine learning algorithms such as Naïve Bayes, to suggest crops based on soil features. Apart from this, the paper implemented a weed detection mechanism using image processing and deep neural networks. The AI-enabled drone will capture real-time video of the elders, and a convolutional neural network (CNN) model trained will identify weed growth. Here the weed's location is sent to the farmer via a mobile application, allowing for manual weed removal, thus preventing crop damage and improving yield. The system proposed consists of environmental sensors that are used to evaluate soil moisture and humidity for monitoring the conditions that may promote weed growth. Here, IoT technologies like Zigbee are utilized for efficient communication between sensor nodes and the cloud database, facilitating data processing and analysis. The model used achieved an accuracy of 90%. Thus, this work offers a user-friendly application that simplifies decision-making for farmers so that agricultural productivity can be boosted by reducing losses due to weeds and enhancing the overall efficiency of crop selection based on real-time soil and environmental data.

The research (Singh et al. 2020) aims to identify the most appropriate environmental conditions such as air humidity, temperature, soil moisture, and light intensity that is optimal for plant development. This work explores the optimization of marigold plant growth by proposing an AI-enabled IoT-based monitoring system. The system proposed various sensors like DHT11, LDR, DS18B20, and soil moisture sensors for capturing environmental conditions and these are connected to Arduino and NodeMCU micro-controllers. The data captured using these sensors is transmitted to cloud using MQTT/HTTP protocols and further stored on the ThingSpeak platform for analysis purpose. Different supervised machine learning algorithms, including Logistic Regression, Gradient Boosting Classier, and Linear Support Vector Classifiers are implemented to analyze the collected data and determine the best growth conditions. This work also implemented image processing techniques using OpenCV and deep neural networks to detect weed growth, thus protecting the marigold plants from nutrient competition. The proposed system is integrated into a Django-based web application. The farmers can input environmental parameters and receive precise crop recommendations using this web application. This approach encourages smart farming practices and raises agricultural output by improving marigold cultivation and offering a scalable framework for other crops also. This paper (Huang et al. 2022) presented a crop growth observation system. It proposes the consolidation of system adaptivity and edge AI to enhance agricultural efficiency. The system exploits artificial intelligence (AI) and wireless sensor networks (WSN) and supports a number of cryptographic functions for decryption of sensor data and it is implemented on FPGA. Convolutional neural networks (CNN) are used for an image fusion technique. This technique blends RGB and deep images taken from Time-of-Flight (ToF) cameras to format using which it is easy to precisely determine Pest and Disease Severity (PDS) levels. These PDS estimators are helpful in identifying target crops. The implementation is performed on the PYNQ-Z2 development board which outperforms performance improvements over traditional CPU and GPU setups, achieving high recognition accuracy and reduced power consumption. With its advanced technical integration, the suggested system aims to maximize agricultural productivity, minimize losses from pests and illnesses, and encourage sustainable farming practices. All things considered, it provides a reliable, effective, and user-friendly solution for smart farming.

3.2 Irrigation Management

This paper (Goap et al. 2018) is about an IoT-based smart system that will predict irrigation requirements in agriculture. The irrigation requirement will be predicted based on sensing data about the ground parameters such as soil moisture, soil temperature, and environmental conditions, along with weather forecast data from the online sources. The system uses sensor nodes to collect data on soil moisture, soil temperature, air temperature, UV radiation, and relative humidity from the crop yield. A machine learning model is trained to process the collected data along with forecasted weather parameters like precipitation, air temperature, humidity, and

UV levels. The system has mainly seven components such as data collection devices, web services for sensor and weather data collection, water motor control, a responsive web interface for monitoring, and IoT-enabled motor pumps. The accuracy of the soil moisture predictions was verified over three weeks of field data. The wireless network is used for the data transmission to the cloud. The real-time visualization and analysis through a web-based decision support system is performed on the cloud. This decision support system helps optimize irrigation strategies. This system has a closed-loop control for automated water supply which makes it an autonomous irrigation scheme.

This study used the Predictive Error Compensated Neural Network (PECNET) model, which is a machine learning technique used to enhance precipitation forecasts for short period for use in precision agriculture (Şentop et al. 2024). The meteorological data from different sources such as the Agricultural Monitoring System of Turkey was used. This model is the combination of neural networks and few error compensation techniques. It was successful to predict hourly precipitation rates. The model was built by considering a variety of parameters, such as air pressure, temperature, rainfall, and humidity.

Eight machine learning models are evaluated in this work (Vij et al. 2020) to predict irrigation water quality (IWQ) across ten parameters: Multiple Linear Regression and Artificial Neural Networks (ANN) (MLR), Adaptive Boosting (AdaBoost), Random Forest (RF), Decision Tree, Support Vector Regression (SVR), k-Nearest Neighbor (kNN), and Stochastic Gradient Descent (SGD). The models examined three hundred samples from nine monitoring stations.

The paper suggested that out of all models except SVR and k-NN accurately predicted most IWQ parameters, and achieved correlation coefficients between 0.56 and 0.99. Here models like ANN, MLR, SGD, and AdaBoost were significantly effective for parameters like Total Dissolved Solids (TDS) and Chloride, while models couldn't predict Magnesium Absorption Ratio (MAR) and Permeability Index (PI) parameters. The findings of this paper significantly highlighted how machine learning may be used to accurately, quickly, and economically forecast irrigation water quality. By providing farmers with the data they require to make irrigation decisions, this enhances agricultural water management and promotes sustainable farming methods.

3.3 Soil Management

This work suggested an AI-based soil mapping model for precision agriculture (Dong et al. 2018). The model is based on VGG16 neural network and identified geo-parcel extremity from high-resolution images. To map the amount of phosphorus available, Random Forest (RF) and deep learning-based Artificial Neural Networks (ANN) algorithms were implemented on soil-based environmental factors. The study worked on 277 data samples and used over 46,000 geo-parcels for model training and validation.

By cutting prediction units by more than 50%, the geo-parcel technique increased prediction efficiency and accuracy. ANN demonstrated marginally better accuracy, while RF provided more stability. The need of accurate environmental data for regulating soil properties was emphasized by this work, which was the pioneer in using geo-parcel mapping in precision agriculture. The worldwide scarcity of clean water has led to the creation of IoT-enabled smart irrigation systems aimed at enhancing water efficiency in precision farming. This investigation (Goap et al. 2018) provides an open-source platform that utilizes sensors to determine the necessary water levels based on weather forecasts, along with soil and environmental factors. With encouraging initial results, the technology, currently being tested on a pilot scale, supports real-time data analysis and automated water management.

Soil moisture, a key element in designing an effective smart irrigation system, is influenced by a variety of environmental conditions, including air temperature, humidity, UV radiation, and soil temperature. Weather forecasts are now much more and credit goes to technological advancements, which also make it possible to more precisely predict changes in soil moisture. This paper proposes a hybrid machine learning-based smart irrigation system based on the Internet of Things (IoT).

3.4 Pest control and Disease Management

The research (Albanese et al. 2021) describes an embedded system that uses deep learning and computer vision technologies to autonomously detect insect infestations in fruit farms. Its main goal is to enable prompt pest detection without requiring human intervention, hence mitigating biotic stress, a major factor in crop yield decline. The device combines a neural accelerator for picture processing and capture inside pheromone traps with a low-power integrated sensing solution. It uses a Raspberry Pi 3 for data collecting and processing management, a Sony IMX219 image sensor, and three machine learning algorithms—a modified LeNet-5, VGG16, and MobileNetV2, all of which are intended to detect pests effectively.

This study (Refaai et al. 2022) analyzes communication protocols for coordinating a fleet of UAVs to investigate how drones can efficiently manage insect infestations. This paper uses feature extraction and transfer learning to assess the efficacy of nine deep neural network models in plant disease detection. SVMs and extreme learning machines (ELMs) are implemented to classify the features that have been extracted, while pretrained models are utilized for feature extraction. The results of the study indicate that deep learning-based feature extraction in conjunction with SVM/ELM classification surpasses transfer learning in performance. The report concludes that the smart farming and precision agriculture was made possible by use of technologies like Internet of Things and remote sensing. These instruments can include satellites and unmanned aerial vehicles (UAVs), which are used to detect pests and notify farmers. Although UAVs have limitations in terms of flying time and battery capacity, they provide more reliability for pest monitoring. Based on simulation results, drones that use distributed search algorithms are more effective at solving pest problems than those that use random searches. The ResNet50 model, when paired with SVM classification, achieves an accuracy of 97.86%. The potential of UAVs and deep learning to enhance precision agriculture's pest management techniques is thus highlighted by this study.

3.5 Supply Chain Management

This study (Carbonneau et al. 2008) examined and applied cutting-edge machine learning methods for predicting skewed demand in agricultural supply chains, including Support Vector Machines (SVM), Recurrent Neural Networks (RNN), and Neural Networks (NN). In this study, these machine learning methods are implemented on dataset and compared with approaches such as naïve forecasting, moving averages, linear regression (MLR), and time-series models. For evaluation purpose, the two datasets were used: one based on a simulated supply chain and another on actual demand data from Canadian Foundries. The results shown in the paper indicate that RNN and SVM perform better than other methods, their forecasting accuracy is not statistically significantly better than that of the MLR model. RNNs are effective at capturing temporal patterns due to their recurrent structure, and SVMs show strong performance on training sets but face challenges with generalization on testing sets.

Here, the other methods like naïve and trend forecasting performed poorly. Even though machine learning models outperformed MLR in accuracy, the differences were not that great, especially when it came to simulated data. Larger improvements were obtained using RNN and SVM for the data. This paper (Baryannis et al. 2019) proposes a supply chain risk prediction framework. It integrates data-driven artificial intelligence (AI) with supply chain risk management (SCRM).The framework proposed can bring together AI experts and supply chain professionals to ensure AI models produce clear and interpretable results for better decision-making in supply chain management This framework is built which can predict delivery delays in a complex aerospace supply chain. The paper discusses about the balance between how easy models are to understand and how well they predict outcomes. This paper states that focusing on interpret ability can reduce accuracy. Decision trees implementation, which are easy to understand, had lower precision and recall than Support Vector Machines (SVM), which were more accurate but harder to interpret. The imbalanced data is balanced using techniques like oversampling, which further improved recall but reduced precision. The authors conclude that interpretable models can offer valuable insights, but a balance is desirable between performances and interpret ability.

This study (Kollia et al. 2021) examines developments in the food supply chain and production by concentrating on three main areas: retail refrigeration energy optimization, greenhouse yield forecast, and food packaging expiration date verification. The deep learning models are implemented to forecast crop yields precisely in order to minimize food loss and better match supply with demand, particularly in controlled greenhouse settings which can improve productivity and security throughout the food supply chain. Deep learning (DL) models, RNNs and LSTMs, have been implemented in greenhouse production for forecasting plant growth and output based on environmental and historical data. It can enable us to be more accurate about harvest scheduling in accordance with market demand. In the food storage industry, DL approaches have effectively reduced energy usage in sizable refrigeration systems, supporting international efforts to cut greenhouse gas emissions.

The DL usage for automatic expiry date identification on food packaging is also highlighted by authors. Automatic detection of expiry date can helpful for merchants and can enhance food safety and save waste. AI use is also implemented and discussed to control energy use during the refrigeration phase, which is important considering the substantial carbon emissions from food storage systems. Overall, the study shows how the food supply chain may be greatly optimized by combining AI and machine learning, hence enhancing sustainability and safety. It urges growers, retailers, and legislators to embrace these technologies broadly.

4 Results and Discussion

This section analyzes the publishers and recent methodologies available in agriculture domain. In the current study, we have considered the IEEE Explore, Springer, Elsevier, Wiley, MDPI, etc., articles to review. Figure 3 and Table 2 shows the illustration of the paper reviewed in various journals and conferences in this book chapter. Also, Fig. 4 and Table 2 shows the book chapter's citation distribution by publisher

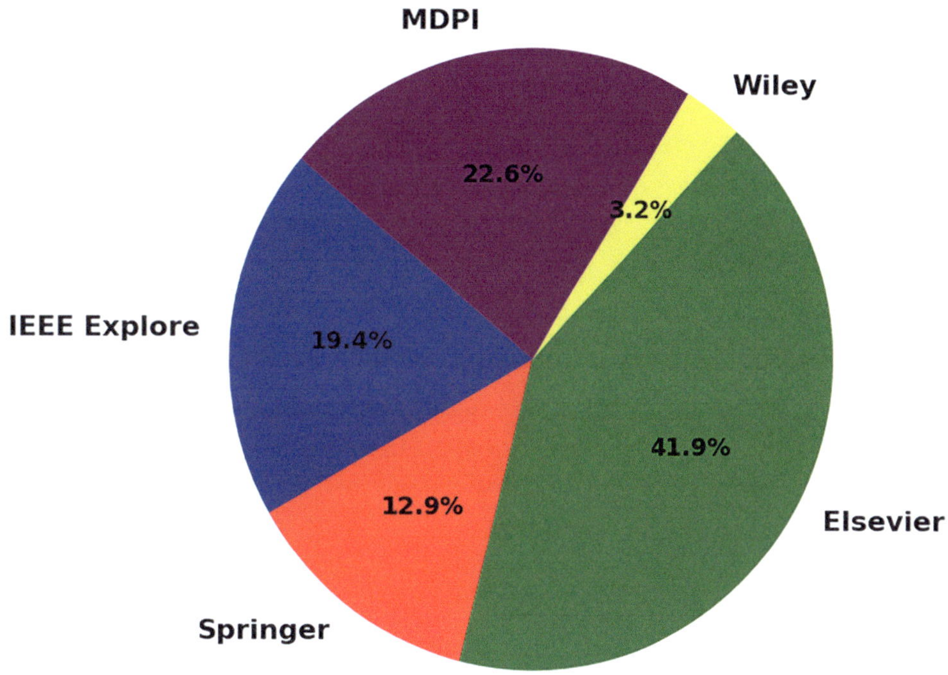

Fig. 3 Distribution of the papers used for reviewing

Table 2 Analysis on paper published according to the publisher's details

Publisher name	References
IEEE explore	Singh et al. (2020), Huang et al. (2022), Şentop et al. (2024), Dong et al. (2018), Albanese et al. (2021), Nagaraja et al. (2019)
Springer	Ramdinthara et al. (2022), Attri et al. (2024), Dasgupta et al. (2020), Bouguettaya et al. (2022)
Elsevier	Jha et al. (2019), Demirel and Kumral (2021), Abiri et al. (2023), Meshram et al. (2021), Attri et al. (2023), Linaza et al. (2021), Dasgupta et al. (2020), Singh et al. (2020), Albanese et al. (2021), Baryannis et al. (2019), Bouguettaya et al. (2022), Tripathi et al. (2022)
Wiley	Refaai et al. (2022)
MDPI	Liakos et al. (2018), Meshram et al. (2021), Araújo et al. (2023), Linaza et al. (2021), Kollia et al. (2021), Albahar (2023), Benos et al. (2021)

shows a significant reliance on Elsevier (41.9%), followed by significant contributions from MDPI and IEEE Explore (22.6%), IEEE Explore (19.4%), Springer (12.9%), and Wiley (3.2%). Elsevier's dominance indicates that the chapter significantly relies on foundational studies, which is consistent with its function as a key source of well-established, high-impact research. With its open-access strategy, MDPI's substantial presence indicates a focus on timeliness and accessibility, which is probably a reflection of recent advancements and multidisciplinary research. While Springer and Wiley's smaller contributions imply selective use for specific insights within the area, IEEE Explore's emphasis on technology and engineering highlights the chapter's involvement with technical and innovative research. By combining both readily available recent studies and traditional, high-impact publications, this citation distribution highlights a well-rounded approach to source selection and offers an in-depth assessment of the state of the field. While the selective usage of Springer and Wiley may offer specialized content that enhances the primary topics, the preponderance of Elsevier, MDPI, and IEEE Explore highlights an emphasis on top-notch, up-to-date, and application-driven research. All things considered, the chapter's academic rigor accessibility, and significance to furthering knowledge in the scientific and technological subjects it addresses are improved by this variety of sources (Table 3).

The analysis of algorithm utilization in research publications on agriculture reveals a variety of methods for choosing machine learning approaches. With four publications, Support Vector Machine (SVM) is the most widely used, probably because of its superiority in classification tasks like yield prediction and crop disease detection. Other algorithms, such as CNN, ANN, *k*-Nearest Neighbor, Grading Boosting, Transfer Learning, RNNs, LSTMs, ANN, Naive Bayes, and Logistic Regression, are equally dispersed and appear in two articles each. This equitable use suggests that scientists are still in the experimental stage, experimenting with diverse approaches to determine which work best for

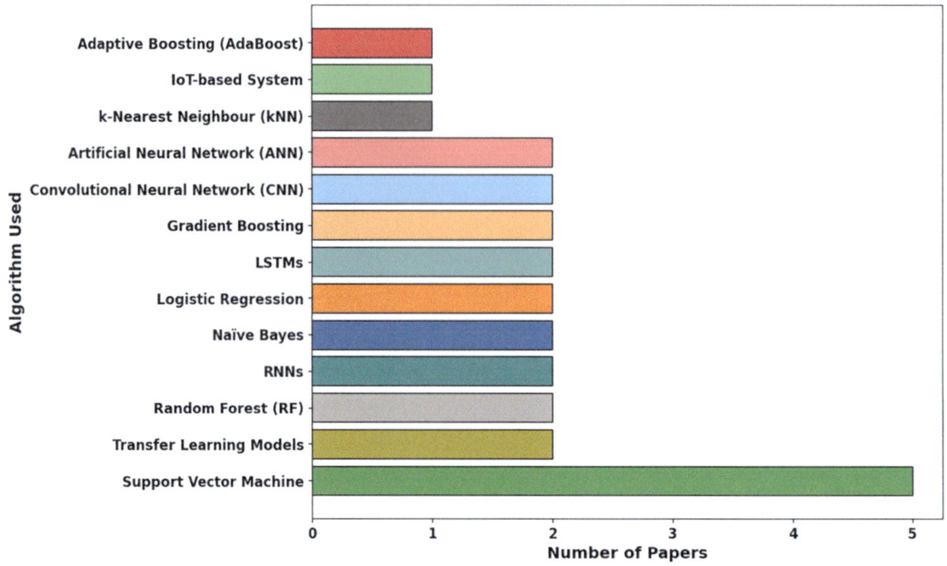

Fig. 4 Number of papers by algorithm used

Table 3 Analysis on methodologies used for solving various agriculture domain problems

Algorithm	References
Naïve Bayes	Dasgupta et al. (2020), Carbonneau et al. (2008)
Convolutional neural network (CNN)	Dasgupta et al. (2020), Huang et al. (2022)
Logistic regression	Singh et al. (2020), Huang et al. (2022)
Gradient boosting	Singh et al. (2020), Huang et al. (2022)
Support vector machine	Singh et al. (2020), Huang et al. (2022), Carbonneau et al. (2008), Baryannis et al. (2019), Kollia et al. (2021)
IoT-based system	Goap et al. (2018)
Transfer Learning models (LeNet, VGG16,MobileNet, ResNet50)	Albanese et al. (2021), Refaai et al. (2022)
RNNs	Carbonneau et al. (2008), Kollia et al. (2021)
LSTMs	Şentop et al. (2024), Kollia et al. (2021)
Adaptive boosting (AdaBoost)	Vij et al. (2020)
Artificial neural network (ANN)	Vij et al. (2020), Dong et al. (2018)
k-Nearest neighbor (*k*NN)	Vij et al. (2020)
Random forest (RF)	Vij et al. (2020), Dong et al. (2018)

various agricultural uses. Although IoT is a developing topic, it may not yet place a strong emphasis on algorithm development or may be coupled with other AI models, as seen by the fact that IoT-based systems and Adaptive Boosting are mentioned in just one publication. Similarly, Adaptive Boosting might have more specialized uses in agriculture. More sophisticated neural network models, like transfer learning, RNNs, and LSTMs, all of which have been discussed in two papers are notably gaining popularity. Theirs inclusion indicates a tendency toward sophisticated methods that are well-suited to managing time-series data, including seasonal crop data or intricate, non-linear agricultural relationships.

5 Conclusion

The traditional farming has been revolutionized into smart farming and precision agriculture by integration of artificial intelligence (AI) and IOT and it leads to sustainable and resource-efficient smart agricultural practices. AI-based solutions use real-time data for agricultural processes such as pest control, crop management, soil monitoring, and irrigation efficiency. This contributes in diminishing the environmental impact and also improves the overall agricultural productivity. By using AI-based analytics, supply chain management can be optimized. This results in reducing food wastage and raised overall profitability. In future, the research should lay stress on increasing small-scale farmers' access to AI. Moreover, in future, the research should address ethical concerns like autonomy and data privacy of the farmer. Research should focus on evaluating the environmental impacts of AI applications for sustainable practices. Finally, to improve agricultural resistance to climate change, AI's predictive powers would need to be developed. By addressing these issues, ethical, cost-effective, and sustainable AI solutions for agriculture will be developed.

References

Abiri R, Rizan N, Balasundram SK, Shahbazi AB, Abdul-Hamid H (2023) Application of digital technologies for ensuring agricultural productivity. Heliyon

Albahar M (2023) A survey on deep learning and its impact on agriculture: challenges and opportunities. Agriculture 13(3):540

Albanese A, Nardello M, Brunelli D (2021) Automated pest detection with DNN on the edge for precision agriculture. IEEE J Emerg Select Topics Circ Syst 11(3):458–467

Araújo SO, Peres RS, Ramalho JC, Lidon F, Barata J (2023) Machine learning applications in agriculture: current trends, challenges, and future perspectives. Agronomy 13(12):2976

Attri I, Awasthi LK, Sharma TP, Rathee P (2023) A review of deep learning techniques used in agriculture. Eco Inform 18:102217

Attri I, Awasthi LK, Sharma TP (2024) Machine learning in agriculture: a review of crop management applications. Multim Tools Appl 83(5):12875–12915

Baryannis G, Dani S, Antoniou G (2019) Predicting supply chain risks using machine learning: the trade-off between performance and interpretability. Futur Gener Comput Syst 1(101):993–1004

Benos L, Tagarakis AC, Dolias G, Berruto R, Kateris D, Bochtis D (2021) Machine learning in agriculture: a comprehensive updated review. Sensors 21(11):3758

Bouguettaya A, Zarzour H, Kechida A, Taberkit AM (2022) Deep learning techniques to classify agricultural crops through UAV imagery: a review. Neural Comput Appl 34(12):9511–9536

Carbonneau R, Laframboise K, Vahidov R (2008) Application of machine learning techniques for supply chain demand forecasting. Eur J Oper Res 184(3):1140–1154

Dasgupta I, Saha J, Venkatasubbu P, Ramasubramanian P (2020) AI crop predictor and weed detector using wireless technologies: a smart application for farmers. Arab J Sci Eng 45:11115–11127

Demirel M, Kumral NA (2021) Artificial intelligence in integrated pest management. In: Artificial intelligence and IoT-based technologies for sustainable farming and smart agriculture. IGI Global, pp 289–313

Dong W, Wu T, Sun Y, Luo J (2018) Digital mapping of soil available phosphorus supported by AI technology for precision agriculture. In: 2018 7th international conference on agro-geoinformatics (agro-geoinformatics). IEEE, pp 1–5

Goap A, Sharma D, Shukla AK, Krishna CR (2018) An IoT based smart irrigation management system using Machine learning and open source technologies. Comput Electron Agric 1(155):41–49

Huang CH, Chen BW, Lin YJ, Zheng JX (2022) Smart crop growth monitoring based on system adaptivity and edge AI. IEEE Access 15(10):64114–64125

Jha K, Doshi A, Patel P, Shah M (2019) A comprehensive review on automation in agriculture using artificial intelligence. Artif Intell Agric 1(2):1–2

Kollia I, Stevenson J, Kollias S (2021) Ai-enabled efficient and safe food supply chain. Electronics 10(11):1223

Kumar AR, Divya TM, Jayasudha BS, Sudha PN (2020) Precision agriculture: a review on its techniques and technologies. Int Res J Mod Eng Technol Sci 2(09):1326–1332

Kumar V, Sharma KV, Kedam N, Patel A, Kate TR, Rathnayake U (2024) A comprehensive review on smart and sustainable agriculture using IoT technologies. Smart Agric Technol 10:100487

Liakos KG, Busato P, Moshou D, Pearson S, Bochtis D (2018) Machine learning in agriculture: a review. Sensors 18(8):2674

Linaza MT, Posada J, Bund J, Eisert P, Quartulli M, Döllner J, Pagani A, Olaizola I, Barriguinha A, Moysiadis T, Lucat L (2021) Data-driven artificial intelligence applications for sustainable precision agriculture. Agronomy 11(6):1227

Liu D, Mishra AK, Ray DK (2020) Sensitivity of global major crop yields to climate variables: a non-parametric elasticity analysis. Sci Total Environ 15(748):141431

Meshram V, Patil K, Meshram V, Hanchate D, Ramkteke SD (2021) Machine learning in agriculture domain: a state-of-art survey. Artif Intell Life Sci 1(1):100010

Nagaraja GS, Soppimath AB, Soumya T, Abhinith A (2019) IoT based smart agriculture management system. In: 2019 4th international conference on computational systems and information technology for sustainable solution (CSITSS). IEEE, pp 1–5

Ramdinthara IZ, Bala PS, Gowri AS (2022) AI-based yield prediction and smart irrigation. Internet of Things Anal Agric 3:113–140

Refaai MR, Dattu VS, Gireesh N, Dixit E, Sandeep CH, Christopher D (2022) [Retracted] Application of IoT-based drones in precision agriculture for pest control. Adv Mater Sci Eng 2022(1):1160258

Şentop MS, Yücel M, Üstündağ BB (2024) AI-based short-term precipitation prediction in precision agriculture. In: 2024 12th international conference on agro-geoinformatics (agro-geoinformatics). IEEE, pp 1–6

Singh R, Srivastava S, Mishra R (2020) AI and IoT based monitoring system for increasing the yield in crop production. In: 2020 international conference on electrical and electronics engineering (ICE3). IEEE, pp 301–305

Sood S, Singh H (2021) Computer vision and machine learning based approaches for food security: a review. Multim Tools Appl 80(18):27973–27999

Sood S, Singh H, Jindal S, Ribeiro-Barros AI, Tevera DS, Goulao LF, Tivana LD (2022) Rust disease classification using deep learning based algorithm: the case of wheat. Food Syst Resilience 24:197

Tripathi P, Kumar N, Rai M, Khan A (2022) Applications of deep learning in agriculture. In: Artificial intelligence applications in agriculture and food quality improvement. IGI Global, pp 17–28

Vij A, Vijendra S, Jain A, Bajaj S, Bassi A, Sharma A (2020) IoT and machine learning approaches for automation of farm irrigation system. Proced Comput Sci 1(167):1250–1257

Wakchaure M, Patle BK, Mahindrakar AK (2023) Application of AI techniques and robotics in agriculture: a review. Artif Intell Life Sci 1(3):100057

Ethical and Privacy Considerations in AI-Driven Wildlife Monitoring

Shalini Kumari and Chander Prabha

Abstract

Artificial Intelligence (AI) has transformed how ecological systems are observed through the monitoring of wildlife, as due to the integration of the theory, it is now possible to collect, analyze, and predict data on a much more significant level. Nevertheless, a high level of ethical and privacy concerns is present in this technological development that requires in-depth scrutiny. This chapter addresses the ethical concerns that are related to wildlife surveillance via AI, including the unintentional harm to animal species, misuse of data, bias in algorithms, and the destruction of natural habitats because of monitoring programs such as AI-enabled drones. Concerns about privacy apply to wildlife and human stakeholders when sensitive information is concerned with endangered species and habitats. The chapter emphasizes that to curb these risks, ethical frameworks, transparency, and accountability during the use of AI need to be strong. It considers current legal and regulatory gaps and points to the need for interdisciplinary cooperation between scientists, policymakers, and local communities to ensure an inclusive governance model is created. The most essential recommendations can be framed into implementing privacy-preserving technologies, algorithmic fairness, and participatory decision-making. These issues can be addressed in a way that AI can be used to facilitate conservation without sacrificing ecological integrity and trust in society. The study will contribute to the discussion on the ethical governance of AI, where practical implications have to be considered to create a balance between technological advancement, the sustainability of the environment, and the sustainability of technology that would be ethical. AI: adaptive, real-time monitoring, community-embedded strategies are identified to warrant fair and sound measures of conservation of wildlife.

Keywords

Artificial intelligence · Wildlife monitoring · Privacy · Deep learning

1 Introduction

The consideration of the element of artificial intelligence (AI) in the wildlife monitoring has become a transformative technology that led to the opening of the path to gather, process, and analyze the data, which is a step forward to the future of studies in the field of ecology and the protection of the whole environment (McClure et al. 2020). However, the combination of AI technology solutions and natural habitat also introduces severe ethical and privacy concerns, which need to be addressed exhaustively and avoided in the future (Isabelle and Westerlund 2022; McClure et al. 2020). It is extremely valuable in conservation, especially when the focal species that was deliberately targeted, such as the elephant, is fundamental to the given environment (Brickson et al. 2023). This is because the ability of AI to analyze massive amounts of data and extract crucial points that would have been missed out on otherwise may decide the future of an environment such as the elephant. Monitoring of wildlife with the help of AI-powered technologies needs a wide ethical framework that should include the potential harm to the wildlife population and the stakeholders as the representatives of the human race (Chakraborty 2024;

S. Kumari · C. Prabha (✉)
Chitkara University Institute of Engineering and Technology, Chitkara University, Rajpura, Punjab, India
e-mail: prabhanice@gmail.com

S. Kumari
e-mail: k.shalini@chitkara.edu.in

K. Subbaraj et al. (eds.), *Computational Bioacoustic Artificial Intelligence*, Sustainable Artificial Intelligence-Powered Applications, https://doi.org/10.1007/978-3-032-06088-4_16

Kuzior et al. 2023). It is worth considering the ethical questions and social consequences of integrating AI in sustainability projects to seek solutions to them (Kuzior et al. 2023). The prospect of seeing the implementation of AI in the form of sustainability programs portends a lot of opportunities to change the way in which we handle environmental issues and manage the available resources (Kuzior et al. 2023). It is also understood that to reap maximum benefit of AI, such arrangements in the form of planning, introduction of accountability, and undertaking of regular checks should be so that it can be made sure that outcomes of the AI technology are making the objective of ecological conservation and that of societal welfare better.

1.1 Ethical Implications of AI in Wildlife Monitoring

Wildlife monitoring with the help of AI leads to a perplexing entanglement of ethical dilemmas that need to be considered thoroughly and actively resolved. Among the most acute, there is the possibility of unintentional damage to animal species due to the implementation of AI in surveillance systems. As an example of how the process could go wrong, one might state that using drones with AI-based facial recognition to monitor the activities of individual animals may disturb natural behavior or cause animals to experience high levels of stress and even abandon their habitats. Also, the process of collecting such sensitive information about wildlife and analyzing it raises doubts regarding the possibility of misuse, such as harnessing the information to create illegal killing or destruction of habitats. Discriminatory results develop as a result of unfair or prejudiced information training the AI systems to create wrong decisions. Consequently, the necessity of clear ethical rules and principles regarding the creation and implementation of AI in wildlife surveillance cannot be overestimated, as specific technologies have to be utilized in the most responsible way and in a way that can reduce the impact on wildlife to a minimum. Interventions requiring the use of AI should be designed with great consideration and an interdisciplinary approach, considering such aspects as design thinking, psychological and sociological considerations, and the economic values so that they possess sustainability but do not pose long-term threats to the environment accidentally (Nishant et al. 2020). The application of AI in sustainability must entail a democratic and participatory way of decision-making to avoid the possible inequities of power (Kuzior et al. 2023). Also, ownership, access, and control of data come into existence owing to the use of AI in wildlife monitoring. The data collected on the population of wildlife are always sensitive to species distribution, behavior, and their habitat usage. Policies ought to be formulated to enable appropriate personalities of the usage of data and accessibility of data, one of which is the assurance of ethical and prudent use of the data. Responsibility in the case of handling methodology should be given to preserve the personally identifiable information and sensitive data. A certain level of transparency must also be incorporated so that people can comprehend the rationale of the choices in matters that affect their lives by creating the factor of trust and acceptability in the process (Kuzior et al. 2023).

In the absence of transparency and accountability, the citizen may have less confidence in the conservation activities and consequently reduce the legitimacy of the activities. In addition, potential problems encountered by AI software, namely algorithm bias, may be the issue that will arise and be developed to track wildlife, and the question of fairness and justice will become an issue at that. Even when AI algorithms are learning from biased data, they will either perpetuate existing disparities or create new surfaces of discrimination. In other words, to give an example, trained AI systems are not expected to identify and monitor species using primarily the information about other segments of the world or other groups, producing misguided conservation efforts. These dangers could be prevented by requesting the open history of algorithms, decision-making procedures, surveillance, removal of inefficiencies, and checking all and any circumstances concerning which AI performances can be influenced (Kandasamy 2024). They ought to embrace the ethical models together with prudent administration in regard to norms, certification, and examination services. By ethical application of AI, inquire into the potential of the AI to enhance wildlife conservation, ensuring it saves the masses of wildlife and the human population as well. Figure 1 shows the ethical considerations of AI in wildlife surveillance.

The guidelines of responsible use of AI dedicated to five main areas: Ethical AI Use, which promotes fairness and responsibility of AI usage; Data Privacy, which highlights the necessity and protection of sensitive wildlife information; Bias Mitigation, which is focused on the reduction of biases inherent in AI algorithms; Transparency, which underlines the importance of openness in AI decision-making; Accountability, which stresses the responsibility of AI systems. The principles offer a direction toward the promotion of trust and integrity of AI technologies. In solving the above ethical dilemma, it is critically important to come up with detailed ethical white papers and codes that would act as a foundation for the development and implementation of AI to monitor wildlife (Santos and Carvalho 2025). These structures should also be generated based on openness, responsibility, fairness, and consideration for both the wild and the human party. Additionally, various stakeholders, including scientists, conservation practitioners, local communities, and policymakers, ought to participate in devising and executing these ethical frameworks in order to accomplish this. Various threats like the potential of biased data, the risk of cyber, and

Fig. 1 Ethical considerations of AI in wildlife surveillance

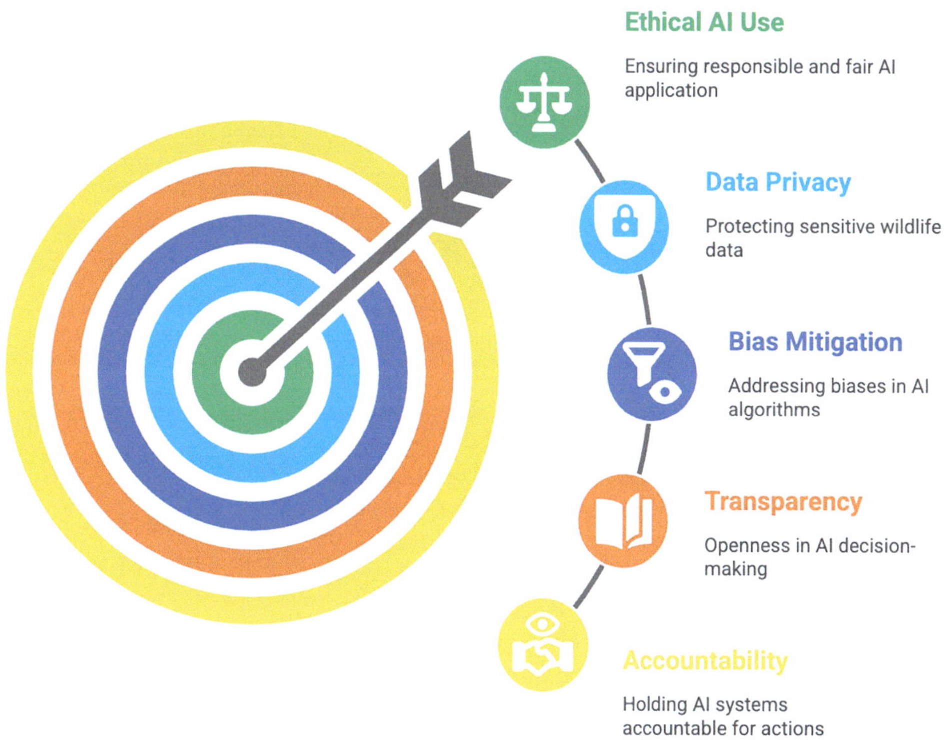

difficulty in measuring the effects of the interventions need to be addressed. The organizations must ensure that those people who developed the AI algorithms and deployed them are accountable for their actions and decisions (Dhopte and Bagde 2023). Besides, the constant monitoring and assessment of AI systems may also help to notice any negative repercussions of this process and eliminate them before they start.

1.2 Deep Learning, Soundscape Ecology, and Species Identification

Deep learning (DL) in wildlife monitoring made it possible to develop new ways of soundscape ecology and species identification, where huge amounts of acoustical and visual data can be collected and analyzed with good accuracy. DL algorithms are able to discriminate against species-specific vocalizations and species-specific behaviors through the use of audio records using technique, such as convolutional neural networks (CNNs), and recurrent neural networks (RNNs) which compiles and trains based on a previous audio recording on audio data, where projects and studies in computer vision use images recorded by camera traps and identify individual animals and classify. Such a development can improve ecological surveillance by automating monotonous activities, revealing unknown trends, and giving real-time information on biodiversity and habitat health. Nevertheless, there are ethical and privacy considerations, including the possible disruptions to wildlife posed by data collection devices such as drones, bias in collected data used as a basis of training models in areas related to conservation, as well as the misuse of sensitive data on species locations. These challenges pose a need to find transparent, accountable, and inclusive structures that value fairness and privacy of data and ecological sustainability. The integration of DL applications in soundscape ecology and species recognition ought to be done through collaboration between researchers, conservationists, and host local communities, to make sure that they are ethical and make a difference in the conservation agenda.

1.3 Significance of the Study

This work can be applicable in legal studies of ethical and privacy issues concerning the use of AI to monitor wildlife, and, generally, it is expected that the outcome of the work will include an inquiry leading to the establishment of a list of guidelines that can support the use of AI as a responsible practice in conservation. Application of AI in criminal justice should be transparent, responsible, and should be within stipulations to avoid discrimination and infringe on civil liberties (Situmeang et al. 2024). This implies ensuring that there is fairness, non-discrimination, and protection of privacy in AI systems (Dehbozorgi et al. 2025; Weiner et al. 2025). The study will be useful in policy formulation and procedures, which can alleviate the dangers of AI to the environmental management process, since it will provide a detailed study of the moral and privacy concerns.

Organization of Chapter: It starts with an Introduction with a brief description of the revolutionary nature of AI in conservation development in connection to the newly raised ethical issues. The Literature Review presents a synthesis of current studies on the technologies used in artificial intelligence in the surveillance of wildlife, ethical concerns, and regulatory loopholes. The Methodology section carries information regarding how the ethical frameworks will be analyzed and what the procedure of data collection and privacy protection will be. Results contain the data on AI performance indicators, identification of ethical problems, and an analysis of privacy risks and are accompanied by visualizations. These results are analyzed in the Discussion, with the focus on the balance between the advantages of technology and the ethical hazards it provokes, and government measures are suggested. Lastly, Future Research Directions state the possible directions of study related to bias mitigation and adaptive AI systems.

2 Literature Review

The emergence of new ways of wildlife management in terms of AI implies both the good and the bad, including: creating issues of ethical concerns and raising questions on the one hand, and on the other hand, it develops an extra option of data collection and analysis. This is because a good data privacy policy should be enacted by a good governance regime, AI calls should become transparent, investments should be made into technology, and collaboration between the relevant stakeholders should be encouraged (Agbozo and Spassov 2018). One should be open about decision-making and applying data when applying AI to human resources (Thakur et al. 2025). The institutions should also have a solution that would ensure that the logics that could discriminate against others are identified and rectified (Thakur et al. 2025). The paper outlines active surveillance, holding someone responsible, and documenting the AI activity to reduce its risks and guarantee that algorithms are impartial and comprehensible (Lepri et al. 2021). One of the key ethical issues is the potential damage to wildlife caused by the application of AI. Consider an example of AI-based surveillance technology to be used on animals, about which they are not aware, and without their permission, or even without troubling the natural behavior of animals, or without harming them. In addition, the AI algorithms could also be involved in making decisions applicable to the process of managing wildlife that are not applied in the best interest of animals, such as culling or relocation programs. And one is not to forget the fact that it has changed too (Kumar and Jyoti 2024). There is also a need to speak about the adverse influence of AI that compromises jobs, privacy violations, and bias, as with the establishment of ethical standards, transparency, and other policies that guarantee the rights and ethical demands of employees. The use of AI should also strengthen and not dilute the welfare of the employees, which is entrenched within the notion of a healthy work-life balance, as much as the personal well-being and growth of the career (Thakur et al. 2025). The other moral point is that AI can increase the already-existing disparities related to the conservation work. They use these biased data to train all these AI algorithms, and they are able to reproduce the biases in the predictions and recommendations. During training AI, one can largely exploit the data on certain people of the planet as compared to others; and, at the conclusion, there may be biased processes of conservation as a result of wrong ID and tracking of the wildlife.

2.1 AI in Wildlife Monitoring: An Overview

The practice of wildlife monitoring is currently changing with the application of AI that provides more sophisticated means of collecting and refining large amounts of data and deriving information (Vatamanu and Tofan 2025). Machine learning and other AI methods, such as computer vision and deep learning, allow conservationists to observe species and characterize habitats and changes to the environment with the accuracy never seen before (Schultz and Seele 2022). The AI is important in analyzing the personalities and behaviors of candidates based on the information on their social media profiles to ensure that they are suitable fashioned to the job (Albaroudi et al. 2024). Video streams and the camera trap photos can be analyzed by using AI algorithms that can identify and detect separate animals, requiring less manual activity in examining captured images. Acoustic monitoring systems that feature the utilization of artificial intelligence are able to detect and categorize the vocalizations of animals, resulting in a useful understanding of the distribution, behavior, and population of the species. AI can enhance the administration of environmental energy by helping to regulate emergent energy through the utilization of predictive analytics and smart grid frameworks (Yadav and Singh 2023). Adverse effects of climate change can also be anticipated through AI: projecting weather issues, reducing carbon emissions, implementing renewable energy, etc. (Patoucha and Radia 2024). AI also improves data analysis because it can pick out patterns and trends that might not be seen by human observers, such as degraded habitats, poaching activity, and invasive species. AI can combine thousands of computers and resources, as well as has the potential to solve big problems, as well as culturally acceptable ways of reducing the

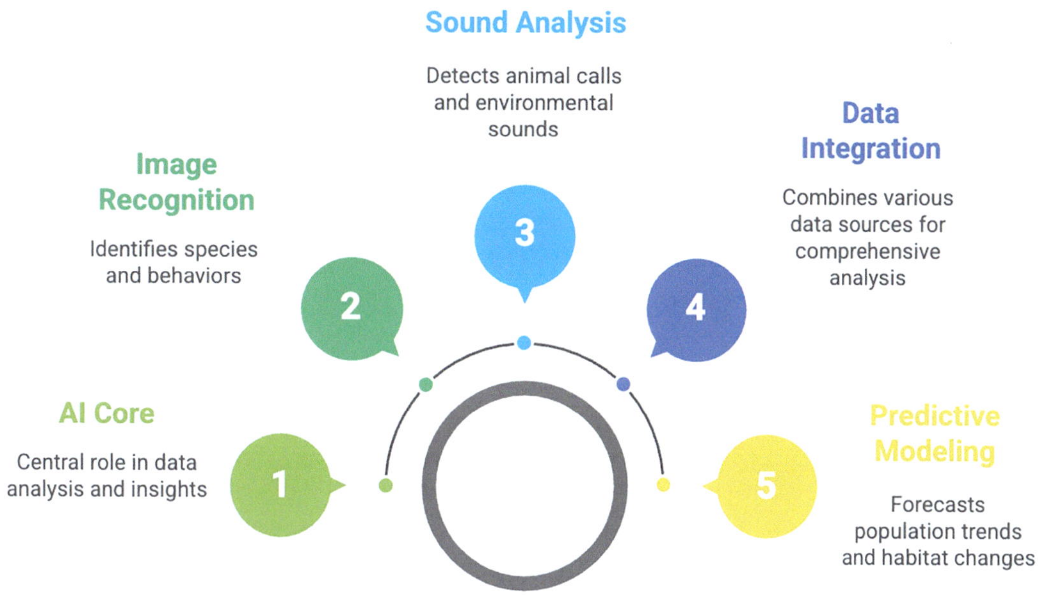

Fig. 2 AI in wildlife monitoring

effects of human activities on natural resources and energy (Nishant et al. 2020). Figure 2 shows the AI in wildlife monitoring.

An approach of using an AI-based technology is powered by animal and environmental monitoring. AI is the main core of the system, and it is essential in the analysis of data and creation of insights. The system combines several information sources, e.g., sound and image data, to produce an in-depth analysis. Sound analysis and image recognition serve to recognize animal calls and environmental sounds, as well as animal behavior and species. Predictive models provide an idea of the trends of population and habitat variations, and this helps in ensuring that there are improved conservation strategies. The set of technologies ensures a comprehensive strategy for ecological monitoring and preservation. AI allows scientists to enhance ecological monitoring, thus facilitating sustainable use of resources and protection of biological diversity, as it makes it easy to track, analyze, and obtain significant data faster due to the automation of tedious tasks. The ability of AI to analyze huge amounts of data, identify patterns, and make estimations helps to implement more efficient conservation policies and the distribution of resources. AI in sustainability enables the achievement of sustainable development goals through the optimization of energy systems, sustainable agriculture, and enhancing environmental monitoring and smart urban planning.

2.2 Ethical Considerations of Artificial Intelligence in Wildlife Surveillance

Moral questions, which are part of the study of AI-based wildlife monitoring, are rather numerous, particularly in regard to the legality and character of the gathering, processing, and use of the information. The issues of data and surveillance security, the possibility of surveillance, and the necessity to increase the technical advancements and ethical responsibilities arise in the implementation of AI systems in environmental observation. An important question is whether it can infringe the privacy of animals, as well as whether animals can be tracked and monitored with the help of AI-equipped surveillance devices without their knowledge, awareness, and consent. Ethical and respectful collection of data pertaining to wildlife should be devised and implemented, and a certain openness and responsibility toward the use of data should emerge. AI has vitality in environmental governance as it encourages organizational and individual behaviors that reduce the amount of energy and natural resources a human activity consumes. There are possibilities that AI can enhance environmental governance, but there are important constraints in this field because machine learning models rely too much on historical data, the human responses to AI-based treatment may be unpredictable, there is an increased risk to cybersecurity, adverse impacts of AI-based applications, and the interference with the effects of measures (Nishant et al. 2020).

2.3 Privacy Concerns in Wildlife Data Collection

The privacy issue is what needs to be deeply protected in the area of concern of using AI in wildlife monitoring, since both animals and human beings will provide rather sensitive information. The reduction of all the problems caused by the disturbances of the wild animal population and subsequent protection of their personal privacy suggest that the process of data collection ought to be carried out in a transparent, ethically agreeable manner (McClure et al. 2020). It also characterizes the machine learning systems, which are likely to come across girths of data where some of this data can be highly confidential and can contain personal information with regards to where endangered animal species can be found or how they live. Loss of data, loss due to data misuse or wrong use of data, or wrong access may lead to highly devastating consequences when dealing with the conservation of the wildlife population. Sound measures of cybersecurity have to be established as a way of forestalling unlawful invasion, information seepage, and mishandling of information collected. The necessity of carrying out data collection in order to save and the moral necessity of saving the privacy of people and animals should be in equilibrium. The thing is that AI systems tend to collect the information in an opaque manner, and thus, "peopling people about the shared or processed information is difficult to do" (Akpobome 2024). Organizations are meant to be transparent in their activities concerning data collection, and they should provide individuals with the possibility of accessing, correcting, and deleting it.

3 Existing Legal and Regulatory Frameworks

Current law and regulation structures might also be insufficient to deal with the specific issues of using AI in wildlife monitoring, and special guidelines and rules will have to be created to deal with the responsible AI implementation (Taddeo et al. 2021). AI provokes questions of liability and accountability in the case of a negative consequence, especially when it comes to autonomous decisions made by algorithms that affect natural resources or wildlife habitats. Lack of properly established laws could delay justice over harm caused by AI by individuals or organizations. According to the authors, polystyrene is used more efficiently because it eliminates the need for some scientific investigators to use properties of substances that are given in nanograms (Montagnani et al. 2024). Currently, the prospects of international policy and legislative work on the issue of liability of AI systems are reduced only to the revision and refinement of the system of liability, but time should be devoted to the achievement of an agreement, because of the complexity of such a problem and the resulting differences in legal traditions applied in different subject jurisdictions in the world (Velasco 2022). The absence of the law needs new legislative regulations and laws that will close the gaps in the current one and make the use of AI in environmental monitoring an ethical and responsible choice. Existing data security and privacy laws, both local and global human rights frameworks, are able to make enough room, through which much of the use and development of the AI systems and technologies could be regulated and restrained (Walsh 2019). Figure 3 shows the legal and regulatory frameworks.

It presents four main categories: Regulatory Guidelines, detailed rules that agencies issue; Core Legal Principles, basic concepts of justice and fairness; Constitutional Laws, basic rights issues and structures of governance; and Statutory Laws, which are specific rules issued by a legislature. All these categories are presented as mutually related to each other and the significance in the area of law. The world is faced with increasing citizen concerns regarding AI, governments are exploring regulatory options, and there exists a risk that other regulatory options will not be given a thorough examination (Percy et al. 2022).

The creation of national plans and policies can be considered a demonstration of concern regarding the opportunities of AI and the necessity to find a balance between innovation, inconsistency, and social implications. The EU leads in terms of the regulation of AI. The regulatory field of

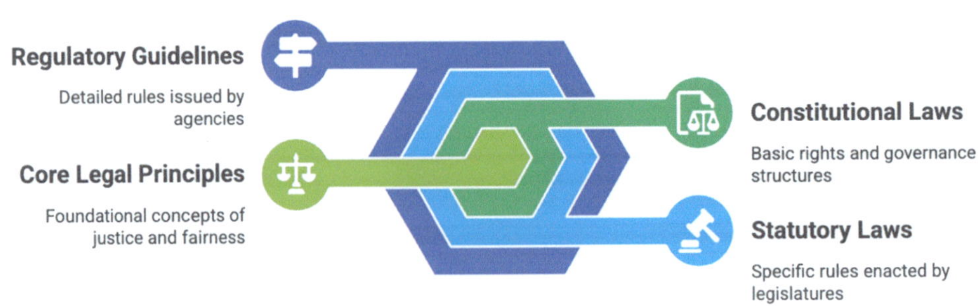

Fig. 3 Legal and regulatory frameworks

AI continues to evolve, with some countries already beginning to implement policymaking concerning AI, despite it still being in its nascent stage. (Montagnani et al. 2024; Stanovsky et al. 2025).

4 Methodology

The scheme of the common ethical principles in using AI can be discovered as a result of a meta-analysis of the provisions of AI governance policies, ethical principles of different public organizations, educational institutions, and corporations, as well as non-governmental organizations, to illustrate a prospective law (Corrêa et al. 2023). The current research proposes a narrower approach to managing research on national government guidelines and offers an extensive analysis that combines different opinions (Saheb and Saheb 2024). It is possible to find systems that can transform the high-level AI ethics systems into practice, and this is being actively designed by numerous organizations and governments (Sanderson et al. 2023). An illustration of this is the OECD AI Principles and UNESCO Recommendation on the Ethics of AI, which gives a detailed explanation of what ethical AI principles are concerning the development and application of AI. This chapter is a review of what can be done to ensure that ethics and governance regarding AI driving are achieved in different nations and regions like the USA, EU, and Asia through reviewing their policies, regulations, and guidelines toward AI driving. It combines the two divergent strategies in order to connect and establish best practices and suggest a complete system of ethical AI regulation in wildlife surveillance (Kulothungan and Gupta 2025). Figure 4 shows the steps to ensure ethical data handling.

A sequential procedure that is undertaken in manipulating data, whereby consideration of ethics is given. It begins with an Ethical Framework that entails the creation of a basis of ethical principles to use throughout the process. Next is Data Collection, where emphasis is placed on the collection of data through acceptable and moral means. The Data Processing step is the third step and entails the process of analyzing and sorting the data to derive insights from it. Lastly, Privacy Protection is the last step, which requires procedures to be put in place to ensure data privacy and confidentiality are ensured. Every action is crucial to the integrity and ethical handling of data in the course of its lifecycle.

4.1 Ethical Framework

The moral problem of algorithmic bias reveals the importance of a responsible and high-performing AI application. The framework provides suggestions on how to govern AI using ethics in wildlife monitoring based on an examination of the ethics, governance, and reduction of biased parts of AI. These comprise establishing guidelines of responsibility, transparency, and equity of the AI system, and the inclusion of techniques for detecting biases in algorithms. It also highlights the importance of multidisciplinary collaboration with

Fig. 4 Steps to ensure ethical data handling

the purpose of ensuring the consideration of moral and social concerns that AI systems have (Walter 2024). To come up with the effective strategies that will promote the use of ethical AI practices, the organization will have to pay a significant focus to the changes in processes and people alongside the rest of the ethical challenges, such as biases, getting effectively addressed (How Do Different Stakeholders Perceive Algorithmic Bias?, n.d.). It is proposed that ethical frameworks can be used to introduce ethical AI frameworks in which the interests of the people involved are met, i.e., ensuring that the best interests of the employees are taken care of and not just the achievement of the institutional end results (Thakur et al. 2025). These frameworks are also largely based on transparency, i.e., the universities must openly explain the workings of AI systems and the decision-making mechanisms (Thakur et al. 2025). Frequent bias audits are also suggested to make sure that everything is fair and the existing biases are not reproduced in AI algorithms (Thakur et al. 2025).

4.2 Data Collection Methods

The usage of new data collection systems, e.g., automated bioacoustics and the use of drones to collect images, introduces new ethical dilemmas, which should be under particular consideration. Sensitive habitats or endangered species are often the main source of data used in wildlife monitoring, and a high level of data collection, storage, and sharing protocols is required to ensure that no harm is brought to the ecological environment. Additionally, wider application of AI in data analysis presupposes algorithmic bias as well as its ability to provide distorted results and inefficient conservation plans. Mitigation of this can only be achieved by the collaboration of programmers and conservationists in the implementation of such strategies that will lead to the minimization of this bias and ensure that the operated AI systems are unbiased and deployed in a fair and transparent manner. Algorithmic bias may have significant consequences on various regions of society, such as education, the job market, and the criminal justice system, and is increasingly challenging to identify and tackle as machine learning algorithms become increasingly sophisticated.

4.3 Data Processing and Analysis

AI in wildlife remark brings several data processing and analysis challenges, such as statistical AI and ML inferences, which demand training data to be representative. The potential implications of AI, including helping law enforcement and having a role in school admissions and hiring decisions, point to its widespread distribution and to the ethical concerns that come with it. These concerns underscore the importance of considering the shortcomings of AI and taking precautions not to rely too much on it in the domains where subjectivity and considerations of ethics and understanding of complex social relations are required.

4.4 Privacy Protection Measures

It is necessary to guard the privacy of individuals and groups affected by the wildlife monitoring operations. Due to the significance of safeguarding personal information and avoiding its misuse, it is essential to perform data security approaches rigorously, such as encryption, access management, and anonymization strategies (Hutami 2024). The indigenous knowledge and traditional practices also need to be respected, and partnerships with the local communities must have the rights and interests of the local communities as a priority. The problem of privacy can be explained by the massive amounts of data collection and processing behind AI technology; the lack of control over allowing unauthorized access to sensitive material, vaguely defined ownership rights over the data, as well as data misuse, are all possible areas of concern (Kandasamy 2024).

5 Results

The privacy of the AI systems is threatened by the fact that their functioning depends on large amounts of data. The issue is concerned with data security and privacy breaches (Hamedani et al. 2025). The way AI fails to effectively understand complex social situations may also cause bias, especially in such potentially controversial areas as recruitment (Albaroudi et al. 2024). Explainability and transparency are essential parts to build confidence and acceptance of AI systems. Figure 5 presents the AI evaluation framework.

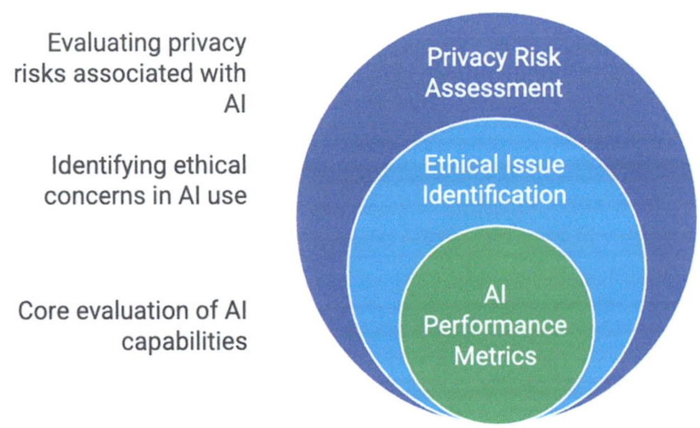

Fig. 5 AI evaluation framework

The correlation of three important elements in the analysis of AI systems: AI Performance Metrics, Ethical Issue Detection, and Privacy Risk Identification. In its essence, performance metrics are used to assess the functionality of AI. Both ethical concerns with the usage of AI and privacy risk assessment are then proposed as important layers, but privacy risk assessment subsumes the ethical concerns on the one hand and the performance of AI on the other. This image shows that the issue of judging AI requires a complex approach, and all three factors, privacy, ethics, and technical performance, should be considered in tandem with each other.

5.1 AI Performance Metrics

The AI performance assessment processes are indispensable to make sure that monitored wildlife is not subject to ineffective, unreliable, and unethical AI systems. Measures such as accuracy, precision, recall, and F1-score are the standard measures determining the performance of the AI algorithms used to discover and categorize wildlife species (Petković 2023). Yet, these indicators cannot be enough to take into consideration the ecology of the situation or the possible biases in the data. To take an extreme case, a model trained to recognize a particular species effectively in a single environment may not do so well in another environment because either of the environmental factors or the quality of the data is different. In addition to that, AI models based on biased inputs can reinforce disparities that exist in society and can give inaccurate or unjust results.

5.2 Ethical Issue Identification

When considered in detail, a proper comprehension of AI ethics requires an elaborate method of defining ethical problems that involves analysis of impacts and values held significant to various stakeholders. The stakeholders need to be engaged so that they can fully understand the ethical implications of the AI technologies, and one of the ways to address potential conflicts is a strategies such as value-sensitive design, which considers the values of the stakeholders in the course of design. The ethics of AI is a complicated issue, and biases should not be neglected, as they do not reduce discrimination.

5.3 Privacy Risk Assessment

A proper privacy risk assessment should be conducted to detect the possible areas of vulnerability and reduce the danger of data breaches or misuse. Data minimization policies, privacy-enhancing technologies, and access restrictions, etc., may contribute to securing sensitive data and make sure that the particular privacy regulations are observed. To achieve responsible and ethical use of AI in wildlife monitoring, it is crucial to address these problems because the use of AI-driven systems will potentially leak sensitive personal data without the user's knowledge (Albaroudi et al. 2024). It is important to conduct research on whether the current safety arrangements for children and people who lack decision-making capacity are in line with this recommendation or whether a different individual may have to take over the responsibility.

5.4 Compliance Analysis

The compliance analysis is necessary to see that AI systems have been dealt with in terms of existing ethical rules, regulations, and industry standards. This involves the evaluation of data governance models, privacy guidelines, and security procedures with a view to determining possible loopholes or areas of non-compliance. Periodic auditing and reviewing are to be performed, so as to ensure that the AI systems do not exceed the potential boundaries and to discuss the potential security measures to be implemented, in order to prevent the ethical/even legal abuse. This is necessary to follow the labor regulations in the context of the use of AI-enabled PA in the employee wellness program to safeguard the rights of the employees and their privacy when collecting and incorporating the wellness information and to provide employees with control over their wellness information and participation in wellness programs and to ensure employees are respected and have autonomy and self-decisions. Comprehensive data privacy and cybersecurity protection should be implemented to ensure the AI system and data on employee wellness remain confidential and not accessible to unauthorized agents.

6 Discussion

The implementation of AI in wildlife monitoring has its prospects and problems, especially in terms of the development of technologies, ethical issues, and privacy and security. Medical data must be included in strict encryption, access control, and constant audit to guard its integrity and manage confidentiality and eliminate unauthorized access or data breach. To encourage trust and a lack of compromise to ethical standards at the time of patient care, they should have full information on the application of AI to their healthcare, especially in complex cases, and the opportunity to give or refuse their consent. Human control is needed, and medical practitioners must be capable of verifying AI recommendations and challenging findings, intervening whenever needed, despite the fact that AI has the potential to enhance decision-making. Ethical frameworks should start off by tackling concerns such as the transparency of algorithms used, privacy of data involved, informed consent, and individual autonomy, and that legislating bodies and regulators should have a responsibility in ensuring that ethical and legal standards are met without hindering innovation and further development in the sector. Principles of ethics and human values are necessary when developing and implementing an AI in the context that healthcare has raised challenging questions about autonomy, dignity, beneficence, and justice. To avoid unintended effects, ethical issues, including bias in the AI algorithm and fairness, should also be addressed. SRAI should be bound by such principles as fairness, equality, and non-discrimination, ensuring that AI systems do not make biased predictions, ensuring fairness and avoiding discrimination and stigmatization.

7 Future Research Directions

A future study should be directed toward establishing strict techniques for assessing the ecological soundness and possible biases of AI systems in various wildlife environments. It also entails coalescing domain knowledge of ecologists, conservation biologists, and the communities in the region to ensure that the AI systems are ethical and achieve conservation goals. Privacy preservation methods such as federated learning and differential privacy should be studied so that we could be able to run joint data analysis and still protect sensitive data. The research in the future can be conducted regarding possible applications of AI in real-time monitoring of patients, precision medicine, and multispectral imaging, so that the possibilities of AI in diagnostics can be expanded. Clinical practice and further research in the future should be aimed at streamlining these AI systems to enhance the accuracy of the diagnosis, reduce decision fatigue, and thus, contribute to physician well-being. In order to preserve patient confidence and achieve better outcomes, further research on ethical problems and the preservation of that human touch in patient care must be conducted, and this pertains to AI usage in healthcare as well. Future studies are necessary to expand our knowledge and improve our understanding of how organizations can effectively apply HR and AI to the healthcare industry because the set of challenges is numerous and complex, and the available amount of organized and evidence-based research on the role of HR and AI in healthcare is minimal.

References

Agbozo E, Spassov K (2018) Establishing efficient governance through data-driven e-Government. 662. https://doi.org/10.1145/3209415.3209419

Akpobome O (2024) The impact of emerging technologies on legal frameworks: a model for adaptive regulation. Int J Res Publ Rev 5(10):5046. https://doi.org/10.55248/gengpi.5.1024.3012

Albaroudi E, Mansouri T, Alameer A (2024) A comprehensive review of AI techniques for addressing algorithmic bias in job hiring [review of a comprehensive review of AI techniques for addressing algorithmic bias in job hiring]. AI 5(1):383. Multidisciplinary Digital Publishing Institute. https://doi.org/10.3390/ai5010019

Brickson L, Vollrath F, Titus AJ (2023) Elephants and algorithms: a review of the current and future role of AI in elephant monitoring [review of elephants and algorithms: a review of the current and future role of AI in elephant monitoring]. arXiv (Cornell University). Cornell University. https://doi.org/10.48550/arxiv.2306.13803

Chakraborty S (2024) Towards a comprehensive assessment of AI's environmental impact. arXiv (Cornell University). https://doi.org/10.48550/arxiv.2405.14004

Corrêa NK, Galvão C, Santos JW, Pino CD, Pinto EP, Barbosa C, Massmann D, Mambrini R, Galvão L, Terem E, de Oliveira N (2023) Worldwide AI ethics: a review of 200 guidelines and recommendations for AI governance [review of worldwide AI ethics: a review of 200 guidelines and recommendations for AI governance]. Patterns 4(10):100857. Elsevier BV. https://doi.org/10.1016/j.patter.2023.100857

Dehbozorgi R, Zangeneh S, Khooshab E, Nia DH, Hanif H, Samian P, Yousefi M, Hashemi F, Vakili M, Jamalimoghadam N, Lohrasebi F (2025) The application of artificial intelligence in the field of mental health: a systematic review [review of The application of artificial intelligence in the field of mental health: a systematic review]. BMC Psychiatry 25(1). BioMed Central. https://doi.org/10.1186/s12888-025-06483-2

Dhopte A, Bagde H (2023) Smart smile: revolutionizing dentistry with artificial intelligence [review of smart smile: revolutionizing dentistry with artificial intelligence]. Cureus, Inc., Cureus. https://doi.org/10.7759/cureus.41227

Hamedani SS, Aslam S, Hamedani SS (2025) AI in business operations: driving urban growth and societal sustainability. Front Artif Intell. https://doi.org/10.3389/frai.2025.1568210

Hutami S (2024) Utilizing technology and artificial intelligence in educational administration to enhance school performance at junior high school. PPSDP Int J Educ 3(2):197. https://doi.org/10.59175/pijed.v3i2.302

International Journal of Recent Technology and Engineering (IJRTE) (2019) International Journal of Recent Technology and Engineering (IJRTE). https://doi.org/10.35940/ijrte.2277-3878

Isabelle DA, Westerlund M (2022) A review and categorization of artificial intelligence-based opportunities in wildlife, ocean and land conservation [review of a review and categorization of artificial intelligence-based opportunities in wildlife, ocean and land conservation]. Sustainability 14(4):1979. Multidisciplinary Digital Publishing Institute. https://doi.org/10.3390/su14041979

Kandasamy UC (2024) Ethical leadership in the age of AI challenges, opportunities and framework for ethical leadership. arXiv (Cornell University). https://doi.org/10.48550/arxiv.2410.18095

Kulothungan V, Gupta D (2025) Towards adaptive AI governance: comparative insights from the U.S., EU, and Asia. https://doi.org/10.48550/ARXIV.2504.00652

Kumar P, Jyoti (2024) Reshaping the library landscape: exploring the integration of artificial intelligence in libraries. IP Indian J Libr Sci Inf Technol 9(1):29. https://doi.org/10.18231/j.ijlsit.2024.005

Kuzior A, Sira M, Brożek P (2023) Ethical implications and social impacts of integrating artificial intelligence into sustainability efforts. Scientific Papers of Silesian University of Technology Organization and Management Series, 2023(176). https://doi.org/10.29119/1641-3466.2023.176.20

Lepri B, Oliver N, Pentland A (2021) Ethical machines: the human-centric use of artificial intelligence [Review of ethical machines: the human-centric use of artificial intelligence]. iScience 24(3):102249. Cell Press. https://doi.org/10.1016/j.isci.2021.102249

McClure EC, Sievers M, Brown CJ, Buelow CA, Ditria EM, Hayes MA, Pearson RM, Tulloch V, Unsworth RKF, Connolly RM (2020) Artificial intelligence meets citizen science to supercharge ecological monitoring [review of artificial intelligence meets citizen science to supercharge ecological monitoring]. Patterns 1(7):100109. Elsevier BV. https://doi.org/10.1016/j.patter.2020.100109

Montagnani ML, Najjar M-C, Davola A (2024) The EU regulatory approach(es) to AI liability, and its application to the financial services market. Comput Law Secur Rev 53:105984. https://doi.org/10.1016/j.clsr.2024.105984

Nishant R, Kennedy M, Corbett J (2020) Artificial intelligence for sustainability: challenges, opportunities, and a research agenda. Int J Inf Manag 53:102104. https://doi.org/10.1016/j.ijinfomgt.2020.102104

Patoucha A, Γαρείου Z (2024) The role of artificial intelligence in environmental sustainability. E3S Web of Conferences 585:11011. https://doi.org/10.1051/e3sconf/202458511011

Percy C, Dragičević S, Sarkar S, Garcez AS. d'Avila (2022) Accountability in AI: from principles to industry-specific accreditation. AI Commun 34(3):181. https://doi.org/10.3233/aic-210080

Petković D (2023) It is not "accuracy vs. explainability"—we need both for trustworthy AI systems. IEEE Trans Technol Soc 4(1):46. https://doi.org/10.1109/tts.2023.3239921

Saheb T, Saheb T (2024) Mapping ethical artificial intelligence policy landscape: a mixed method analysis. Sci Eng Ethics 30(2). https://doi.org/10.1007/s11948-024-00472-6

Sanderson C, Douglas D, Lu Q, Schleiger E, Whittle J, Lacey J, Newnham G, Hajkowicz S, Robinson C, Hansen D (2023) AI ethics principles in practice: perspectives of designers and developers. IEEE Trans Technol Soc 4(2):171. https://doi.org/10.1109/tts.2023.3257303

Santos MRC, Carvalho LC (2025) AI-driven participatory environmental management: Innovations, applications, and future prospects. J Environ Manag 373:123864. https://doi.org/10.1016/j.jenvman.2024.123864

Schultz MD, Seele P (2022) Towards AI ethics' institutionalization: knowledge bridges from business ethics to advance organizational AI ethics. AI Ethics 3(1):99. https://doi.org/10.1007/s43681-022-00150-y

Situmeang SMT, Harliyanto R, Zulkarnain PD, Mahdi U, Nugroho T (2024) The role of artificial intelligence in criminal justice. Glob Int J Innov Res 2(8):1966. https://doi.org/10.59613/global.v2i8.264

Stanovsky G, Keydar R, Perl G, Habba E (2025) Beyond benchmarks: on the false promise of AI regulation. arXiv (Cornell University). https://doi.org/10.48550/arxiv.2501.15693

Taddeo M, Tsamados A, Cowls J, Floridi L (2021) Artificial intelligence and the climate emergency: opportunities, challenges, and recommendations. One Earth 4(6):776. https://doi.org/10.1016/j.oneear.2021.05.018

Thakur RA, Talukdar M, Iyer S, Liu L-C, Chen C-L, Singh A (2025) AI and employee well-being: assessing the ethical implications of AI-driven human resource practices in Indian Universities

Vatamanu AF, Tofan M (2025) Integrating artificial intelligence into public administration: challenges and vulnerabilities. Adm Sci 15(4):149. https://doi.org/10.3390/admsci15040149

Velasco C (2022) Cybercrime and artificial intelligence. an overview of the work of international organizations on criminal justice and the international applicable instruments. ERA Forum 23(1):109. https://doi.org/10.1007/s12027-022-00702-z

Walsh T (2019) The effective and ethical development of artificial intelligence: an opportunity to improve our wellbeing. https://apo.org.au/sites/default/files/resource-files/2019-07/apo-nid250306.pdf

Walter Y (2024) Managing the race to the moon: Global policy and governance in artificial intelligence regulation—a contemporary overview and an analysis of socioeconomic consequences. Discov Artif Intell 4(1). https://doi.org/10.1007/s44163-024-00109-4

Weiner E, Dankwa-Mullan I, Nelson WA, Hassanpour S (2025) Ethical challenges and evolving strategies in the integration of artificial intelligence into clinical practice. PLOS Dig Health 4(4). https://doi.org/10.1371/journal.pdig.0000810

Yadav M, Singh G (2023) Environmental sustainability with artificial intelligence. EPRA Int J Multi Res (IJMR) 213. https://doi.org/10.36713/epra13325

MIX
Papier aus verantwortungsvollen Quellen
Paper from responsible sources
FSC® C105338

If you have any concerns about our products,
you can contact us on
ProductSafety@springernature.com

In case Publisher is established outside the EU,
the EU authorized representative is:
Springer Nature Customer Service Center GmbH
Europaplatz 3, 69115 Heidelberg, Germany

Printed by Libri Plureos GmbH
in Hamburg, Germany